continued on back

Robust Statistics

Robust Statistics

PETER J. HUBER

Professor of Statistics
Harvard University
Cambridge, Massachusetts

John Wiley & Sons

New York • Chichester • Brisbane • Toronto

Copyright © 1981 by John Wiley & Sons, Inc.

All rights reserved. Published simultaneously in Canada.

Reproduction or translation of any part of this work
beyond that permitted by Sections 107 or 108 of the
1976 United States Copyright Act without the permission
of the copyright owner is unlawful. Requests for
permission or further information should be addressed to
the Permissions Department, John Wiley & Sons, Inc.

Library of Congress Cataloging in Publication Data:

Huber, Peter J
  Robust statistics.

  (Wiley series in probability and mathematical
statistics)
  "A Wiley-Interscience publication."
  Includes index.
  1. Robust statistics. I. Title.

QA276.H785   519.5   80-18627
ISBN 0-471-41805-6

Printed in the United States of America

10 9 8 7 6 5 4 3 2

Preface

The present monograph is the first systematic, book-length exposition of robust statistics. The technical term "robust" was coined only in 1953 (by G. E. P. Box), and the subject matter acquired recognition as a legitimate topic for investigation only in the mid-sixties, but it certainly never was a revolutionary new concept. Among the leading scientists of the late nineteenth and early twentieth century, there were several practicing statisticians (to name but a few: the astronomer S. Newcomb, the astrophysicist A. Eddington, and the geophysicist H. Jeffreys), who had a perfectly clear, operational understanding of the idea; they knew the dangers of long-tailed error distributions, they proposed probability models for gross errors, and they even invented excellent robust alternatives to the standard estimates, which were rediscovered only recently. But for a long time theoretical statisticians tended to shun the subject as being inexact and "dirty." My 1964 paper may have helped to dispel such prejudices. Amusingly (and disturbingly), it seems that lately a kind of bandwagon effect has evolved, that the pendulum has swung to the other extreme, and that "robust" has now become a magic word, which is invoked in order to add respectability.

This book gives a solid foundation in robustness to both the theoretical and the applied statistician. The treatment is theoretical, but the stress is on concepts, rather than on mathematical completeness. The level of presentation is deliberately uneven: in some chapters simple cases are treated with mathematical rigor; in others the results obtained in the simple cases are transferred by analogy to more complicated situations (like multiparameter regression and covariance matrix estimation), where proofs are not always available (or are available only under unrealistically severe assumptions). Also selected numerical algorithms for computing robust estimates are described and, where possible, convergence proofs are given.

Chapter 1 gives a general introduction and overview; it is a must for every reader. Chapter 2 contains an account of the formal mathematical background behind qualitative and quantitative robustness, which can be skipped (or skimmed) if the reader is willing to accept certain results on faith. Chapter 3 introduces and discusses the three basic types of estimates (M-, L-, and R-estimates), and Chapter 4 treats the asymptotic minimax

v

theory for location estimates; both chapters again are musts. The remaining chapters branch out in different directions and are fairly independent and self-contained; they can be read or taught in more or less any order.

The book does not contain exercises—I found it hard to invent a sufficient number of problems in this area that were neither trivial nor too hard—so it does not satisfy some of the formal criteria for a textbook. Nevertheless I have successfully used various stages of the manuscript as such in graduate courses.

The book also has no pretensions of being encyclopedic. I wanted to cover only those aspects and tools that I personally considered to be the most important ones. Some omissions and gaps are simply due to the fact that I currently lack time to fill them in, but do not want to procrastinate any longer (the first draft for this book goes back to 1972). Others are intentional. For instance, adaptive estimates were excluded because I would now prefer to classify them with nonparametric rather than with robust statistics, under the heading of nonparametric efficient estimation. The so-called Bayesian approach to robustness confounds the subject with admissible estimation in an *ad hoc* parametric supermodel, and still lacks reliable guidelines on how to select the supermodel and the prior so that we end up with something robust. The coverage of L- and R-estimates was cut back from earlier plans because they do not generalize well and get awkward to compute and to handle in multiparameter situations.

A large part of the final draft was written when I was visiting Harvard University in the fall of 1977; my thanks go to the students, in particular to P. Rosenbaum and Y. Yashizoe, who then sat in my seminar course and provided many helpful comments.

P. J. Huber

Cambridge, Massachusetts
July 1980

Contents

Robust Statistics

CHAPTER 1

Generalities

1.1 WHY ROBUST PROCEDURES?

Statistical inferences are based only in part upon the observations. An equally important base is formed by prior assumptions about the underlying situation. Even in the simplest cases, there are explicit or implicit assumptions about randomness and independence, about distributional models, perhaps prior distributions for some unknown parameters, and so on.

These assumptions are not supposed to be exactly true—they are mathematically convenient rationalizations of an often fuzzy knowledge or belief. As in every other branch of applied mathematics, such rationalizations or simplifications are vital, and one justifies their use by appealing to a vague continuity or stability principle: a minor error in the mathematical model should cause only a small error in the final conclusions.

Unfortunately, this does not always hold. During the past decades one has become increasingly aware that some of the most common statistical procedures (in particular, those optimized for an underlying normal distribution) are excessively sensitive to seemingly minor deviations from the assumptions, and a plethora of alternative "robust" procedures have been proposed.

The word "robust" is loaded with many—sometimes inconsistent—connotations. We use it in a relatively narrow sense: for our purposes, *robustness signifies insensitivity to small deviations from the assumptions.*

Primarily, we are concerned with *distributional robustness*: the shape of the true underlying distribution deviates slightly from the assumed model (usually the Gaussian law). This is both the most important case and the best understood one. Much less is known about what happens when the other standard assumptions of statistics are not quite satisfied and about the appropriate safeguards in these other cases.

The following example, due to Tukey (1960), shows the dramatic lack of distributional robustness of some of the classical procedures.

1

Example 1.1 Assume that we have a large, randomly mixed batch of n "good" and "bad" observations x_i of the same quantity μ. Each single observation with probability $1-\varepsilon$ is a "good" one, with probability ε a "bad" one, where ε is a small number. In the former case x_i is $\mathfrak{N}(\mu, \sigma^2)$, in the latter $\mathfrak{N}(\mu, 9\sigma^2)$. In other words all observations have the same mean, but the errors of some are increased by a factor of 3.

Equivalently, we could say that the x_i are independent, identically distributed with the common underlying distribution

$$F(x) = (1-\varepsilon)\Phi\left(\frac{x-\mu}{\sigma}\right) + \varepsilon\Phi\left(\frac{x-\mu}{3\sigma}\right), \qquad (1.1)$$

where

$$\Phi(x) = \frac{1}{\sqrt{2\pi}} \int_{-\infty}^{x} e^{-y^2/2} \, dy \qquad (1.2)$$

is the standard normal cumulative.

Two time-honored measures of scatter are the mean absolute deviation

$$d_n = \frac{1}{n} \sum |x_i - \bar{x}| \qquad (1.3)$$

and the mean square deviation

$$s_n = \left[\frac{1}{n} \sum (x_i - \bar{x})^2\right]^{1/2}. \qquad (1.4)$$

There was a dispute between Eddington (1914, p. 147) and Fisher (1920, footnote on p. 762) about the relative merits of d_n and s_n. Eddington advocated the use of the former: "This is contrary to the advice of most textbooks; but it can be shown to be true." Fisher seemingly settled the matter by pointing out that for normal observations s_n is about 12% more efficient than d_n.

Of course, the two statistics measure different characteristics of the error distribution. For instance, if the errors are exactly normal, s_n converges to σ, while d_n converges to $\sqrt{2/\pi}\,\sigma \approx 0.80\sigma$. So we must be precise about how their performances are to be compared; we use the asymptotic relative efficiency (ARE) of d_n relative to s_n, defined as follows:

$$ARE(\varepsilon) = \lim_{n \to \infty} \frac{var(s_n)/(Es_n)^2}{var(d_n)/(Ed_n)^2}$$

$$= \frac{\left[\dfrac{3(1+80\varepsilon)}{(1+8\varepsilon)^2} - 1 \right] \Big/ 4}{\dfrac{\pi(1+8\varepsilon)}{2(1+2\varepsilon)^2} - 1}.$$

The results are summarized in the Exhibit 1.1.1.

The result is disquieting: just 2 bad observations in 1000 suffice to offset the 12% advantage of the mean square error, and $ARE(\varepsilon)$ reaches a maximum value greater than 2 at about $\varepsilon = 0.05$.

This is particularly unfortunate since in the physical sciences typical "good data" samples appear to be well modeled by an error law of the form (1.1) with ε in the range between 0.01 and 0.1. (This does not imply that these samples contain between 1% and 10% gross errors, although this is very often true; the above law (1.1) may just be a convenient description of a slightly longer-tailed than normal distribution.) Thus it becomes painfully clear that the naturally occurring deviations from the idealized model are large enough to render meaningless the traditional asymptotic optimality theory: in practice we should certainly prefer d_n to s_n, since it is better for all ε between 0.002 and 0.5.

ε	$ARE(\varepsilon)$
0	0.876
0.001	0.948
0.002	1.016
0.005	1.198
0.01	1.439
0.02	1.752
0.05	2.035
0.10	1.903
0.15	1.689
0.25	1.371
0.5	1.017
1.0	0.876

Exhibit 1.1.1 Asymptotic efficiency of mean absolute relative to mean square deviation. From Huber (1977b), with permission of the publisher.

To avoid misunderstandings, we should hasten to emphasize what is *not* implied here. First, the above does not imply that we advocate the use of the mean absolute deviation (there are still better estimates of scale). Second, some people have argued that the example is unrealistic insofar as the "bad" observations will stick out as outliers, so any conscientious statistician will do something about them before calculating the mean square error. This is beside the point; outlier rejection followed by the mean square error might very well beat the performance of the mean absolute error, but we are concerned here with the behavior of the *unmodified* classical estimates.

The example clearly has to do with longtailedness: lengthening the tails of the underlying distribution explodes the variance of s_n (d_n is much less affected). Shortening the tails, on the other hand, produces quite negligible effects on the distributions of the estimates. (It may impair the absolute efficiency by decreasing the asymptotic Cramér-Rao bound, but the latter is so unstable under small changes of the distribution that this effect cannot be taken very seriously.)

The sensitivity of classical procedures to longtailedness is typical and not limited to this example. As a consequence "distributionally robust" and "outlier resistant," although conceptually distinct, are practically synonymous notions. Any reasonable, formal or informal, procedure for rejecting outliers will prevent the worst.

We might therefore ask whether robust procedures are needed at all; perhaps a two-step approach would suffice:

(1) First clean the data by applying some rule for outlier rejection.

(2) Then use classical estimation and testing procedures on the remainder.

Would these steps do the same job in a simpler way?
Unfortunately they will not, for the following reasons:

(1) It is rarely possible to separate the two steps cleanly; for instance, in multiparameter regression problems outliers are difficult to recognize unless we have reliable, robust estimates for the parameters.

(2) Even if the original batch of observations consists of normal observations interspersed with some gross errors, the cleaned data will not be normal (there will be statistical errors of both kinds, false rejections and false retentions), and the situation is even worse when the original batch derives from a genuine nonnormal distribution, instead of from a gross-error framework. Therefore the classical normal theory is not applicable to cleaned samples, and the actual performance of such a two-step procedure may be more difficult to work out than that of a straight robust procedure.

(3) It is an empirical fact that the best rejection procedures do not quite reach the performance of the best robust procedures. The latter apparently are superior because they can make a smooth transition between full acceptance and full rejection of an observation.

1.2 WHAT SHOULD A ROBUST PROCEDURE ACHIEVE?

We are adopting what might be called an "applied parametric viewpoint": we have a parametric model, which hopefully is a good approximation to the true underlying situation, but we cannot and do not assume that it is exactly correct. Therefore any statistical procedure should possess the following desirable features:

(1) It should have a reasonably good (optimal or nearly optimal) efficiency at the assumed model.

(2) It should be robust in the sense that small deviations from the model assumptions should impair the performance only slightly, that is, the latter (described, say, in terms of the asymptotic variance of an estimate, or of the level and power of a test) should be close to the nominal value calculated at the model.

(3) Somewhat larger deviations from the model should not cause a catastrophe.

If asymptotic performance criteria are used, some care is needed. In particular, the convergence should be uniform over a neighborhood of the model, or there should be at least a one-sided uniform bound, because otherwise we cannot guarantee robustness for any finite n, no matter how large n is. This point has often been overlooked in the past.

It should be emphasized once more that the occurrence of gross errors in a small fraction of the observations is to be regarded as a small deviation, and that, in view of the extreme sensitivity of some classical procedures, a primary goal of robust procedures is to safeguard against gross errors.

The literature contains many other explicit and implicit goals for robust procedures, for example, high asymptotic *relative efficiency* (relative to some classical reference procedures), or high *absolute efficiency*, and this either for completely arbitrary (sufficiently smooth) underlying distributions, or for a specific parametric family.

However, it seems to me that these goals are secondary in importance, and they should never be allowed to take precedence over the above-mentioned three.

Robust, Nonparametric, and Distribution-Free

Traditionally, robust procedures have been classified together with non-parametric and distribution-free ones. In our view the three notions have very little overlap.

A procedure is called *nonparametric* if it is supposed to be used for a broad, not-parametrized set of underlying distributions. For instance, the sample mean and the sample median are *the* nonparametric estimates of the population mean and median, respectively. Although nonparametric the sample mean is highly sensitive to outliers and therefore very non-robust. In the relatively rare cases where one is *specifically* interested in estimating the true population mean, there is little choice except to pray and use the sample mean.

A test is called *distribution-free* if the probability of falsely rejecting the null hypothesis is the same for all possible underlying continuous distributions (optimal robustness of validity). The typical examples are the two-sample rank tests for testing equality between distributions. Most distribution-free tests happen to have a reasonably stable power and thus also a good robustness of total performance. But this seems to be a fortunate accident, since distribution-freeness does not imply anything about the behavior of the power function.

Estimates derived from a distribution-free test are sometimes also called distribution-free, but this is a misnomer: the stochastic behavior of point estimates is intimately connected with the power (not the level) of the parent tests and depends on the underlying distribution. The only exceptions are interval estimates derived from rank tests: for example the interval between two specified sample quantiles catches the true median with a fixed probability (but still the distribution of the length of this interval depends on the underlying distribution).

Robust methods, as conceived in this book, are much closer to the classical parametric ideas than to nonparametric or distribution-free ones. They are destined to work with parametric models; the only differences are that the latter are no longer supposed to be literally true, and that one is also trying to take this into account in a formal way.

In accordance with these ideas, we intend to standardize robust esti-mates such that they are *consistent estimates* of the unknown parameters *at the idealized model*. Because of robustness they will not drift too far away if the model is only approximately true. Outside of the model we then may *define* the parameter to be estimated in terms of the limiting value of the estimate—for example, if we use the sample median, then the natural estimand is the population median, and so on.

Adaptive Procedures

Stein (1956) discovered the possibility of devising nonparametric efficient tests and estimates. Later, several authors, in particular Takeuchi (1971), Beran (1974), Sacks (1975), and Stone (1975), described specific location estimates that are asymptotically efficient for all sufficiently smooth symmetric densities. Since we may say that these estimates adapt themselves to the underlying distribution, they have become known under the name of *adaptive procedures*. See also the review article by Hogg (1974).

At one time, it almost seemed as if the ultimate goal of robust estimation were to construct fully efficient adaptive estimates.

However, the connection between adaptivity and robustness is far from clear. The crucial point is that in robustness the emphasis rests much more on safety than on efficiency. The behavior of adaptive procedures under asymmetry is practically unexplored. For extremely large samples, where at first blush adaptive estimates look particularly attractive, the statistical variability of the estimate falls below its potential bias (caused by asymmetric contamination and the like), and robustness would therefore suggest to move toward a less efficient estimate, namely the sample median, that minimizes bias (see Section 4.2). We therefore prefer to follow Stein's original terminology and to classify adaptive estimates not under robustness, but under the heading of efficient nonparametric procedures.

Resistant Procedures

A statistical procedure is called *resistant* (see Mosteller and Tukey, 1977, p. 203) if the value of the estimate (or test statistic) is insensitive to small changes in the underlying *sample* (small changes in all, or large changes in a few of the values). The underlying distribution does not enter at all. This notion is particularly appropriate for (exploratory) data analysis and is of course conceptually distinct from robustness. However, in view of Hampel's theorem (Section 2.6), the two notions are for all practical purposes synonymous.

1.3 QUALITATIVE ROBUSTNESS

In this section we motivate and give a formal definition of qualitative asymptotic robustness. For statistics representable as a functional T of the empirical distribution, qualitative robustness is essentially equivalent to weak(-star) continuity of T, and for the sake of clarity we first discuss this particular case.

Many of the most common test statistics and estimators depend on the sample $(x_1, \ldots, x_n)$ only through the empirical distribution function

$$F_n(x) = n^{-1} \sum 1_{\{x_i < x\}}, \tag{3.1}$$

or, for more general sample spaces, through the empirical measure

$$F_n = n^{-1} \sum \delta_{x_i} \tag{3.2}$$

where δ_x stands for the pointmass 1 at x, that is, we can write

$$T_n(x_1, \ldots, x_n) = T(F_n) \tag{3.3}$$

for some functional T defined (at least) on the space of empirical measures. Often T has a natural extension to (a subspace of) the space $\mathfrak{M}$ of all probability measures on the sample space. For instance, if the limit in probability exists, put

$$T(F) = \lim_{n \to \infty} T(F_n), \tag{3.4}$$

where F is the true underlying common distribution of the observations. If a functional T satisfies (3.4), it is called *consistent* at F.

Example 3.1   *The Test Statistic of the Neyman-Pearson Lemma*   The most powerful tests between two densities p_0 and p_1 are based on a statistic of the form

$$\int \psi(x) F_n(dx) = \frac{1}{n} \sum \psi(x_i) \tag{3.5}$$

with

$$\psi(x) = \log \frac{p_1(x)}{p_0(x)}. \tag{3.6}$$

Example 3.2   The maximum likelihood estimate of θ for an assumed underlying family of densities $f(x, \theta)$ is a solution of

$$\int \psi(x, \theta) F_n(dx) = 0 \tag{3.7}$$

with

$$\psi(x, \theta) = \frac{\partial}{\partial \theta} \log f(x, \theta).$$ (3.8)

Example 3.3 The α-trimmed mean can be written as

$$\bar{X}_\alpha = \frac{1}{1 - 2\alpha} \int_\alpha^{1-\alpha} F_n^{-1}(t) \, dt.$$ (3.9)

Example 3.4 The so-called Hodges-Lehmann estimate is one-half of the median of the convolution square

$$\frac{1}{2} \mathrm{med}(F_n * F_n).$$ (3.10)

(NOTE: this is the median of all n^2 pairwise means $(x_i + x_j)/2$; the more customary versions use only the pairs $i < j$ or $i \leq j$, but are asymptotically equivalent.)

Assume now that the sample space is Euclidean, or more generally, a complete, separable metrizable space. We claim that, in this case, the natural robustness (more precisely, resistance) requirement for a statistic of the form (3.3) is that T should be continuous with respect to the weak(-star) topology. By definition this is the weakest topology in the space $\mathfrak{M}$ of all probability measures such that the map

$$F \to \int \psi \, dF$$ (3.11)

from $\mathfrak{M}$ into $\mathbb{R}$ is continuous whenever ψ is bounded and continuous. The converse is also true: if a linear functional of the form (3.11) is weakly continuous, then ψ must be bounded and continuous; see Chapter 2 for details.

The motivation behind this claim is the following basic resistance requirement. Take a linear statistic of the form (3.5) and make a small change in the sample, that is, make either small changes in all of the observations x_i (rounding, grouping) or large changes in a few of them (gross errors, blunders). If ψ is bounded and continuous, then this will result in a small change of $T(F_n) = \int \psi \, dF_n$. But if ψ is not bounded, then a single, strategically placed gross error can completely upset $T(F_n)$. If ψ is not continuous, and if F_n happens to put mass onto discontinuity points, then small changes in many of the x_i may produce a large change in $T(F_n)$.

We conclude from this that our vague, intuitive notion of resistance or robustness should be made precise as follows: a linear functional T is robust everywhere if and only if (iff) the corresponding ψ is bounded and continuous, that is, iff T is weakly continuous.

We could take this last property as our definition and call a (not necessarily linear) statistical functional T robust if it is weakly continuous.

But, following Hampel (1971), we prefer to adopt a slightly more general definition.

Let the observations x_i be independent identically distributed, with common distribution F, and let (T_n) be a sequence of estimates or test statistics $T_n = T_n(x_1, \ldots, x_n)$. Then this sequence is called *robust at* $F = F_0$ if the sequence of maps of distributions

$$F \rightarrow \mathcal{L}_F(T_n) \tag{3.12}$$

is equicontinuous at F_0, that is, if we take a suitable distance function d_* in the space $\mathfrak{M}$ of probability measures, metrizing the weak topology, and assume that, for each $\varepsilon > 0$, there is a $\delta > 0$ and an $n_0 > 0$ such that, for all F and all $n \geqslant n_0$,

$$d_*(F_0, F) \leqslant \delta \Rightarrow d_*\big(\mathcal{L}_{F_0}(T_n), \mathcal{L}_F(T_n)\big) \leqslant \varepsilon. \tag{3.13}$$

If the sequence (T_n) derives from a functional $T_n = T(F_n)$, then it is shown in Section 2.6 that this definition is essentially equivalent to weak continuity of T.

Note the close formal analogy between this definition of robustness and stability of ordinary differential equations; let $y_x(\cdot)$ be the solution with initial value $y(0) = x$ of the differential equation

$$\frac{dy}{dt} = f(t, y).$$

Then we have stability at $x = x_0$ if, for all $\varepsilon > 0$, there is a $\delta > 0$ such that, for all x and all $t \geqslant 0$,

$$d(x_0, x) \leqslant \delta \Rightarrow d\big(y_{x_0}(t), y_x(t)\big) \leqslant \varepsilon.$$

1.4 QUANTITATIVE ROBUSTNESS

For several reasons it may be useful to describe quantitatively how greatly a small change in the underlying distribution F changes the distribution

$\mathcal{L}_F(T_n)$ of an estimate or test statistic $T_n = T_n(x_1, \ldots, x_n)$. A few crude and simple numerical quantifiers might be more effective than a very detailed description.

To fix the idea assume that $T_n = T(F_n)$ derives from a functional T. In most cases of practical interest, T_n is then consistent:

$$T_n \rightarrow T(F), \quad \text{in probability}, \tag{4.1}$$

and asymptotically normal

$$\mathcal{L}_F\{\sqrt{n}\,[T_n - T(F)]\} \rightarrow \mathfrak{N}(0, A(F, T)). \tag{4.2}$$

Then it is convenient to discuss the quantitative large sample robustness of T in terms of the behavior of its asymptotic bias $T(F) - T(F_0)$ and asymptotic variance $A(F, T)$ in some neighborhood $\mathcal{P}_\varepsilon(F_0)$ of the model distribution F_0.

For instance, $\mathcal{P}_\varepsilon$ might be a *Lévy neighborhood*

$$\mathcal{P}_\varepsilon(F_0) = \{F \,|\, \forall t, \, F_0(t - \varepsilon) - \varepsilon \leqslant F(t) \leqslant F_0(t + \varepsilon) + \varepsilon\}, \tag{4.3}$$

or a *contamination "neighborhood"*

$$\mathcal{P}_\varepsilon(F_0) = \{F \,|\, F = (1 - \varepsilon)F_0 + \varepsilon H, \, H \in \mathfrak{M}\} \tag{4.4}$$

(the latter is not a neighborhood in the topological sense). Equation 4.4 is also called the *gross error model*.

The two most important characteristics then are the maximum bias

$$b_1(\varepsilon) = \sup_{F \in \mathcal{P}_\varepsilon} |T(F) - T(F_0)| \tag{4.5}$$

and the maximum variance

$$v_1(\varepsilon) = \sup_{F \in \mathcal{P}_\varepsilon} A(F, T). \tag{4.6}$$

We often consider a restricted supremum of $A(F, T)$ also, assuming that F varies only over some slice of $\mathcal{P}_\varepsilon$ where $T(F)$ stays constant, for example, only over the set of symmetric distributions.

Unfortunately, the above approach to the problem is conceptually inadequate; we should like to establish that, for sufficiently large n, our estimate T_n behaves well for *all* $F \in \mathcal{P}_\varepsilon$. A description in terms of b_1 and v_1 would allow us to show that, for each *fixed* $F \in \mathcal{P}_\varepsilon$, T_n behaves well for sufficiently large n. The distinction involves an interchange in the order of quantifiers and is fundamental, but has been largely ignored in the literature.

A better approach is as follows. Let $M(F, T_n)$ be the median of $\mathcal{L}_F[T_n - T(F_0)]$ and let $Q_t(F, T_n)$ be a normalized t-quantile range of $\mathcal{L}_F(\sqrt{n}\, T_n)$, where, for any distribution G, the normalized t-quantile range is defined as

$$Q_t = \frac{G^{-1}(1-t) - G^{-1}(t)}{\Phi^{-1}(1-t) - \Phi^{-1}(t)}, \tag{4.7}$$

Φ being the standard normal cumulative. The value of t is arbitrary, but fixed, say $t = 0.25$ (interquartile range) or $t = 0.025$ (95% range, which is convenient in view of the traditional 95% confidence intervals). For a normal distribution, Q_t coincides with the standard deviation; therefore Q_t^2 is sometimes called pseudo-variance.

Then define the maximum asymptotic bias and variance, respectively, as

$$b(\varepsilon) = \lim_n \sup_{F \in \mathcal{P}_\varepsilon} |M(F, T_n)|, \tag{4.8}$$

$$v(\varepsilon) = \lim_n \sup_{F \in \mathcal{P}_\varepsilon} Q_t(F, T_n)^2. \tag{4.9}$$

THEOREM 4.1 If b_1 and v_1 are well-defined, we have $b(\varepsilon) \geqslant b_1(\varepsilon)$ and $v(\varepsilon) \geqslant v_1(\varepsilon)$.

Proof Let $T(F_0) = 0$ for simplicity and assume that T_n is consistent: $T(F_n) \to T(F)$. Then $\lim_n M(F, T_n) = T(F)$, and we have the following obvious inequality, valid for any $F \in \mathcal{P}_\varepsilon$:

$$b(\varepsilon) = \lim_n \sup_{F \in \mathcal{P}_\varepsilon} |M(F, T_n)| \geqslant \lim_n |M(F, T_n)| = |T(F)|;$$

hence

$$b(\varepsilon) \geqslant \sup_{F \in \mathcal{P}_\varepsilon} |T(F)| = b_1(\varepsilon).$$

Similarly, if $\sqrt{n}\,[T_n - T(F)]$ has a limiting normal distribution, we have $\lim_n Q_t(F, T_n)^2 = A(F, T)$, and $v(\varepsilon) \geqslant v_1(\varepsilon)$ follows in the same fashion as above. ∎

The quantities b and v are awkward to handle, so we usually work with b_1 and v_1 instead. We are then, however, obliged to check whether, for the particular $\mathcal{P}_\varepsilon$ and T under consideration, we have $b_1 = b$ and $v_1 = v$. Fortunately, this is usually true.

THEOREM 4.2 If $\mathcal{P}_\varepsilon$ is the Lévy neighborhood, then $b(\varepsilon) \leqslant b_1(\varepsilon+0) = \lim_{\eta \downarrow \varepsilon} b_1(\eta)$.

Proof According to the Glivenko-Cantelli theorem, we have $\sup |F_n(x) - F(x)| \to 0$ in probability, uniformly in F. Thus for any $\delta > 0$, the probability of $F_n \in \mathcal{P}_\delta(F)$, and hence of $F_n \in \mathcal{P}_{\varepsilon+\delta}(F_0)$, will tend to 1, uniformly in F for $F \in \mathcal{P}_\varepsilon(F_0)$. Hence $b(\varepsilon) \leqslant b_1(\varepsilon + \delta)$ for all $\delta > 0$. ∎

Note that, for the above types of neighborhoods, $\mathcal{P}_1 = \mathfrak{M}$ is the set of all probability measures on the sample space, so $b(1)$ is the worst possible value of b (usually ∞). We define the *asymptotic breakdown point* of T at F_0 as

$$\varepsilon^* = \varepsilon^*(F_0, T) = \sup\{\varepsilon | b(\varepsilon) < b(1)\}. \qquad (4.10)$$

Roughly speaking, the breakdown point gives the limiting fraction of bad outliers the estimator can cope with. In many cases ε^* does not depend on F_0, and it is often the same for all the usual choices for $\mathcal{P}_\varepsilon$.

Example 4.1 The breakdown point of the α-trimmed mean is $\varepsilon^* = \alpha$. (This is intuitively obvious; for a formal derivation see Section 3.3.)

Similarly we may also define an asymptotic variance breakdown point

$$\varepsilon^{**} = \varepsilon^{**}(F_0, T) = \sup\{\varepsilon | v(\varepsilon) < v(1)\}, \qquad (4.11)$$

but this is a much less useful notion.

1.5 INFINITESIMAL ASPECTS

What happens if we add one more observation with value x to a very large sample? Its suitably normed limiting influence on the value of an estimate or test statistic $T(F_n)$ can be expressed as

$$IC(x, F, T) = \lim_{s \to 0} \frac{T((1-s)F + s\delta_x) - T(F)}{s} \qquad (5.1)$$

where δ_x denotes the pointmass 1 at x. The above quantity, considered as a function of x, has been introduced by Hampel (1968, 1974) under the name *influence curve (IC)* or *influence function*, and is perhaps the most useful heuristic tool of robust statistics. It is treated in more detail in Section 2.5.

If T is sufficiently regular, it can be linearized near F in terms of $IC(x, F, T)$; if G is near F, then the leading terms of a Taylor expansion are

$$T(G) = T(F) + \int IC(x, F, T)[G(dx) - F(dx)] + \cdots . \qquad (5.2)$$

We have

$$\int IC(x, F, T)F(dx) = 0, \qquad (5.3)$$

and, if we substitute the empirical distribution F_n for G in the above expansion, we obtain

$$\sqrt{n}\,(T(F_n) - T(F)) = \sqrt{n}\,\int IC(x, F, T)F_n(dx) + \cdots$$

$$= \frac{1}{\sqrt{n}} \sum IC(x_i, F, T) + \cdots . \qquad (5.4)$$

By the central limit theorem, the leading term on the right-hand side is asymptotically normal with mean 0, if the x_i are independent with common distribution F. Since it is often true (but not easy to prove) that the remaining terms are asymptotically negligible, $\sqrt{n}\,[T(F_n) - T(F)]$ is then asymptotically normal with mean 0 and variance

$$A(F, T) = \int IC(x, F, T)^2 F(dx). \qquad (5.5)$$

Thus the influence function has two main uses. First, it allows us to assess the relative influence of individual observations toward the value of an estimate or test statistic. If it is unbounded, an outlier might cause trouble. Its maximum absolute value,

$$\gamma^* = \sup_x |IC(x, F, T)|, \qquad (5.6)$$

has been called *gross error sensitivity* by Hampel. It is related to the maximum bias (4.5): take the gross error model (4.4), then, approximately,

$$T(F) - T(F_0) \cong \varepsilon \int IC(x, F_0, T)H(dx). \qquad (5.7)$$

Hence

$$b_1(\varepsilon) = \sup|T(F) - T(F_0)| \cong \varepsilon\gamma^*. \tag{5.8}$$

However some risky and possibly illegitimate interchanges of suprema and passages to the limit are involved here. We give two examples later (Section 3.5) where

(1) $\gamma^* < \infty$ but $b_1(\varepsilon) = \infty$ for all $\varepsilon > 0$.
(2) $\gamma^* = \infty$ but $\lim b(\varepsilon) = 0$ for $\varepsilon \to 0$.

Second, the influence curve allows an immediate and simple, heuristic assessment of the asymptotic properties of an estimate, since it allows us to guess an explicit formula (5.5) for the asymptotic variance (which then has to be proved rigorously by other means).

There are several finite sample and/or difference quotient versions of (5.1); the most important ones are the *sensitivity curve* (Tukey 1970) and the *jackknife* (Quenouille 1956, Tukey 1958, Miller 1964, 1974). We obtain the sensitivity curve if we replace F by F_{n-1} and s by $1/n$ in (5.1):

$$SC_{n-1}(x) = \frac{T\left(\frac{n-1}{n}F_{n-1} + \frac{1}{n}\delta_x\right) - T(F_{n-1})}{\frac{1}{n}}$$

$$= n\left[T_n(x_1, \ldots, x_{n-1}, x) - T_{n-1}(x_1, \ldots, x_{n-1})\right].$$

The jackknife is defined as follows. Consider an estimate $T_n(x_1, \ldots, x_n)$, which is essentially the "same" across different sample sizes (for instance, assume that it is a functional of the empirical distribution). Then the *ith jackknifed pseudo-value* is, by definition,

$$T_{ni}^* = nT_n - (n-1)T_{n-1}(x_1, \ldots, x_{i-1}, x_{i+1}, \ldots, x_n).$$

For example, if T_n is the sample mean, then $T_{ni}^* = x_i$. This is an approximation to $IC(x_i)$; more precisely, if we substitute F_n for F and $-1/(n-1)$ for s in (5.1), we obtain

$$\frac{T\left(\frac{n}{n-1}F_n - \frac{1}{n-1}\delta_{x_i}\right) - T(F_n)}{-\frac{1}{n-1}}$$

$$= (n-1)\left[T_n - T_{n-1}(x_1, \ldots, x_{i-1}, x_{i+1}, \ldots, x_n)\right]$$

$$= T_{ni}^* - T_n.$$

If T_n is a consistent estimate of θ, whose bias has the asymptotic expansion

$$E(T_n - \theta) = \frac{a_1}{n} + \frac{a_2}{n^2} + O\left(\frac{1}{n^3}\right),$$

then

$$T_n^* = \frac{1}{n} \sum_i T_{ni}^*$$

has a smaller bias:

$$E(T_n^* - \theta) = -\frac{a_2}{n^2} + O\left(\frac{1}{n^3}\right).$$

Example 5.1   If $T_n = 1/n \sum (x_i - \bar{x})^2$, then

$$T_{ni}^* = \frac{n}{n-1}(x_i - \bar{x})^2$$

and

$$T_n^* = \frac{1}{n-1} \sum (x_i - \bar{x})^2.$$

Tukey (1958) pointed out that

$$\frac{1}{n(n-1)} \sum (T_{ni}^* - T_n^*)^2$$

(a finite sample version of 5.5) usually is a good estimator of the variance of T_n. (It can also be used as an estimate of the variance of T_n^*, but it is better matched to T_n.)

> WARNING   In some cases, namely when the influence function $IC(x; F, T)$ does not depend smoothly on F, the jackknife is in trouble and may yield a variance that is worse than useless. This happens, in particular, for estimates that are based on a small number of order statistics, like the median.

1.6   OPTIMAL ROBUSTNESS

In Section 1.4 we introduced some quantitative measures of robustness. They certainly are not the only ones. But, as we defined robustness to mean insensitivity with regard to small deviations from the assumptions, any quantitative measure of robustness must somehow be concerned with the maximum degradation of performance possible for an ε-deviation from the assumptions. An *optimally robust* procedure then minimizes this maximum degradation and hence will be a minimax procedure of some kind. As we have considerable freedom in how we quantize performance and ε-deviations, we also have a host of notions of optimal robustness, of various usefulness, and of various mathematical manageability.

Exact, finite sample minimax results are available for two simple, but important special cases: the first corresponds to a robustification of the Neyman-Pearson lemma, and the second yields interval estimates of location. They are treated in Chapter 10. While the resulting tests and estimates are quite simple, the approach does not generalize well. In particular, it does not seem possible to obtain explicit, finite-sample results when there are nuisance parameters (e.g., when scale is unknown).

If we use asymptotic performance criteria (like asymptotic variances), we obtain *asymptotic minimax estimates*, treated in Chapters 4 to 6. These asymptotic theories work well only if there is a high degree of symmetry (left-right symmetry, translation invariance, etc.), but they are able to cope with nuisance parameters. By a fortunate accident some of the asymptotic minimax estimates, although derived under quite different assumptions, coincide with certain finite-sample minimax estimates; this gives a strong heuristic support for using asymptotic optimality criteria.

Multiparameter regression, and the estimation of *covariance matrices* possess enough symmetries that the above asymptotic optimality results are transferable (Chapters 7 and 8). However the value of this transfer is somewhat questionable because of the fact that in practice the number of observations per parameter tends to be uncomfortably low. Other, design-related dangers, like leverage points, may become more important than distributional robustness itself.

In problems lacking invariance, for instance in the general one-parameter estimation problem, Hampel (1968) has proposed optimizing robustness by minimizing the asymptotic variance at the model, subject to a bound on the gross-error sensitivity γ^* defined by (5.6). This approach is technically the simplest one, but it has some conceptual drawbacks; reassuringly, it again yields the same estimates as those obtained by the exact, finite-sample minimax approach when the latter is applicable. For details, see Section 11.1.

1.7 COMPUTATION OF ROBUST ESTIMATES

In many practical applications of (say) the method of least squares, the actual setting up and solving of the least squares equations occupies only a small fraction of the total length of the computer program. We should therefore strive for robust algorithms that can easily be patched into existing programs, rather than for comprehensive robust packages.

This is in fact possible. Technicalities are discussed in Chapter 7; the salient ideas are as follows.

Assume we are doing a least squares fit on observations y_i, yielding fitted values $\hat{y}_i$, and residuals $r_i = y_i - \hat{y}_i$. Let s_i be some estimate of the standard error of y_i (or, even better, of the standard error of r_i).

We metrically Winsorize the observations y_i and replace them by pseudo-observations y_i^*:

$$y_i^* = y_i, \qquad \text{if } |r_i| \leqslant cs_i,$$

$$= \hat{y}_i - cs_i, \qquad \text{if } r_i < -cs_i,$$

$$= \hat{y}_i + cs_i, \qquad \text{if } r_i > cs_i. \tag{7.1}$$

The constant c regulates the amount of robustness; good choices are in the range between 1 and 2, say $c = 1.5$.

Then use the pseudo-observations y_i^* to calculate new fitted values $\hat{y}_i$ (and new s_i), and iterate to convergence.

If all observations are equally accurate, the classical estimate of the variance of a single observation would be

$$s^2 = \frac{1}{n-p} \sum r_i^2, \tag{7.2}$$

and we can then estimate the standard error of the residual r_i by $s_i = \sqrt{1 - h_i}\, s$, where h_i is the ith diagonal element of $H = X(X^TX)^{-1}X^T$.

If we use modified residuals $r_i^* = y_i^* - \hat{y}_i$ instead of the r_i, we clearly would underestimate scale; we can correct this bias (to a zero order approximation) by putting

$$s^2 = \frac{\dfrac{1}{n-p} \sum r_i^{*2}}{\left(\dfrac{m}{n}\right)^2}, \tag{7.3}$$

where $n-p$ is the number of observations minus the number of parameters, and m is the number of unmodified observations ($y_i^* = y_i$).

It is evident that this procedure deflates the influence of outliers. Moreover there are versions of this procedure that are demonstrably convergent; they converge to a reasonably well-understood M-estimate.

These ideas yield a completely general recipe to robustize almost any procedure: we first "clean" the data by pulling outliers towards their fitted values in the manner of (7.1), refitting iteratively until convergence is obtained. Then we apply the procedure in question to the pseudo-observations y_i^*. Compare Huber (1979) and Kleiner et al. (1979) for specific examples.

CHAPTER 2

The Weak Topology and its Metrization

2.1 GENERAL REMARKS

This chapter attempts to give a more or less self-contained account of the formal mathematics underlying qualitative and quantitative robustness. It can be skipped by a reader who is willing to accept a number of results on faith: the more important ones are quoted and explained in an informal, heuristic fashion at the appropriate places elsewhere in this book.

The principal background references for this chapter are Prohorov (1956) and Billingsley (1968); some details on Polish spaces are most elegantly treated in Neveu (1964).

2.2 THE WEAK TOPOLOGY

Ordinarily, our sample space Ω is a finite dimensional Euclidean space. Somewhat more generally, we assume throughout this chapter that Ω is a Polish space, that is, a topological space whose topology is metrizable by some metric d, such that Ω is complete and separable (i.e., contains a countable dense subset). Let $\mathfrak{M}$ be the space of all probability measures on $(\Omega, \mathfrak{B})$, where $\mathfrak{B}$ is the Borel-σ-algebra (i.e., the smallest σ-algebra containing the open subsets of Ω). By $\mathfrak{M}'$ we denote the set of finite signed measures on $(\Omega, \mathfrak{B})$, that is, the linear space generated by $\mathfrak{M}$. We use capital latin italic letters for the elements of Ω; if $\Omega = \mathbb{R}$ is the real line, we use the same letter F for both the measure and the associated distribution function, with the convention that $F(\cdot)$ denotes the distribution function, $F\{\cdot\}$ the set function: $F(x) = F\{(-\infty, x)\}$.

It is well known that every measure $F \in \mathfrak{M}$ is regular in the sense that any Borel set $B \in \mathfrak{B}$ can be approximated in F-measure by compact sets C

from below and by open sets G from above:

$$\sup_{C \subset B} F\{C\} = F\{B\} = \inf_{G \supset B} F\{G\}. \tag{2.1}$$

Compare, for example, Neveu (1964).

The weak(-star) topology in $\mathfrak{M}$ is the weakest topology such that, for every bounded continuous function ψ, the map

$$F \to \int \psi \, dF \tag{2.2}$$

from $\mathfrak{M}$ into $\mathbb{R}$ is continuous.

Let L be a linear functional on $\mathfrak{M}$ (or, more precisely, the restriction to $\mathfrak{M}$ of a linear functional on $\mathfrak{M}'$).

LEMMA 2.1 A linear functional L is weakly continuous on $\mathfrak{M}$ iff it can be represented in the form

$$L(F) = \int \psi \, dF \tag{2.3}$$

for some bounded continuous function ψ.

Proof Evidently, every functional representable in this way is linear and weakly continuous on $\mathfrak{M}$. Conversely, assume that L is weakly continuous and linear. Put

$$\psi(x) = L(\delta_x)$$

where δ_x denotes the measure putting a pointmass 1 at x. Then, because of linearity, (2.3) holds for all F with finite support. Clearly, whenever x_n is a sequence of points converging to x, then $\delta_{x_n} \to \delta_x$ weakly; hence

$$\psi(x_n) = L(\delta_{x_n}) \to L(\delta_x) = \psi(x),$$

and ψ must be continuous. If ψ should be unbounded, say $\sup \psi(x) = \infty$, then choose a sequence of points such that $\psi(x_n) \geq n^2$, and let (with an arbitrary x_0)

$$F_n = \left(1 - \frac{1}{n}\right)\delta_{x_0} + \frac{1}{n}\delta_{x_n}.$$

Clearly, $F_n \to \delta_{x_0}$ weakly, but $L(F_n) = \psi(x_0) + (1/n)[\psi(x_n) - \psi(x_0)]$ diverges.

This contradicts the assumed continuity of L; hence ψ must be bounded. Furthermore the measures with finite support are dense in $\mathfrak{M}$ (for every $F \in \mathfrak{M}$ and every finite set $\{\psi_1, \ldots, \psi_n\}$ of bounded continuous functions, we can easily find a measure F^* with finite support such that $\int \psi_i \, dF^* - \int \psi_i \, dF$ is arbitrarily small simultaneously for all i); hence the representation (2.3) holds for all $F \in \mathfrak{M}$. ∎

LEMMA 2.2 The following statements are equivalent:

(1) $F_n \to F$ weakly.
(2) $\liminf F_n\{G\} \geqslant F\{G\}$ for all open sets G.
(3) $\limsup F_n\{A\} \leqslant F\{A\}$ for all closed sets A.
(4) $\lim F_n\{B\} = F\{B\}$ for all Borel sets with F-null boundary (i.e., $F\{\mathring{B}\} = F\{B\} = F\{\bar{B}\}$, where $\mathring{B}$ denotes the interior and $\bar{B}$ the closure of B.

Proof We show (1)⟹(2)⇄(3)⟹(4)⟹(1).

Equivalence of (2) and (3) is obvious, and we now show that they imply (4).

If B has F-null boundary, then it follows from (2) and (3) that

$$\liminf F_n\{\mathring{B}\} \geqslant F\{\mathring{B}\} = F\{B\} = F\{\bar{B}\} \geqslant \limsup F_n\{\bar{B}\}.$$

As

$$F_n\{\mathring{B}\} \leqslant F_n\{B\} \leqslant F_n\{\bar{B}\},$$

(4) follows.

We now show (1)⟹(2). Let $\varepsilon > 0$, let G be open, and let $A \subset G$ be a closed set such that $F\{A\} \geqslant F\{G\} - \varepsilon$ (remember that F is regular). By Urysohn's lemma [cf. Kelley (1955)] there is a continuous function ψ satisfying $1_A \leqslant \psi \leqslant 1_G$. Hence (1) implies

$$\liminf F_n\{G\} \geqslant \lim \int \psi \, dF_n = \int \psi \, dF \geqslant F\{A\} \geqslant F\{G\} - \varepsilon.$$

Since ε was arbitrary, (2) follows.

It remains to show (4)⟹(1). It suffices to verify $\int \psi \, dF_n \to \int \psi \, dF$ for positive ψ, say $0 \leqslant \psi \leqslant M$; thus we can write

$$\int \psi \, dF_n = \int_0^M F_n\{\psi > t\} \, dt. \tag{2.4}$$

For almost all t, $\{\psi > t\}$ is an open set with F-null boundary. Hence the integrand in (2.4) converges to $F\{\psi > t\}$ for almost all t, and (1) now follows from the dominated convergence theorem. ∎

COROLLARY 2.3 On the real line, weak convergence $F_n \to F$ holds iff the sequence of distribution functions converges at every continuity point of F.

Proof If F_n converges weakly, then (4) implies at once convergence at the continuity points of F. Conversely, if F_n converges at the continuity points of F, then a straightforward monotonicity argument shows that

$$F(x) = F(x-0) \leqslant \liminf F_n(x) \leqslant \limsup F_n(x+0) \leqslant F(x+0), \quad (2.5)$$

where $F(x+0)$ and $F(x-0)$ denote the left and right limits of F at x, respectively. We now verify (2). Every open set G is a disjoint union of open intervals (a_i, b_i); thus

$$F_n(G) = \sum \left[F_n(b_i) - F_n(a_i + 0) \right].$$

Fatou's lemma now yields, in view of (2.5),

$$\liminf F_n(G) \geqslant \sum \liminf \left[F_n(b_i) - F_n(a_i + 0) \right]$$

$$\geqslant \sum \left[F(b_i) - F(a_i + 0) \right] = F(G). \qquad ∎$$

DEFINITION 2.4 A subset $\mathcal{S} \subset \mathfrak{M}$ is called tight if, for every $\varepsilon > 0$, there is a compact set $K \subset \Omega$ such that, for all $F \in \mathcal{S}$, $F\{K\} \geqslant 1 - \varepsilon$.

In particular, every finite subset is tight [this follows from regularity (2.1)].

LEMMA 2.5 A subset $\mathcal{S} \subset \mathfrak{M}$ is tight iff, for every $\varepsilon > 0, \delta > 0$, there is a finite union

$$B = \bigcup_i B_i$$

of δ-balls, $B_i = \{y \mid d(x_i, y) < \delta\}$, such that, for all $F \in \mathcal{S}$, $F(B) \geqslant 1 - \varepsilon$.

Proof If $\mathcal{S}$ is tight, then the existence of such a finite union of δ-balls follows easily from the fact that every compact set $K \subset \Omega$ can be covered by a finite union of open δ-balls.

Conversely, given $\varepsilon > 0$, choose, for every natural number k, a finite union $B_k = \cup_{i=1}^{n_k} B_{ki}$ of $1/k$-balls B_{ki}, such that, for all $F \in \mathcal{S}$, $F(B_k) \geqslant 1 - \varepsilon 2^{-k}$.

Let $K = \cap B_k$, then evidently $F(K) \geqslant 1 - \Sigma \varepsilon 2^{-k} = 1 - \varepsilon$. We claim that K is compact. As K is closed it suffices to show that every sequence (x_n) with $x_n \in K$ has an accumulation point (for Polish spaces, sequential compactness implies compactness). For each k, $B_{k1}, \ldots, B_{kn_k}$ form a finite cover of K; hence it is possible to inductively choose sets B_{ki_k} such that, for all m, $A_m = \cap_{k \leqslant m} B_{ki_k}$ contains infinitely many members of the sequence (x_n). Thus if we pick a subsequence $x_{n_m} \in A_m$, it will be a Cauchy sequence, $d(x_{n_m}, x_{n_l}) \leqslant 2/\min(m, l)$, and, since Ω is complete, it converges. ■

THEOREM 2.6 (Prohorov)   A set $S \subset \mathfrak{M}$ is tight iff its weak closure is weakly compact.

Proof   In view of Lemma 2.2(3) a set is tight iff its weak closure is, so it suffices to prove the theorem for weakly closed sets $S \subset \mathfrak{M}$.

Let $\mathcal{C}$ be the space of bounded continuous functions on Ω. We rely on Daniell's theorem [see Neveu (1964), Proposition II.7.1], according to which a positive, linear functional L on $\mathcal{C}$, satisfying $L(1) = 1$, is induced by a probability measure F: $L(\psi) = \int \psi \, dF$ for some $F \in \mathfrak{M}$ iff $\psi_n \downarrow 0$ (pointwise) implies $L(\psi_n) \downarrow 0$.

Let $\mathcal{L}$ be the space of positive linear functionals on $\mathcal{C}$, satisfying $L(1) \leqslant 1$, topologized by the topology of pointwise convergence on $\mathcal{C}$. Then $\mathcal{L}$ is compact, and S can be identified with a subspace $S \subset \mathcal{L}$ in a natural way. Evidently, S is compact iff it is closed as a subspace of $\mathcal{L}$.

Now assume that S is tight. Let $L \in \mathcal{L}$ be in the closure of S; we want to show that $L(\psi_n) \downarrow 0$ for every monotone decreasing sequence $\psi_n \downarrow 0$ of bounded continuous functions. Without loss of generality we can assume $0 \leqslant \psi_n \leqslant 1$. Let $\varepsilon > 0$ and let K be such that, for all $F \in S$, $F(K) \geqslant 1 - \varepsilon$. The restriction of ψ_n to the compact set K converges not only pointwise but uniformly, say $\psi_n \leqslant \varepsilon$ on K for $n \geqslant n_0$. Thus for all $F \in S$ and all $n \geqslant n_0$,

$$0 \leqslant \int \psi_n \, dF = \int_K \psi_n \, dF + \int_{K^c} \psi_n \, dF$$

$$\leqslant \int_K \varepsilon \, dF + \int_{K^c} 1 \, dF \leqslant 2\varepsilon.$$

It follows that $0 \leqslant L(\psi_n) \leqslant 2\varepsilon$; hence $\lim L(\psi_n) = 0$, since ε was arbitrary. Thus L is induced by a probability measure; hence it lies in S (which by assumption is a weakly closed subset of $\mathfrak{M}$), and thus S is compact (S being closed in $\mathcal{L}$).

Conversely, assume that S is compact, and let $\psi_n \in \mathcal{C}$ and $\psi_n \downarrow 0$. Then $\int \psi_n \, dF \downarrow 0$ pointwise on the compact set S; thus, also uniformly, $\sup_{F \in S} \int \psi_n \, dF \downarrow 0$. We now choose ψ_n as follows. Let $\delta > 0$ be given. Let (x_n) be a dense sequence in Ω, and by Urysohn's lemma, let φ_i be a continuous function with values between 0 and 1, such that $\varphi_i(x)=0$ for $d(x_i, x) \leqslant \delta/2$ and $\varphi_i(x)=1$ for $d(x_i, x) \geqslant \delta$. Put $\psi_n(x)=\inf\{\varphi_i(x)|i \leqslant n\}$. Then $\psi_n \downarrow 0$ and $\psi_n \geqslant 1_{A_n^c}$, where A_n is the union of the δ-balls around x_i, $i=1,\ldots, n$. Hence $\sup_{F \in \mathcal{S}} F(A_n^c) \leqslant \sup_{F \in \mathcal{S}} \int \psi_n \, dF \downarrow 0$, and the conclusion follows from Lemma 2.5.

2.3   LÉVY AND PROHOROV METRICS

We now show that the space $\mathfrak{M}$ of probability measures on a Polish space Ω, topologized by the weak topology, is itself a Polish space, that is complete separable metrizable.

For the real line $\Omega = \mathbb{R}$, the most manageable metric metrizing $\mathfrak{M}$ is the so-called Lévy distance.

DEFINITION 3.1   The Lévy distance between two distribution functions F and G is

$$d_L(F, G) = \inf\{\varepsilon | \forall x \; F(x-\varepsilon)-\varepsilon \leqslant G(x) \leqslant F(x+\varepsilon)+\varepsilon\}.$$

LEMMA 3.2 d_L is a metric.

Proof   We have to verify (1) $d_L(F,G) \geqslant 0, d_L(F,G)=0$ iff $F=G$; (2) $d_L(F,G)=d_L(G, F)$; (3) $d_L(F, H) \leqslant d_L(F,G)+d_L(G, H)$. All of this is immediate.   ∎

NOTE $\sqrt{2} \; d_L(F, G)$ is the maximum distance between the graphs of F and G, measured along a 45°-direction (see Exhibit 2.3.1).

THEOREM 3.3   The Lévy distance metrizes the weak topology.

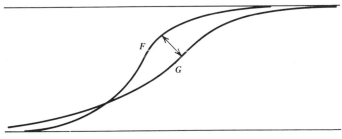

Exhibit 2.3.1

Proof In view of Lemma 2.3 it suffices to show that convergence of $F_n \to F$ at the continuity points of F and $d_L(F, F_n) \to 0$ are equivalent. (1) Assume $d_L(F, F_n) \to 0$. If x is a continuity point of F, then $F(x \pm \varepsilon) \pm \varepsilon \to F(x)$ as $\varepsilon \to 0$; hence F_n converges at x. (2) Assume that $F_n \to F$ at the continuity points of F. Let $x_0 < x_1 < \cdots < \cdots < x_N$ be continuity points of F such that $F(x_0) < \varepsilon/2$, $F(x_N) > 1 - \varepsilon/2$, and that $x_{i+1} - x_i < \varepsilon$. Let n_0 be so large that, for all i and all $n \geqslant n_0$, $|F_n(x_i) - F(x_i)| < \varepsilon/2$. Then for $x_{i-1} \leqslant x \leqslant x_i$,

$$F_n(x) \leqslant F_n(x_i) < F(x_i) + \frac{\varepsilon}{2} \leqslant F(x + \varepsilon) + \varepsilon.$$

This bound obviously also holds for $x < x_0$ and for $x > x_N$. In the same way we establish $F_n(x) \geqslant F(x - \varepsilon) - \varepsilon$. ∎

For general Polish sample spaces Ω, the weak topology in $\mathfrak{M}$ can be metrized by the so-called *Prohorov distance*. Conceptually, this is the most attractive metric; however, it is not very manageable for actual calculations. We need a few preparatory definitions.

For any subset $A \subset \Omega$, we define the *closed δ-neighborhood* of A as

$$A^\delta = \left\{ x \in \Omega \mid \inf_{y \in A} d(x, y) \leqslant \delta \right\}. \tag{3.1}$$

LEMMA 3.4 For any arbitrary set A, we have

$$A^\delta = \overline{A}^\delta = \overline{\overline{A}^\delta} = \overline{A^\delta}$$

(where an overbar denotes closure). In particular, A^δ is closed.

Proof It suffices to show

$$\overline{\overline{A}^\delta} \subset A^\delta.$$

Let

$$\eta > 0 \quad \text{and} \quad x \in \overline{\overline{A}^\delta}.$$

Then we can successively find $y \in \overline{A}^\delta$, $z \in \overline{A}$, and $t \in A$, such that $d(x, y) < \eta$, $d(y, z) < \delta + \eta$, and $d(z, t) < \eta$. Thus $d(x, t) < \delta + 3\eta$, and, since η was arbitrary, $x \in A^\delta$. ∎

Let $G \in \mathfrak{M}$ be a fixed probability measure, and let $\varepsilon, \delta > 0$. Then the set

$$\mathscr{P}_{\varepsilon, \delta} = \left\{ F \in \mathfrak{M} \mid F\{A\} \leqslant G\{A^\delta\} + \varepsilon \text{ for all } A \in \mathscr{B} \right\} \tag{3.2}$$

is called a *Prohorov neighborhood* of G. Often we assume $\varepsilon = \delta$.

DEFINITION 3.5   The Prohorov distance between two members $F, G \in \mathfrak{M}$ is

$$d_p(F, G) = \inf\{\varepsilon > 0 \mid F\{A\} \leqslant G\{A^\varepsilon\} + \varepsilon \text{ for all } A \in \mathfrak{B}\}.$$

We have to show that this is a metric. First, we show that it is symmetric in F and G; this follows immediately from the following lemma.

LEMMA 3.6   If $F\{A\} \leqslant G\{A^\delta\} + \varepsilon$ for all $A \in \mathfrak{B}$, then $G\{A\} \leqslant F\{A^\delta\} + \varepsilon$ for all $A \in \mathfrak{B}$.

Proof   Let $\delta' > \delta$ and insert $A = B^{\delta'c}$ into the premiss (here superscript c denotes complementation). This yields $G\{B^{\delta'c\delta c}\} \leqslant F\{B^{\delta'}\} + \varepsilon$. We now show that $B \subset B^{\delta'c\delta c}$, or, which is the same, $B^{\delta'c\delta} \subset B^c$. Assume $x \in B^{\delta'c\delta}$, then $\exists y \notin B^{\delta'}$ with $d(x, y) \leqslant \delta'$; thus $x \notin B$, because otherwise $d(x, y) > \delta'$. It follows that $G(B) \leqslant F\{B^{\delta'}\} + \varepsilon$. Since $B^\delta = \cap_{\delta' > \delta} B^{\delta'}$, the assertion of the lemma follows.

We now show that $d_p(F, G) = 0$ implies $F = G$. Since $\cap_{\varepsilon > 0} A^\varepsilon = \bar{A}$, it follows from $d_p(F, G) = 0$ that $F\{A\} \leqslant G\{A\}$ and $G\{A\} \leqslant F\{A\}$ for all closed sets A; this implies $F = G$ (remember that all our measures are regular). To prove the triangle inequality, assume $d_p(F, G) \leqslant \varepsilon$ and $d_p(G, H) \leqslant \delta$, then $F\{A\} \leqslant G\{A^\varepsilon\} + \varepsilon \leqslant H\{(A^\varepsilon)^\delta\} + \varepsilon + \delta$. Thus it suffices to verify $(A^\varepsilon)^\delta \subset A^{\varepsilon + \delta}$, which is a simple consequence of the triangle inequality for d. ■

THEOREM 3.7 (Strassen)   The following two statements are equivalent:

(1) $F\{A\} \leqslant G\{A^\delta\} + \varepsilon$ for all $A \in \mathfrak{B}$.
(2)   There are (dependent) random variables X and Y with values in Ω, such that $\mathcal{L}(X) = F$, $\mathcal{L}(Y) = G$, and $P\{d(X, Y) \leqslant \delta\} \geqslant 1 - \varepsilon$.

Proof   As $\{X \in A\} \subset \{Y \in A^\delta\} \cup \{d(X, Y) > \delta\}$, (1) is an immediate consequence of (2). The proof of the converse is contained in a famous paper of Strassen [(1965), p. 436 ff]. ■

NOTE 1   In the above theorem we may put $\delta = 0$. Then, since F and G are regular, (1) is equivalent to the assumption that the difference in total variation between F and G satisfies $d_{TV}(F, G) = \sup_{A \in \mathfrak{B}} |F(A) - G(A)| \leqslant \varepsilon$. In this case Strassen's theorem implies that there are two random variables X and Y with marginal distributions F and G, respectively, such that $P(X \neq Y) \leqslant \varepsilon$. However, the total variation distance does not metrize the weak topology.

NOTE 2   If G is the idealized model and F is the true underlying distribution, such that $d_p(F, G) \leqslant \varepsilon$, then Strassen's theorem shows that we can always assume that there is an ideal (but unobservable) random variable Y with $\mathcal{L}(Y) = G$, and an observable X with $\mathcal{L}(X) = F$, such that $P\{d(X, Y) \leqslant \varepsilon\}$ $\geqslant 1 - \varepsilon$, that is, the Prohorov distance provides both for small errors occurring with large probability, and for large errors occurring with low probability, in a very explicit, quantitative fashion.

THEOREM 3.8   The Prohorov metric metrizes the weak topology in $\mathfrak{M}$.

Proof   Let $P \in \mathfrak{M}$ be fixed. Then a basis for the neighborhood system of P in the weak topology is furnished by the sets of the form

$$\left\{ Q \in \mathfrak{M} \,\middle|\, \left| \int \varphi_i \, dQ - \int \varphi_i \, dP \right| < \varepsilon, i = 1, \ldots, k \right\}, \tag{3.3}$$

where the φ_i are bounded continuous functions. In view of Lemma 2.2 there are three other bases for this neighborhood system, namely: those furnished by the sets

$$\left\{ Q \in \mathfrak{M} \,\middle|\, Q(G_i) > P(G_i) - \varepsilon, i = 1, \ldots, k \right\} \tag{3.4}$$

where the G_i are open; those furnished by the sets

$$\left\{ Q \in \mathfrak{M} \,\middle|\, Q(A_i) < P(A_i) + \varepsilon, i = 1, \ldots, k \right\} \tag{3.5}$$

where the A_i are closed; and those furnished by the sets

$$\left\{ Q \in \mathfrak{M} \,\middle|\, |Q(B_i) - P(B_i)| < \varepsilon, i = 1, \ldots, k \right\} \tag{3.6}$$

where the B_i have P-null boundary.

We first show that each neighborhood of the form (3.5) contains a Prohorov neighborhood. Assume that P, ε, and a closed set A are given. Clearly, we can find a δ, $0 < \delta < \varepsilon$, such that $P(A^\delta) < P(A) + \frac{1}{2} \varepsilon$. If $d_p(P, Q)$ $< \frac{1}{2} \delta$, then

$$Q(A) < P(A^\delta) + \tfrac{1}{2} \delta < P(A) + \varepsilon.$$

It follows that (3.5) contains a Prohorov neighborhood. In order to show the converse, let $\varepsilon > 0$ be given. Choose $\delta < \frac{1}{2} \varepsilon$. In view of Lemma 2.5 there is a finite union of sets A_i with diameter $< \delta$ such that

$$P\left(\bigcup_{i=1}^{k} A_i \right) > 1 - \delta.$$

We can choose the A_i to be disjoint and to have P-null boundaries. If $\mathcal{C}$ is the (finite) class of unions of A_i, then every element of $\mathcal{C}$ has a P-null boundary. By (3.6) there is a weak neighborhood $\mathcal{U}$ of P such that

$$\mathcal{U} = \{ Q \mid |Q(B) - P(B)| < \delta \text{ for } B \in \mathcal{C} \}.$$

We now show that $d_P(P, Q) < \varepsilon$ if $Q \in \mathcal{U}$. Let $B \in \mathcal{B}$ be an arbitrary set, and let A be the union of the sets A_i that intersect B. Then

$$B \subset A \cup \left[\bigcup_1^k A_i \right]^c \quad \text{and} \quad A \subset B^\delta,$$

and hence

$$P(B) < P(A) + \delta < Q(A) + 2\delta < Q(B^\delta) + 2\delta.$$

The assertion follows. ■

THEOREM 3.9 $\mathcal{M}$ is a Polish space.

Proof   It remains to show that $\mathcal{M}$ is separable and complete. We already noted (proof of Lemma 2.1) that the measures with finite support are dense in $\mathcal{M}$. Now let $\Omega_0 \subset \Omega$ be a countable dense subset; then it is easy to see that already the countable set $\mathcal{M}_0$ is dense in $\mathcal{M}$, where $\mathcal{M}_0$ consists of the measures whose finite support is contained in Ω_0 and that have rational masses. This establishes separability. Now let $\{P_n\}$ be a Cauchy sequence in $\mathcal{M}$. Let $\varepsilon > 0$ be given, and chose n_0 such that $d_P(P_n, P_m) \leqslant \varepsilon/2$ for $m, n \geqslant n_0$, that is $P_m(A) \leqslant P_n(A^{\varepsilon/2}) + \varepsilon/2$. The finite sequence $\{P_m\}_{m < n_0}$ is tight, so by Lemma 2.5 there is a finite union B of $\varepsilon/2$-balls such that $P_m(B) \geqslant 1 - \varepsilon/2$ for $m \leqslant n_0$. But then $P_n(B^{\varepsilon/2}) \geqslant P_{n_0}(B) - \varepsilon/2 \geqslant 1 - \varepsilon$, and, since $B^{\varepsilon/2}$ is contained in a finite union of ε-balls (with the same centers as the balls forming B), we conclude from Lemma 2.5 that the sequence $\{P_n\}$ is tight. Hence it has an accumulation point in $\mathcal{M}$ (which by necessity is unique). ■

2.4   THE BOUNDED LIPSCHITZ METRIC

The weak topology can also be metrized by other metrics. An interesting one is the so-called *bounded Lipschitz metric* d_{BL}. Assume that the distance function d in Ω is bounded by 1 {if necessary, replace it by $d(x, y)/[1 + d(x, y)]$}. Then define

$$d_{BL}(F, G) = \sup \left| \int \psi \, dF - \int \psi \, dG \right|, \tag{4.1}$$

where the supremum is taken over all functions ψ satisfying the Lipschitz condition

$$|\psi(x) - \psi(y)| \leqslant d(x, y). \tag{4.2}$$

LEMMA 4.1 d_{BL} is a metric.

Proof The only nontrivial part is to show that $d_{BL}(F, G) = 0$ implies $F = G$. Clearly, it implies $\int \psi \, dF = \int \psi \, dG$ for all functions satisfying the Lipschitz condition $|\psi(x) - \psi(y)| \leqslant cd(x, y)$ for some c. In particular, let $\psi(x) = (1 - cd(x, A))^+$, with $d(x, A) = \inf\{d(x, y) | y \in A\}$; then $|\psi(x) - \psi(y)| < cd(x, y)$ and $1_A \leqslant \psi \leqslant 1_{A^{1/c}}$. Let $c \to \infty$, then it follows that $F(A) = G(A)$ for all closed sets A; hence $F = G$. ∎

Also for this metric an analogue of Strassen's theorem holds [first proved by Kantorovič and Rubinstein (1958) in a special case].

THEOREM 4.2 The following two statements are equivalent:

(1) $d_{BL}(F, G) \leqslant \varepsilon$.
(2) *There are random variables X and Y with $\mathcal{L}(X) = F$, and $\mathcal{L}(Y) = G$, such that*

$$Ed(X, Y) \leqslant \varepsilon.$$

Proof (2)⇒(1) is trivial:

$$\left| \int \psi \, dF - \int \psi \, dG \right| = |E\psi(X) - E\psi(Y)| \leqslant E|\psi(X) - \psi(Y)| \leqslant Ed(X, Y).$$

To prove the reverse implication, we first assume that Ω is a finite set. Then the assertion is, essentially, a particular case of the Kuhn-Tucker (1951) theorem, but a proof from scratch may be more instructive. Assume that the elements of Ω are numbered from 1 to n; then the probability measures F and G are represented by n-tuples $(f_1, \ldots, f_n)$ and $(g_1, \ldots, g_n)$ of real numbers, and we are looking for a probability on $\Omega \times \Omega$, represented by a matrix u_{ij}. Thus we attempt to minimize

$$Ed(X, Y) = \sum d_{ij} u_{ij} \tag{4.3}$$

under the side conditions

$$u_{ij} \geqslant 0,$$

$$\sum_i u_{ij} = g_j, \tag{4.4}$$

$$\sum_j u_{ij} = f_i,$$

where d_{ij} satisfies

$$d_{ij} \geqslant 0, \qquad d_{ii} = 0,$$

$$d_{ij} = d_{ji}, \tag{4.5}$$

$$d_{ik} \leqslant d_{ij} + d_{jk}.$$

There exist matrices u_{ij} satisfying the side conditions, for example $u_{ij} = f_i g_j$, and it follows from a simple compactness argument that there is a solution to our minimum problem. With the aid of Lagrange multipliers λ_i and μ_j, it can be turned into an unconstrained minimum problem: minimize

$$\sum (d_{ij} - \lambda_i - \mu_j) u_{ij} \tag{4.6}$$

on the orthant

$$u_{ij} \geqslant 0.$$

At the minimum (which we know to exist) we must have the following implications:

$$u_{ij} > 0 \quad \Rightarrow \quad d_{ij} = \lambda_i + \mu_j, \tag{4.7}$$

$$u_{ij} = 0 \quad \Rightarrow \quad d_{ij} \geqslant \lambda_i + \mu_j, \tag{4.8}$$

because otherwise (4.6) could be decreased through a suitable small change in some of the u_{ij}. We note that (4.4), (4.7), and (4.8) imply that the minimum value η of (4.3) satisfies

$$\eta = \sum d_{ij} u_{ij} = \sum (\lambda_i + \mu_j) u_{ij} = \sum \lambda_i f_i + \sum \mu_i g_i. \tag{4.9}$$

Assume for the moment that $\mu_i = -\lambda_i$ for all i [this would follow from (4.7) if $u_{ii} > 0$ for all i]. Then (4.7) and (4.8) show that λ satisfies the Lipschitz condition $|\lambda_i - \lambda_j| \leqslant d_{ij}$, and (4.9) now gives $\eta \leqslant \varepsilon$; thus assertion (2) of the theorem holds.

In order to establish $\mu_i = -\lambda_i$, for a fixed i, assume first that both $f_i > 0$ and $g_i > 0$. Then both the ith row and the ith column u_{ij} must contain a strictly positive element. If $u_{ii} > 0$, then (4.7) implies $\lambda_i + \mu_i = d_{ii} = 0$, and we are finished. If $u_{ii} = 0$, then there must be a $u_{ij} > 0$ and a $u_{ki} > 0$. Therefore

$$\lambda_i + \mu_j = d_{ij},$$

$$\lambda_k + \mu_i = d_{ki},$$

and the triangle inequality gives

$$\lambda_k + \mu_j \leqslant d_{kj} \leqslant d_{ki} + d_{ij} = \lambda_k + \mu_i + \lambda_i + \mu_j;$$

hence

$$0 \leqslant \mu_i + \lambda_i \leqslant d_{ii} = 0,$$

and thus $\lambda_i + \mu_i = 0$.

In the case $f_i = g_i = 0$ there is nothing to prove (we may drop the ith point from consideration).

The most troublesome case is when just one of f_i and g_i is 0, say $f_i > 0$ and $g_i = 0$. Then $u_{ki} = 0$ for all k, and $\lambda_k + \mu_i \leqslant d_{ki}$, but μ_i is not uniquely determined in general; in particular, note that its coefficient in (4.9) then is 0. So we increase μ_i until, for the first time, $\lambda_k + \mu_i = d_{ki}$ for some k. If $k = i$, we are finished. If not, there must be some j for which $u_{ij} > 0$ since $f_i > 0$; thus $\lambda_i + \mu_j = d_{ij}$, and we can repeat the argument with the triangle inequality from before.

This proves the theorem for finite sets Ω.

We now show that it holds whenever the support of F and G is finite, say $\{x_1, \ldots, x_n\}$. In order to do this, it suffices to show that any function ψ defined on the set $\{x_1, \ldots, x_n\}$ and satisfying the Lipschitz condition $|\psi(x_i) - \psi(x_j)| \leqslant d(x_i, x_j)$ can be extended to a function satisfying the Lipschitz condition everywhere in Ω.

Let $x_1, x_2, \ldots$ be a dense sequence in Ω, and assume inductively that ψ is defined and satisfies the Lipschitz condition on $\{x_1, \ldots, x_n\}$. Then ψ will satisfy it on $\{x_1, \ldots, x_{n+1}\}$ iff $\psi(x_{n+1})$ can be defined such that

$$\psi(x_{n+1}) \in \left\{ \max_{1 \leqslant i \leqslant n} \left[\psi(x_i) - d(x_i, x_{n+1}) \right], \min_{1 \leqslant i \leqslant n} \left[\psi(x_i) + d(x_i, x_{n+1}) \right] \right\}$$

$$(4.10)$$

It suffices to show that the interval in question is not empty, that is, for all $i, j \leqslant n$,

$$\psi(x_i) - d(x_i, x_{n+1}) \leqslant \psi(x_j) + d(x_j, x_{n+1}),$$

or equivalently,

$$\psi(x_i) - \psi(x_j) \leqslant d(x_i, x_{n+1}) + d(x_j, x_{n+1}),$$

and this is obviously true in view of the triangle inequality.

Thus it is possible to extend the definition of ψ to a dense set, and from there, by uniform continuity, to the whole of Ω.

For general measures F and G, the theorem now follows from a straightforward passage to the limit, as follows.

First, we show that, for every $\delta > 0$ and every F, there is a measure F^* with finite support such that $d_{BL}(F, F^*) < \delta$. In order to see this, find first a compact $K \subset \Omega$ such that $F(K) > 1 - \delta/2$, cover K by a finite number of disjoint sets $U_1, \ldots, U_n$ with diameter $< \delta/2$, put $U_0 = K^c$, and select points $x_i \in U_i$, $i = 0, \ldots, n$. Define F^* with support $\{x_0, \ldots, x_n\}$ by $F^*\{x_i\} = F\{U_i\}$. Then, for any ψ satisfying the Lipschitz condition, we have

$$\left| \int \psi \, dF^* - \int \psi \, dF \right| \leqslant \sum \left[\max_{U_i} \psi(x) - \min_{U_i} \psi(x) \right] F^*\{x_i\}$$

$$\leqslant F^*\{x_0\} + \sum_{i>0} \frac{\delta}{2} F^*\{x_i\} < \delta.$$

Thus we can approximate F and G by F^* and G^*, respectively, such that the starred measures have finite support, and

$$d_{BL}(F^*, G^*) < \varepsilon + 2\delta.$$

Then find a measure P^* on $\Omega \times \Omega$ with marginals F^* and G^* such that

$$\int d(X, Y) \, dP^* < \varepsilon + 2\delta.$$

If we take a sequence of δ values converging to 0, then the corresponding sequence P^* clearly is tight in the space of probability measures on $\Omega \times \Omega$, and the marginals converge weakly to F and G, respectively. Hence there is a weakly convergent subsequence of the P^*, whose limit P then satisfies (2). This terminates the proof of the theorem. ∎

COROLLARY 4.3   For all $F, G \in \mathfrak{M}$, we have

$$d_P(F, G)^2 \leqslant d_{BL}(F, G) \leqslant 2 d_P(F, G).$$

In particular, d_P and d_{BL} define the same topology.

Proof   For any probability measure P on $\Omega \times \Omega$, we have

$$\int d(X, Y) \, dP \leqslant \varepsilon P\{d(X, Y) \leqslant \varepsilon\} + P\{d(X, Y) > \varepsilon\}$$

$$= \varepsilon + (1 - \varepsilon) P\{d(X, Y) > \varepsilon\}.$$

If $d_P(F, G) \leqslant \varepsilon$, we can (by Theorem 3.8) choose P so that this is bounded by $\leqslant \varepsilon + (1 - \varepsilon)\varepsilon \leqslant 2\varepsilon$, which establishes $d_{BL} \leqslant 2d_P$. On the other hand Markov's inequality gives

$$P\{(d(X, Y) > \varepsilon\} \leqslant \frac{\int d(X, Y) \, dP}{\varepsilon} \leqslant \varepsilon$$

if $d_{BL}(F, G) \leqslant \varepsilon^2$; thus $d_P^2 < d_{BL}$. ∎

Some Further Inequalities

The *total variation distance*

$$d_{TV}(F, G) = \sup_{A \in \mathfrak{B}} |F\{A\} - G\{A\}| \tag{4.11}$$

and, on the real line, the *Kolmogorov distance*

$$d_K(F, G) = \sup |F(x) - G(x)| \tag{4.12}$$

do not generate the weak topology, but they possess other convenient properties. In particular, we have the inequalities

$$d_L \leqslant d_P \leqslant d_{TV}, \tag{4.13}$$

$$d_L \leqslant d_K \leqslant d_{TV}. \tag{4.14}$$

Proof (4.13, 4.14) The defining equation for the Prohorov distance

$$d_P(F, G) = \inf\{\varepsilon | \forall A \in \mathfrak{B}, F\{A\} \leqslant G\{A^\varepsilon\} + \varepsilon\} \tag{4.15}$$

is turned into a definition of the Lévy distance if we decrease the range of conditions to sets A of the form $(-\infty, x]$ and $[x, \infty)$. It is turned into a definition of the total variation distance if we replace A^ε by A and thus make the condition harder to fulfill. This again can be converted into a definition of Kolmogorov distance if we restrict the range of A to sets $(-\infty, x]$ and $[x, \infty)$. Finally, if we increase A on the right-hand side of the inequality in (4.15) and replace it by A^ε, we decrease the infimum and obtain the Lévy distance. ∎

2.5 FRÉCHET AND GÂTEAUX DERIVATIVES

Assume that d_* is a metric [or pseudo-metric—we shall not actually need $d_*(F, G) = 0 \Rightarrow F = G$], in the space $\mathfrak{M}$ of probability measures, that:

(1) Is compatible with the weak topology in the sense that $\{F | d_*(G, F) < \varepsilon\}$ is open for all $\varepsilon > 0$.

(2) Is compatible with the affine structure of $\mathfrak{M}$: let $F_t = (1-t)F_0 + tF_1$, then $d_*(F_t, F_s) = O(|t-s|)$.

The "usual" distance functions metrizing the weak topology of course satisfy the first condition; they also satisfy the second, but this has to be checked in each case.

In the case of the Lévy metric we note that

$$|F_t(x) - F_s(x)| = |t-s| \, |F_1(x) - F_0(x)| \leqslant |t-s|;$$

hence $d_K(F_t, F_s) \leqslant |t-s|$ and, *a fortiori*,

$$d_L(F_t, F_s) \leqslant |t-s|.$$

In the case of the Prohorov metric, we have, similarly,

$$|F_t\{A\} - F_s\{A\}| = |t-s| \cdot |F_1\{A\} - F_0\{A\}| \leqslant |t-s|;$$

hence

$$d_P(F_t, F_s) \leqslant |t-s|.$$

In the case of the bounded Lipschitz metric, we have, for any ψ satisfying the Lipschitz condition,

$$\left| \int \psi \, dF_t - \int \psi \, dF_s \right| = |t-s| \cdot \left| \int \psi \, dF_1 - \int \psi \, dF_0 \right|.$$

Let $\bar\psi = \sup \psi(x)$, and $\underline\psi = \inf \psi(x)$, then $\bar\psi - \underline\psi \leqslant \sup d(x, y) \leqslant 1$; thus $\int \psi \, dF_1 - \int \psi \, dF_0 \leqslant \int \bar\psi \, dF_1 - \int \underline\psi \, dF_0 \leqslant 1$. It follows that $d_{BL}(F_t, F_s) \leqslant |t-s|$.

We say that a statistical functional T is *Fréchet differentiable* at F if it can be approximated by a linear functional L (defined on the space of finite signed measures) such that, for all G,

$$|T(G) - T(F) - L(G-F)| = o\left[d_*(F, G) \right]. \tag{5.1}$$

Of course $L = L_F$ depends on the base point F.

It is easy to see that L is (essentially) unique: if L_1 and L_2 are two such linear functionals, then their difference satisfies

$$|(L_1 - L_2)(G-F)| = o\left[d_*(F, G) \right],$$

and, in particular, with $F_t = (1-t)F + tG$, we obtain

$$|(L_1 - L_2)(F_t - F)| = t|(L_1 - L_2)(G - F)|$$

$$= o(d_*(F, F_t)) = o(t);$$

hence $L_1(G-F) = L_2(G-F)$ for all G.

It follows that L is uniquely determined on the space of finite signed measures of total mass 0, and we may arbitrarily standardize it by putting $L(F) = 0$.

If T were defined not just one some convex set, but in a full, open neighborhood of F in some linear space, weak continuity of T at F together with (5.1) would imply that L is continuous in G at $G = F$, and, since L is linear, it then would follow that L is continuous everywhere.

Unfortunately, this is not so, and thus a somewhat more roundabout approach appears necessary.

We note first that, if we define

$$\psi(x) = L(\delta_x - F), \tag{5.2}$$

then, by linearity of L,

$$L(G - F) = \int \psi \, dG \tag{5.3}$$

for all G with finite support.

In particular, with $F_t = (1-t)F + tG$, we then obtain

$$|T(F_t) - T(F) - L(F_t - F)| = |(F_t) - T(F) - tL(G - F)|$$

$$= \left| T(F_t) - T(F) - t \int \psi \, dG \right|$$

$$= o(d_*(F, F_t)) = o(t). \tag{5.4}$$

Assume that T is continuous at F; then $d_*(F, F_t) = O(t)$ implies that $|T(F_t) - T(F)| = o(1)$ uniformly in G. A comparison with the preceding formula (5.4) yields that ψ must be bounded.

We may rewrite (5.4) as

$$\left| \int \psi \, dG - \frac{T(F_t) - T(F)}{t} \right| = o(1),$$

holding uniformly in G. Now, if T is continuous in a neighborhood of F, and if $G_n \to G$ weakly, then

$$F_{n,t} = (1-t)F + tG_n \to F_t \quad \text{weakly.}$$

Since t can be chosen arbitrarily small, we must have $\int \psi \, dG_n \to \int \psi \, dG$. In particular, by letting $G_n = \delta_{x_n}$, with $x_n \to x$, we obtain that ψ must be continuous. If G is an arbitrary probability measure, and the G_n are approximations to G with finite support, then the same argument shows that $\int \psi \, dG_n$ converges simultaneously to $\int \psi \, dG$ (since ψ is bounded and continuous) and to $L(G-F)$; hence $L(G-F) = \int \psi \, dG_n$ holds for all $G \in \mathfrak{M}$. Thus we have proved the following proposition.

PROPOSITION 5.1 If T is weakly continuous in a neighborhood of F and Fréchet differentiable at F, then its Fréchet derivative at F is a weakly continuous linear functional, and it is representable as

$$L(G-F) = \int \psi_F \, dG \tag{5.5}$$

with ψ_F bounded and continuous, and $\int \psi_F \, dF = 0$.

Unfortunately the concept of Fréchet differentiability appears to be too strong: in too many cases, the Fréchet derivative does not exist, and even if it does, the fact is difficult to establish.

About the weakest concept of differentiability is the *Gâteaux derivative* [in the statistical literature it has usually been called the Volterra derivative, but this happens to be a misnomer, cf. Reeds (1976)]. We say that a functional T is Gâteaux differentiable at F if there is a linear functional $L = L_F$ such that, for all $G \in \mathfrak{M}$,

$$\lim_{t \to 0} \frac{T(F_t) - T(F)}{t} = L_F(G-F), \tag{5.6}$$

with

$$F_t = (1-t)F + tG.$$

Clearly, if T is Fréchet differentiable, it is also Gâteaux differentiable, and the two derivatives L_F agree. We usually assume in addition that the Gâteaux derivative L_F is representable by a measurable function ψ_F, conveniently standardized such that $\int \psi_F \, dF = 0$:

$$L_F(G-F) = \int \psi_F \, dG.$$

[Note that there are discontinuous linear functionals that cannot be represented as integrals with respect to measurable function ψ, e.g., $L(F)$ = sum of the jumps $F(x+0) - F(x-0)$ of the distribution function F.]

If we put $G = \delta_x$, (5.6) gives the value of $\psi_F(x)$; following Hampel (1968, 1974b) we write

$$IC(x; F, T) = \lim_{t \to 0} \frac{T(F_t) - T(F)}{t}, \tag{5.7}$$

where $F_t = (1 - t)F + \delta_x$ and IC stands for influence curve.

The Gâteaux derivative is, after all, nothing but the ordinary derivative of the real valued function $T(F_t)$ with respect to the real parameter t. If we integrate the derivative of an absolutely continuous function, we get back the function; in this particular case we obtain the useful identity

$$T(F_1) - T(F_0) = \int_0^1 \int IC(x; F_t, T) \, d(F_1 - F_0) \, dt. \tag{5.8}$$

Proof   We have

$$T(F_1) - T(F_0) = \int_0^1 \frac{d}{dt} T(F_t) \, dt.$$

Now

$$\frac{d}{dt} T(F_t) = \lim_{h \to 0} \frac{T(F_{t+h}) - T(F_t)}{h}$$

and, since

$$F_{t+h} = \left(1 - \frac{h}{1-t}\right) F_t + \frac{h}{1-t} F_1,$$

we obtain, provided the Gâteaux derivative exists at F_t,

$$\frac{d}{dt} T(F_t) = \frac{1}{1-t} \int IC(x; F_t, T) \, d(F_1 - F_t)$$

$$= \int IC(x; F_t, T) \, d(F_1 - F_0). \qquad \blacksquare$$

If the empirical distribution F_n converges to the true one at the rate $n^{-1/2}$,

$$d_*(F, F_n) = O_P(n^{-1/2}), \tag{5.9}$$

and if T has a Fréchet derivative at F, (5.1) and (5.5) allow a one-line asymptotic normality proof

$$\sqrt{n}\left[T(F_n)-T(F)\right]=\sqrt{n}\int\psi_F dF_n+\sqrt{n}\,o\left[d_*(F,F_n)\right]$$

$$=\frac{1}{\sqrt{n}}\sum\psi_F(x_i)+o_P(1);$$

hence the left-hand side is asymptotically normal with mean 0 and variance $\int\psi_F^2 dF$.

For the Lévy distance, (5.9) is true; this follows at once from (4.14) and the well-known properties of the Kolmogorov-Smirnov test statistic. Unfortunately (5.9) is false for both the Prohorov and the bounded Lipschitz distance, as soon as F has sufficiently long tails [rational tails $F\{|X|>t\}\sim t^{-k}$ for some k, suffice for (5.9) to fail].

Proof The idea behind the proof is that, for long-tailed distributions F, the extreme order statistics are widely scattered, so, if we surrounded them by ε-neighborhoods, we catch very little of the mass of F. To be specific assume that $F(x)=|x|^{-k}$ for large negative x, let δ be a small positive number to be fixed later, let $m=n^{1/2+\delta}$, and let $A=\{x_{(1)},\dots,x_{(m)}\}$ be the set of the m leftmost order statistics. Put $\varepsilon=\frac{1}{2}n^{-1/2+\delta}$. We intend to show that, for large n,

$$F_n\{A\}-\varepsilon\geqslant F\{A^\varepsilon\};$$

hence $d_P(F,F_n)\geqslant\varepsilon$. We have $F_n\{A\}=m/n=2\varepsilon$, so it suffices to show $F\{A^\varepsilon\}\leqslant\varepsilon$. We only sketch the calculations.

We have, approximately,

$$F\{A^\varepsilon\}\lesssim\sum_1^m 2\varepsilon f(x_{(i)}),$$

where f is the density of F. Now $x_{(i)}$ can be represented as $F^{-1}(u_{(i)})$, where $u_{(i)}$ is the ith order statistic from the uniform distribution on $(0,1)$, and $f(F^{-1}(t))=kt^{(k+1)/k}$. Thus

$$F\{A^\varepsilon\}\lesssim 2\varepsilon k\sum_1^m(u_{(i)})^{(k+1)/k}.$$

Since $u_{(i)} \approx i/(n+1)$, we can approximate the right-hand side

$$\approx 2\varepsilon k \sum_1^m \left(\frac{i}{n+1}\right)^{(k+1)/k}$$

$$\approx 2\varepsilon k n \int_0^{m/n} t^{(k+1)/k} \, dt$$

$$= \frac{2\varepsilon k n}{2+1/k} \left(\frac{m}{n}\right)^{2+1/k}$$

$$= \frac{2k}{2+1/k} \frac{1}{2} n^{-1/2+\delta} n (n^{-1/2+\delta})^{2+1/k}$$

$$= \frac{k}{2+1/k} n^{-(1/2)-(1/2k)+[3+(1/k)]\delta}.$$

If we choose δ sufficiently small, this is of a smaller order of magnitude than ε. ∎

Compare also Kersting (1978).

On the other hand (5.9) is true for d_P and d_{BL} if F is the uniform distribution on a finite interval, but it fails again for the uniform distribution on the unit cube in three or more dimensions [see Dudley (1969) for details].

It seems we are in trouble here because of a phenomenon that has been colorfully called the "curse of dimensionality"; the higher the dimension, the more empty space there is, and it becomes progressively more difficult to relate the coarse and sparse empirical measure to the true one.

Mere Gâteaux differentiability does not suffice to establish asymptotic normality [unless we also have higher order derivatives, cf. von Mises (1937, 1947), who introduced the concept of differentiable statistical functionals, and Filippova (1962)]. The most promising intermediate approach seems to be the one by Reeds (1976), which is based on the notion of compact differentiability (Averbukh and Smolyanov 1967, 1968).

2.6   HAMPEL'S THEOREM

We recall Hampel's definition of qualitative asymptotic robustness (cf. Section 1.3).

Let the observations x_i be independent, with common distribution F, and let $T_n = T_n(x_1, \ldots, x_n)$ be a sequence of estimates or test statistics with values in $\mathbb{R}^k$. This sequence is called robust at $F = F_0$ if the sequence of maps of distributions

$$F \to \mathcal{L}_F(T_n)$$

is equicontinuous at F_0, that is, if, for every $\varepsilon > 0$, there is a $\delta > 0$ and an n_0 such that, for all F and all $n \geqslant n_0$,

$$d_*(F_0, F) \leqslant \delta \Rightarrow d_*\big(\mathcal{L}_{F_0}(T_n), \mathcal{L}_F(T_n)\big) \leqslant \varepsilon. \tag{6.1}$$

Here, d_* is any metric generating the weak topology; to be specific we work with the Lévy metric for F and the Prohorov metric for $\mathcal{L}(T_n)$.

Assume that $T_n = T(F_n)$ derives from a functional T, which is defined on some weakly open subset of $\mathfrak{M}$.

PROPOSITION 6.1 If T is weakly continuous at F, then $\{T_n\}$ is consistent at F, in the sense that $T_n \to T(F)$ in probability and almost surely.

Proof It follows from the Glivenko-Cantelli theorem and (4.14) that, in probability and almost surely,

$$d_L(F, F_n) \leqslant d_K(F, F_n) \to 0;$$

hence $F_n \to F$ weakly, and thus $T(F_n) \to T(F)$. ∎

The following is a variant of somewhat more general results first proved by Hampel (1971).

THEOREM 6.2 Assume that $\{T_n\}$ is consistent in a neighborhood of F_0. Then T is continuous at F_0 iff $\{T_n\}$ is robust at F_0.

Proof Assume first that T is continuous at F_0. We can write

$$d_P\big(\mathcal{L}_{F_0}(T_n), \mathcal{L}_F(T_n)\big) \leqslant d_P\big(\delta_{T(F_0)}, \mathcal{L}_{F_0}(T_n)\big) + d_P\big(\delta_{T(F_0)}, \mathcal{L}_F(T_n)\big),$$

where $\delta_{T(F_0)}$ denotes the degenerate law concentrated at $T(F_0)$. Thus robustness at F_0 is proved if we can show that, for each $\varepsilon > 0$, there is a $\delta > 0$ and an n_0, such that $d_L(F_0, F) \leqslant \delta$ implies

$$d_P\big(\delta_{T(F_0)}, \mathcal{L}_F(T(F_n))\big) \leqslant \tfrac{1}{2}\varepsilon, \qquad \text{for } n \geqslant n_0.$$

It follows from the easy part of Strassen's theorem (Theorem 3.7) that this last inequality holds if we can show

$$P_F\{d(T(F_0), T(F_n)) \leq \tfrac{1}{2}\varepsilon\} \geq 1 - \tfrac{1}{2}\varepsilon.$$

But, since T is continuous at F_0, there is a $\delta > 0$ such that $d_L(F_0, F) \leq 2\delta$ implies $d(T(F_0), T(F)) \leq \tfrac{1}{2}\varepsilon$, so it suffices to show

$$P_F\{d_L(F_0, F_n) \leq 2\delta\} \geq 1 - \tfrac{1}{2}\varepsilon.$$

We note that Glivenko-Cantelli convergence is uniform in F: for each $\delta > 0$ and $\varepsilon > 0$, there is an n_0 such that, for all F and all $n \geq n_0$,

$$P_F\{d_L(F, F_n) \leq \delta\} \geq 1 - \tfrac{1}{2}\varepsilon.$$

But, since $d_*(F_0, F_n) \leq d_*(F_0, F) + d_*(F, F_n)$, we have established robustness at F_0.

Conversely, assume that $\{T_n\}$ is robust at F_0. We note that, for degenerate laws δ_x, which put all mass on a single point x, Prohorov distance degenerates to the ordinary distance: $d_P(\delta_x, \delta_y) = d(x, y)$.

Since T_n is consistent for each F in some neighborhood of F_0, we have $d_P(\delta_{T(F)}, \mathcal{L}_F(T_n)) \to 0$. Hence (6.1) implies, in particular,

$$d_L(F_0, F) \leq \delta \Rightarrow d_P\big(\delta_{T(F_0)}, \delta_{T(F)}\big) = d(T(F_0), T(F)) \leq \varepsilon.$$

It follows that T is continuous at F_0. ∎

CHAPTER 3

The Basic Types of Estimates

3.1 GENERAL REMARKS

This chapter introduces three basic types of estimates (M, L, and R) and discusses their qualitative and quantitative robustness properties. They correspond, respectively, to maximum likelihood type estimates, linear combinations of order statistics, and estimates derived from rank tests.

For reasons discussed in more detail near the end of Section 3.5, the emphasis is on the first type, the M-estimates: they are the most flexible ones, and they generalize straightforwardly to multiparameter problems, even though (or, perhaps, because) they are not automatically scale invariant and have to be supplemented for practical applications by an auxiliary estimate of scale (see Chapters 6 ff).

3.2 MAXIMUM LIKELIHOOD TYPE ESTIMATES (M-ESTIMATES)

Any estimate T_n, defined by a minimum problem of the form

$$\sum \rho(x_i; T_n) = \min!, \tag{2.1}$$

or by an implicit equation

$$\sum \psi(x_i; T_n) = 0, \tag{2.2}$$

where ρ is an arbitrary function, $\psi(x; \theta) = (\partial/\partial\theta)\rho(x; \theta)$, is called an M-estimate [or maximum likelihood type estimate; note that the choice $\rho(x; \theta) = -\log f(x; \theta)$ gives the ordinary ML estimate].

We are particularly interested in location estimates

$$\sum \rho(x_i - T_n) = \min! \tag{2.3}$$

43

or

$$\sum \psi(x_i - T_n) = 0. \tag{2.4}$$

This last equation can be written equivalently as

$$\sum w_i \cdot (x_i - T_n) = 0 \tag{2.5}$$

with

$$w_i = \frac{\psi(x_i - T_n)}{x_i - T_n}; \tag{2.6}$$

this gives a formal representation of T_n as a weighted mean

$$T_n = \frac{\sum w_i x_i}{\sum w_i} \tag{2.7}$$

with weights depending on the sample.

Remark   The functional version of (2.1) may cause trouble: we cannot in general define $T(F)$ to be a value of t that minimizes

$$\int \rho(x; t) F(dx). \tag{2.8}$$

Note, for instance, that the median corresponds to $\rho(x; t) = |x - t|$, but that

$$\int |x - t| F(dx) \equiv \infty \tag{2.9}$$

identically in t unless F has a finite first absolute moment. There is a simple remedy: replace $\rho(x; t)$ by $\rho(x; t) - \rho(x; t_0)$ for some fixed t_0, that is, in the case of the median, minimize

$$\int (|x - t| - |x|) F(dx) \tag{2.10}$$

instead of (2.9).

The functional derived from (2.2), defining $T(F)$ by

$$\int \psi(x; T(F)) F(dx) = 0, \tag{2.11}$$

does not suffer from this difficulty, but it may have more solutions [corresponding to local minima of (2.8)].

Influence Function of M-Estimates

To calculate the influence function of an M-estimate, we insert $F_t = (1-t)F + tG$ for F into (2.11) and take the derivative with respect to t at $t=0$. In detail, if we put for short

$$\dot{T} = \lim_{t \to 0} \frac{T(F_t) - T(F)}{t},$$

we obtain, by differentiation of the defining equation (2.11),

$$\dot{T} \int \frac{\partial}{\partial \theta} \psi(x; T(F)) F(dx) + \int \psi(x; T(F))(dG - dF) = 0. \qquad (2.12)$$

For the moment we do not worry about regularity conditions. We recall from (2.5.7) that, for $G = \delta_x$, $\dot{T}$ gives the value of the influence function at x, so, by solving (2.12) for $\dot{T}$, we obtain

$$IC(x; F, T) = \frac{\psi(x; T(F))}{-\int (\partial/\partial\theta)\psi(x; T(F))F(dx)}. \qquad (2.13)$$

In other words the influence function of an M-estimate is proportional to ψ.

In the special case of a location problem, $\psi(x; \theta) = \psi(x - \theta)$, we obtain

$$IC(x; F, T) = \frac{\psi[x - T(F)]}{\int \psi'[x - T(F)] F(dx)}. \qquad (2.14)$$

We conclude from this in a heuristic way that $\sqrt{n}\,[T_n - T(F)]$ is asymptotically normal with mean 0 and variance

$$A(F, T) = \int IC(x; F, T)^2 F(dx). \qquad (2.15)$$

However, this must be checked by a rigorous proof.

Asymptotic Properties of M-Estimates

A fairly simple and straightforward theory is possible if $\psi(x; \theta)$ is monotone in θ; more general cases are treated in Chapter 6.

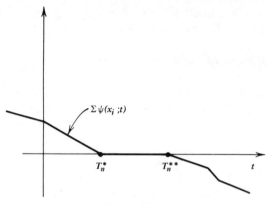

Exhibit 3.2.1

Assume that $\psi(x; \theta)$ is measurable in x and decreasing (i.e., nonincreasing) in θ, from strictly positive to strictly negative values. Put

$$T_n^* = \sup\left\{ t \mid \sum_1^n \psi(x_i; t) > 0 \right\},$$

$$T_n^{**} = \inf\left\{ t \mid \sum_1^n \psi(x_i; t) < 0 \right\}. \qquad (2.16)$$

Clearly, $-\infty < T_n^* \leqslant T_n^{**} < \infty$, and any value T_n satisfying $T_n^* \leqslant T_n \leqslant T_n^{**}$ can serve as our estimate. Exhibit 3.2.1 may help with the interpretation of T_n^* and T_n^{**}.

Note that

$$\{T_n^* < t\} \subset \left\{ \sum \psi(x_i; t) \leqslant 0 \right\} \subset \{T_n^* \leqslant t\},$$

$$\{T_n^{**} < t\} \subset \left\{ \sum \psi(x_i; t) < 0 \right\} \subset \{T_n^{**} \leqslant t\}. \qquad (2.17)$$

Hence

$$P\{T_n^* < t\} = P\left\{ \sum \psi(x_i; t) \leqslant 0 \right\},$$

$$P\{T_n^{**} < t\} = P\left\{ \sum \psi(x_i; t) < 0 \right\}, \qquad (2.18)$$

at the continuity points t of the left-hand side.

The distribution of the customary midpoint estimate $\frac{1}{2}(T_n^* + T_n^{**})$ is somewhat difficult to work out, but the randomized estimate T_n, which selects one of T_n^* or T_n^{**} at random with equal probability, has an explicitly expressible distribution function

$$P\{T_n < t\} = \tfrac{1}{2} P\left\{ \sum \psi(x_i; t) \leqslant 0 \right\} + \tfrac{1}{2} P\left\{ \sum \psi(x_i; t) < 0 \right\}. \qquad (2.19)$$

It follows that the *exact distributions* of T_n^*, T_n^{**}, and T_n can be calculated from the convolution powers of $\mathcal{L}(\psi(x; t))$.

Asymptotic approximations can be found by expanding $G_n = \mathcal{L}(\sum_1^n \psi(x_i; t))$ into an asymptotic series.

We may take the traditional Edgeworth expansion

$$G_n(x) \sim \Phi(x) + \varphi(x) \left[\frac{1}{\sqrt{n}} R_3(x) + \frac{1}{n} R_4(x) + \cdots \right]. \qquad (2.20)$$

However, this gives a somewhat poor approximation in the tails, that is, precisely in the region where we are most interested. Therefore it is preferable to use so-called saddlepoint techniques and recenter the distributions at the point of interest. Thus if we have independent random variables Y_i with density $f(x)$ and would like to determine the distribution G_n of $Y_1 + \cdots + Y_n$ at the point t, we replace the original density f by a conjugate density f_t:

$$f_t(z) = c_t e^{a_t z} f(t + z), \qquad (2.21)$$

where c_t and a_t are chosen such that this is a probability density with expectation 0. See Daniels (1954).

Later Hampel (1973b) noticed that the principal error term of the saddlepoint method seems to reside in the normalizing constant (standardizing the total mass of G_n to 1), so it would be advantageous not to expand G_n or its density g_n, but $g_n'/g_{n'}$ and then to determine the normalizing constant by numerical integration. Thus his procedure can be summarized as follows. Determine the second and the normalized third moment of f_t:

$$\sigma_t^2 = \int z^2 f_t(z)\, dz,$$

$$\lambda_{3,t} = \frac{\int z^3 f_t(z)\, dz}{\sigma_t^3}. \qquad (2.22)$$

Then we get from the first two terms of the Edgeworth expansion, calculated at $x = 0$, that

$$\frac{g_n'}{g_n} \sim -na_t - \frac{\lambda_{3,t}}{2\sigma_t}. \qquad (2.23)$$

From this we obtain g_n and G_n by two numerical integrations and one exponentiation; the integration constant must be determined such that G_n has total mass 1. It turns out that the first integration can be done explicitly:

$$\log g_n(t) \sim -nc_t - \log \sigma_t + \text{const}. \qquad (2.24)$$

This method appears to give fantastically accurate approximations down to very small sample sizes ($n = 3$ or 4). See Field and Hampel (1980) for details.

We now turn to the *limiting distribution* of T_n. Put

$$\lambda(t) = \lambda(t, F) = E_F \psi(X, t). \qquad (2.25)$$

If λ exists and is finite for at least one value of t, then it exists and is monotone (although not necessarily finite) for all t. This follows at once from the remark that $\psi(X; t) - \psi(X; s)$ is positive for $t \leqslant s$ and hence has a well defined expectation (possibly $+\infty$).

PROPOSITION 2.1   Assume that there is a t_0 such that $\lambda(t) > 0$ for $t < t_0$ and $\lambda(t) < 0$ for $t > t_0$. Then both T_n^* and T_n^{**} converge in probability and almost surely to t_0.

Proof   This follows easily from (2.18) and the weak (strong) law of large numbers applied to $(1/n)\sum \psi(x_i; t_0 \pm \varepsilon)$.   ■

COROLLARY 2.2   If $T(F)$ is uniquely defined, then T_n is consistent at F: $T_n \to T(F)$ in probability and almost surely.

Note that $\lambda(s; F) = \lambda(t; F)$ implies $\psi(x; s) = \psi(x; t)$ a.e. $[F]$, so for many purposes $\lambda(t)$ furnishes a more convenient parameterization than t itself. If λ is continuous, then Proposition 2.1 can be restated by saying that $\lambda(T_n)$ is a consistent estimate of 0; this holds also if λ vanishes on a nondegenerate interval. Also other aspects of the asymptotic behavior of T_n are best studied through that of $\lambda(T_n)$. Since λ is monotone decreasing, we have, in particular,

$$\{-\lambda(T_n) < -\lambda(t)\} \subset \{T_n < t\} \subset \{T_n \leqslant t\} \subset \{-\lambda(T_n) \leqslant -\lambda(t)\}. \qquad (2.26)$$

We now plan to show that $\sqrt{n}\ \lambda(T_n)$ is asymptotically normal under the following.

ASSUMPTIONS

(A-1) $\psi(x, t)$ is measurable in x and monotone decreasing in t.

(A-2)   There is at least one t_0 for which $\lambda(t_0) = 0$.

Let Γ_0 be the set of t-values for which $\lambda(t) = 0$.

(A-3) λ is continuous in a neighborhood of Γ_0.

(A-4) $\sigma(t)^2 = E_F[\psi(X; t)^2] - \lambda(t, F)^2$ is finite, nonzero, and continuous in a neighborhood of Γ_0. Put $\sigma_0 = \sigma(t_0)$.

Asymptotically, all T_n, $T_n^* \leqslant T_n \leqslant T_n^{**}$, show the same behavior; formally, we work with T_n^*.

Let y be an arbitrary real number. With the aid of (A-3) define a sequence t_n, for sufficiently large n, such that $y = -\sqrt{n}\ \lambda(t_n)$. Put

$$Y_{ni} = \frac{\psi(x_i; t_n) - \lambda(t_n)}{\sigma(t_n)}. \tag{2.27}$$

The Y_{ni}, $1 \leqslant i \leqslant n$, are independent, identically distributed random variables with expectation 0 and variance 1. We have, in view of (2.18) and (2.26),

$$P\{-\sqrt{n}\ \lambda(T_n^*) < y\} = P\{T_n^* < t_n\}$$

$$= P\left\{\frac{1}{\sqrt{n}}\ \sum Y_{ni} \leqslant \frac{y}{\sigma(t_n)}\right\} \tag{2.28}$$

if $y/\sqrt{n}$ is a continuity point of the distribution of $\lambda(T_n^*)$, that is, for almost all y.

LEMMA 2.3   When $n \to \infty$, then

$$P\left\{\frac{1}{\sqrt{n}}\ \sum Y_{ni} < z\right\} \to \Phi(z)$$

uniformly in z.

Proof   We have to verify Lindeberg's condition, which in our case reads: for every $\varepsilon > 0$,

$$E\{Y_{ni}^2; |Y_{ni}| > \sqrt{n}\ \varepsilon\} \to 0$$

as $n \to \infty$. Since λ and σ are continuous, this is equivalent to: for every $\varepsilon > 0$, as $n \to \infty$,

$$E\{\psi(x; t_n)^2; |\psi(x; t_n)| > \sqrt{n}\ \varepsilon\} \to 0.$$

Thus it suffices to show that the family of random variables $(\psi(x; t_n))_{n > n_0}$ is uniformly integrable [cf. Neveu (1964), p. 48]. But, since ψ is monotone,

$$\psi(X; s)^2 \leqslant \psi(X; s_0)^2 + \psi(X; s_1)^2$$

for $s_0 \leqslant s \leqslant s_1$; hence the family in view of (A-4) is majorized by an integrable random variable, and thus is uniformly integrable. ∎

In view of (2.28) we thus have the following theorem.

THEOREM 2.4   Under assumptions (A-1) to (A-4)

$$P\left\{-\sqrt{n}\ \lambda(T_n) < y\right\} - \Phi\left(\frac{y}{\sigma_0}\right) \to 0 \tag{2.29}$$

uniformly in y. In other words $\sqrt{n}\ \lambda(T_n)$ is asymptotically normal $N(0, \sigma_0^2)$.

Proof   It only remains to show that the convergence is uniform. This is clearly true for any bounded y-interval $[-y_0, y_0]$, so, if $\varepsilon > 0$ is given and if we choose y_0 so large that $\Phi(-y_0/\sigma_0) < \varepsilon/2$ and n_0 so large that (2.29) is $< \varepsilon/2$ for all $n \geqslant n_0$ and all $y \in [-y_0, y_0]$, it follows that (2.29) must be $< \varepsilon$ for all y. ∎

COROLLARY 2.5   If λ has a derivative $\lambda'(t_0) < 0$, then $\sqrt{n}\ (T_n - t_0)$ is asymptotically normal with mean 0 and variance $\sigma_0^2/(\lambda'(t_0))^2$.

Proof   In this case

$$t_n = t_0 - \frac{y}{\sqrt{n}\ \lambda'(t_0)} + o\left(\frac{1}{\sqrt{n}}\right),$$

so the corollary follows from a comparison between (2.28) and (2.29). ∎

If we compare this with the heuristically derived expression (2.15), we notice that the latter is correct *only if we can interchange the order of integration and differentiation in the denominator of* (2.13), that is, if

$$\lambda'(t; F) = \frac{\partial}{\partial t} \int \psi(x; t) F(dx) = \int \frac{\partial}{\partial t} \psi(x; t) F(dx)$$

at $t = T(F)$.

To illustrate some of the issues, we take the location case, $\psi(x; t) = \psi(x - t)$. If F has a smooth density, we can write

$$\lambda(t; F) = \int \psi(x - t) f(x) \, dx = \int \psi(x) f(x + t) \, dx;$$

thus

$$\lambda'(t; F) = \int \psi(x) f'(x + t) \, dx$$

may be well behaved even if ψ is not differentiable.

If $F = (1 - \varepsilon) G + \varepsilon \delta_{x_0}$ is a mixture of a smooth distribution and a point-mass, we have

$$\lambda(t; F) = (1 - \varepsilon) \int \psi(x - t) g(x) \, dx + \varepsilon \psi(x_0 - t)$$

and

$$\lambda'(t, F) = (1 - \varepsilon) \int \psi(x) g'(x + t) \, dx - \varepsilon \psi'(x_0 - t).$$

Hence if ψ' is discontinuous and happens to have a jump at the point $x_0 - T(F)$, then the left-hand side and the right-hand side derivatives of λ at $t = T(F)$ exist but are different. As a consequence $\sqrt{n}\,[T_n - T(F)]$ has a nonnormal limiting distribution: it is pieced together from the left half and the right half of two normal distributions with different standard deviations.

Hitherto we have been concerned with a fixed underlying distribution F. From the point of view of robustness, such a result is of limited use; we would really like to have the convergence in Theorem 2.4 be uniform with respect to F in some neighborhood of the model distribution F_0. For this we need more stringent regularity conditions.

For instance, let us assume that $\psi(x; t)$ is bounded and continuous as a function of x and that the map $t \to \psi(\cdot; t)$ is continuous for the topology of uniform convergence. Then $\lambda(t; F)$ and $\sigma(t; F)$ depend continuously on both t and F. With the aid of the Berry-Esseen theorem, it is then possible to put a bound on (2.29) that is uniform in F [cf. Feller (1966), p. 515 ff].

Of course, this does not yet suffice to make the asymptotic variance of $\sqrt{n}\,[T_n - T(F)]$,

$$A(F, T) = \frac{\sigma(T(F); F)^2}{(\lambda'(T(F); F))^2}, \tag{2.30}$$

continuous as a function of F.

Quantitative and Qualitative Robustness of M-Estimates

We now calculate the maximum bias b_1 (see Section 1.4) for M-estimates. Specifically, we consider the location case, $\psi(x; t) = \psi(x - t)$, with a monotone increasing ψ, and for $\mathcal{P}_\varepsilon$ we take a Lévy neighborhood (the results for Prohorov neighborhoods happen to be the same). For simplicity we assume that the target value is $T(F_0) = 0$.

Put

$$b_+(\varepsilon) = \sup\{T(F)|d_L(F_0, F) \leqslant \varepsilon\} \qquad (2.31)$$

and

$$b_-(\varepsilon) = \inf\{T(F)|d_L(F_0, F) \leqslant \varepsilon\}; \qquad (2.32)$$

then

$$b_1(\varepsilon) = \max\{b_+(\varepsilon), -b_-(\varepsilon)\}. \qquad (2.33)$$

In view of Theorems 1.4.1 and 1.4.2, we have $b_1(\varepsilon) = b(\varepsilon)$ at the continuity points of b_1.

As before we let

$$\lambda(t; F) = \int \psi(x - t) F(dx).$$

We note that λ is decreasing in t, and that it increases if F is made stochastically larger [see, e.g., Lehmann (1959), p. 74, Lemma 2(i)]. The solution $t = T(F)$ of $\lambda(t; F) = 0$ is not necessarily unique; we have $T^*(F) \leqslant T(F) \leqslant T^{**}(F)$ with

$$T^*(F) = \sup\{t|\lambda(t; F) > 0\},$$

$$T^{**}(F) = \inf\{t|\lambda(t; F) < 0\}, \qquad (2.34)$$

and we are concerned with the worst possible choice of $T(F)$ when we determine b_+ and b_-.

The stochastically largest member of the set $d_L(F_0, F) \leqslant \varepsilon$ is the (improper) distribution F_1 (it puts mass ε at $+\infty$):

$$F_1(x) = (F_0(x - \varepsilon) - \varepsilon)^+; \qquad (2.35)$$

that is,

$$F_1(x) = 0, \qquad\qquad \text{for } x \leqslant x_0 + \varepsilon$$

$$= F_0(x - \varepsilon) - \varepsilon, \qquad \text{for } x > x_0 + \varepsilon,$$

with x_0 satisfying

$$F_0(x_0) = \varepsilon.$$

We gloss over some (inessential) complications that arise in the discontinuous case, when ε does not belong to the set of values of F_0.

Thus

$$\lambda(t; F) \leqslant \lambda(t; F_1) = \int_{x_0}^{\infty} \psi(x - t + \varepsilon) F_0(dx) + \varepsilon \psi(\infty), \qquad (2.36)$$

and

$$b_+(\varepsilon) = \inf\{t \mid \lambda(t; F_1) < 0\}. \qquad (2.37)$$

The other quantity $b_-(\varepsilon)$ is calculated analogously; in the important special case where F_0 is symmetric and ψ is an odd function, we have, of course,

$$b_1(\varepsilon) = b_+(\varepsilon) = -b_-(\varepsilon).$$

We conclude that $b_+(\varepsilon) < b_+(1) = \infty$, provided $\psi(+\infty) < \infty$ and

$$\lim_{t \to \infty} \lambda(t; F_1) = (1 - \varepsilon)\psi(-\infty) + \varepsilon\psi(+\infty) < 0. \qquad (2.38)$$

Thus in order to avoid breakdown on the right-hand side, we should have $\varepsilon/(1 - \varepsilon) < -\psi(-\infty)/\psi(+\infty)$. If we also take the left-hand side into account, we obtain that the breakdown point is

$$\varepsilon^* = \frac{\eta}{1 + \eta} \qquad (2.39)$$

with

$$\eta = \min\left\{ -\frac{\psi(-\infty)}{\psi(+\infty)}, -\frac{\psi(+\infty)}{\psi(-\infty)} \right\}, \qquad (2.40)$$

and that it reaches its best possible value $\varepsilon^* = \frac{1}{2}$ if $\psi(-\infty) = -\psi(+\infty)$. If ψ is unbounded, we have $\varepsilon^* = 0$.

The continuity properties of T are also easy to establish. Put

$$\|\psi\| = \psi(+\infty) - \psi(-\infty); \tag{2.41}$$

then (2.36) implies

$$\lambda(t+\varepsilon; F_0) - \|\psi\|\varepsilon \leqslant \lambda(t; F) \leqslant \lambda(t-\varepsilon; F_0) + \|\psi\|\varepsilon.$$

Hence if ψ is bounded and $\lambda(t; F_0)$ has a unique zero at $t = T(F_0)$, then $T(F) \to T(F_0)$ as $\varepsilon \to 0$, and T thus is continuous at F_0. On the other hand if ψ is unbounded, or if the zero of $\lambda(t; F_0)$ is not unique, then T cannot be continuous at F_0, as we can easily verify.

We summarize these results in a theorem.

THEOREM 2.6 Let ψ be a monotone increasing, but not necessarily continuous, function that takes values of both signs. Then the M-estimator T of location, defined by $\int \psi(x - T(F)) F(dx) = 0$, is weakly continuous at F_0 iff ψ is bounded and $T(F_0)$ is unique. The breakdown point ε^* is given by (2.39) and (2.40) and reaches its maximal value $\varepsilon^* = \frac{1}{2}$ whenever $\psi(-\infty) = -\psi(+\infty)$.

Example 2.1 The median, corresponding to $\psi(x) = \text{sign}(x)$, is a continuous functional at every F_0 whose median is uniquely defined.

Example 2.2 If ψ is bounded and strictly monotone, then the corresponding M-estimate is everywhere continuous.

If ψ is not monotone, the situation is much more complicated. To be specific take

$$\psi(x) = \sin(x), \qquad \text{for } -\pi \leqslant x \leqslant \pi$$

$$= 0, \qquad\qquad \text{elsewhere}$$

(an estimate proposed by D. F. Andrews). Then $\Sigma \psi(x_i - T_n)$ has many distinct zeros in general, and even vanishes identically for large absolute values of T_n. Two possibilities for narrowing down the choice of solutions are:

(1) Take the absolute minimum of $\Sigma \rho(x_i - T_n)$, with

$$\rho(x) = 1 - \cos(x), \qquad \text{for } -\pi \leqslant x \leqslant \pi$$

$$= 2, \qquad\qquad \text{elsewhere.}$$

(2) Take the solution nearest to the sample median.

For computational reasons we prefer (2) or a variant thereof; start an iterative root-finding procedure at the sample median and accept whatever root it converges to.

In case (2), the procedure inherits the high breakdown point $\varepsilon^* = \frac{1}{2}$ from the median.

Consistency and asymptotic normality of M-estimates are treated again in Sections 6.2 and 6.3.

3.3   LINEAR COMBINATIONS OF ORDER STATISTICS (L-ESTIMATES)

Consider a statistic that is a linear combination of order statistics, or more generally, of some function h of them:

$$T_n = \sum_{i=1}^{n} a_{ni} h(x_{(i)}). \qquad (3.1)$$

We assume that the weights are generated by a (signed) measure M on $(0, 1)$:

$$a_{ni} = \frac{1}{2} M\left\{\left(\frac{i-1}{n}, \frac{i}{n}\right)\right\} + \frac{1}{2} M\left\{\left[\frac{i-1}{n}, \frac{i}{n}\right]\right\}. \qquad (3.2)$$

(This choice preserves the total mass, $\sum_i a_{ni} = M\{(0, 1)\}$, and symmetry of the coefficients, if M is symmetric about $t = \frac{1}{2}$.)

Then $T_n = T(F_n)$ derives from the functional

$$T(F) = \int h(F^{-1}(s)) M(ds). \qquad (3.3)$$

We have exact equality $T_n = T(F_n)$ if we regularize the integrand at its discontinuity points and replace it by

$$\tfrac{1}{2} h(F_n^{-1}(s-0)) + \tfrac{1}{2} h(F_n^{-1}(s+0)), \qquad (3.4)$$

but only asymptotic equivalence if we do not care. Here, the inverse of any distribution function F is defined in the usual way as

$$F^{-1}(s) = \inf\{x \mid F(x) \geqslant s\} \qquad 0 < s < 1. \qquad (3.5)$$

Influence Function of *L*-Estimates

It is now a matter of plain calculus to find the influence function $IC(x; F, T)$ of T: insert $F_t = (1 - t)F + tG$ into (3.3), and take the derivative with respect to t at $t = 0$, for $G = \delta_x$.

We begin with the derivative of $T_s = F_t^{-1}(s)$, that is, of the s-quantile. If we differentiate the identity

$$F_t(F_t^{-1}(s)) = s \tag{3.6}$$

with respect to t at $t = 0$, we obtain

$$G(F^{-1}(s)) - F(F^{-1}(s)) + f(F^{-1}(s))\dot{T}_s = 0, \tag{3.7}$$

or

$$\dot{T}_s = \frac{s - G(F^{-1}(s))}{f(F^{-1}(s))}. \tag{3.8}$$

If $G = \delta_x$ is the pointmass 1 at x, this gives the value of the influence function of T_s:

$$IC(x; F, T_s) = \frac{s - 1}{f(F^{-1}(s))}, \qquad \text{for} \quad x < F^{-1}(s)$$

$$= \frac{s}{f(F^{-1}(s))}, \qquad \text{for} \quad x > F^{-1}(s). \tag{3.9}$$

Quite clearly, these calculations make sense only if F has a nonzero finite derivative f at $F^{-1}(s)$, but then they are legitimate.

By the chain rule for differentiation, the influence function of $h(T_s)$ is

$$IC(x; F, h(T_s)) = IC(x; F, T_s)h'(T_s), \tag{3.10}$$

and that of T itself then is

$$IC(x; F, T) = \int IC(x; F, h(T_s))M(ds)$$

$$= \int \frac{sh'(F^{-1}(s))}{f(F^{-1}(s))}M(ds) - \int_{F(x)}^1 \frac{h'(F^{-1}(s))}{f(F^{-1}(s))}M(ds).$$

$$\tag{3.11}$$

Of course, the legitimacy of taking the derivative under the integral sign in (3.3) must be checked in each particular case.

If M has a density m, it may be more convenient to write (3.11) as

$$IC(x; F, T) = \int_{-\infty}^{x} h'(y) m(F(y)) \, dy - \int_{-\infty}^{\infty} (1 - F(y)) h'(y) m(F(y)) \, dy.$$

(3.12)

This can be easily remembered through its derivative:

$$\frac{d}{dx} IC(x; F, T) = h'(x) m(F(x)).$$
(3.13)

The last two formulas also hold if F does not have a density. This can easily be seen by starting from an alternative version of (3.3):

$$T(F) = \int h(F^{-1}(s)) m(s) \, ds = \int h(y) m(F(y)) F(dy)$$

$$= -\int h'(y) M(F(y)) \, dy.$$
(3.14)

If we now insert F_s and differentiate, we obtain (3.12). Of course, here also the legitimacy of the integration by parts and of the differentiation under the integral sign must be checked but, for the "usual" h and m, this does not present a problem.

Example 3.1 For the median ($s = \frac{1}{2}$) we have

$$IC(x; F, T_{1/2}) = \frac{-1}{2f(F^{-1}(\frac{1}{2}))} \qquad \text{for } x < F^{-1}(\tfrac{1}{2})$$

$$= \frac{1}{2f(F^{-1}(\frac{1}{2}))} \qquad \text{for } x > F^{-1}(\tfrac{1}{2}).$$
(3.15)

Example 3.2 If $T(F) = \Sigma \beta_i F^{-1}(s_i)$, then $IC(x; F, T)$ has jumps of size $\beta_i / f(F^{-1}(s_i))$ at the points $x = F^{-1}(s_i)$.

Example 3.3 The α-trimmed mean corresponds to $h(x) = x$ and

$$m(s) = \frac{1}{1 - 2\alpha}, \qquad \text{for } \alpha < s < 1 - \alpha$$

$$= 0, \qquad\qquad \text{otherwise;}$$
(3.16)

thus

$$T(F) = \frac{1}{1-2\alpha} \int_{\alpha}^{1-\alpha} F^{-1}(s)\,ds. \qquad (3.17)$$

Note that the α-trimmed mean $T(F_n)$, as defined by (3.17), has the following property: if αn is an integer, then αn observations are removed from each end of the sample and the mean of the rest is taken. If it is not an integer, say $\alpha n = \lfloor \alpha n \rfloor + p$, then $\lfloor \alpha n \rfloor$ observations are removed from each end, and the next observations $x_{(s[\alpha n]+1)}$ and $x_{(n-[\alpha n])}$ are given the reduced weight $1-p$.

The influence function of the α-trimmed mean is, according to (3.12),

$$IC(x) = \frac{1}{1-2\alpha}[F^{-1}(\alpha) - W(F)], \qquad \text{for } x < F^{-1}(\alpha)$$

$$= \frac{1}{1-2\alpha}[x - W(F)], \qquad \text{for } F^{-1}(\alpha) \leqslant x \leqslant F^{-1}(1-\alpha)$$

$$= \frac{1}{1-2\alpha}[F^{-1}(1-\alpha) - W(F)], \qquad \text{for } x > F^{-1}(1-\alpha). \qquad (3.18)$$

Here W is the functional corresponding to the so-called α-Winsorized mean:

$$W(F) = \int_{\alpha}^{1-\alpha} F^{-1}(s)\,ds + \alpha F^{-1}(\alpha) + \alpha F^{-1}(1-\alpha)$$

$$= (1-2\alpha)T(F) + \alpha F^{-1}(\alpha) + \alpha F^{-1}(1-\alpha). \qquad (3.19)$$

Clearly, there will be trouble if the corner points $F^{-1}(\alpha)$ and $F^{-1}(1-\alpha)$ are not uniquely determined (i.e. if F^{-1} has jumps there).

Example 3.4 The α-Winsorized mean (3.19) has the influence curve

$$IC(x; F, W)$$

$$= F^{-1}(\alpha) - \frac{\alpha}{f(F^{-1}(\alpha))} - C(F), \qquad \text{for } x < F^{-1}(\alpha)$$

$$= x - C(F), \qquad \text{for } F^{-1}(\alpha) < x < F^{-1}(1-\alpha)$$

$$= F^{-1}(1-\alpha) + \frac{\alpha}{f(F^{-1}(1-\alpha))} - C(F), \qquad \text{for } x > F^{-1}(1-\alpha),$$

$$(3.20)$$

with

$$C(F) = W(F) - \frac{\alpha^2}{f(F^{-1}(f(F^{-1}(\alpha)))} - \frac{\alpha^2}{f(F^{-1}(1-\alpha))}. \quad (3.21)$$

Thus the influence curve has jumps at $F^{-1}(\alpha)$ and $F^{-1}(1-\alpha)$.

The α-Winsorized mean corresponds to: replace the values of the αn leftmost observations by that of $x_{(\alpha n+1)}$, the values of the αn rightmost observations by that of $x_{(n-\alpha n)}$, and take the mean of this modified sample. The heuristic idea behind this proposal is that we did not want to "throw away" the αn leftmost and rightmost observations as in the trimmed mean, but wanted only to reduce their influences to those of a more moderate order statistic. This exemplifies how unreliable our intuition can be; we know now from looking at the influence functions that the trimmed mean *does not* throw away all of the information sitting in the discarded observations, but that it *does* exactly what the Winsorized mean was *supposed to* do!

Quantitative and Qualitative Robustness of L-Estimates

We now calculate the maximum bias b_1 (see Section 1.4) for L-estimates. To fix the idea assume that $h(x) = x$ and that M is a positive measure with total mass 1. Clearly, the resulting functional then corresponds to a location estimate; if F_{aX+b} denotes the distribution of the random variable $aX+b$, we have

$$T(F_{aX+b}) = aT(F_X) + b, \quad \text{for} \quad a \geq 0. \quad (3.22)$$

It is rather evident that T cannot be continuous if the support of M (i.e., the smallest closed set with total mass 1) contains 0 or 1. Let α be the largest real number such that $[\alpha, 1-\alpha]$ contains the support of M; then, also evidently, the breakdown point satisfies $\varepsilon^* \leq \alpha$. We now show that $\varepsilon^* = \alpha$.

Assume that the target value is $T(F_0) = 0$, let $0 < \varepsilon < \alpha$, and define b_+, b_- as in (2.31) and (2.32). Then with F_1 as in (2.35), we have

$$b_+(\varepsilon) = \int F_1^{-1}(s) M(ds) = \varepsilon + \int_\alpha^{1-\alpha} F_0^{-1}(s+\varepsilon) M(ds),$$

and symmetrically,

$$b_-(\varepsilon) = -\varepsilon + \int_\alpha^{1-\alpha} F_0^{-1}(s-\varepsilon) M(ds),$$

and $b_1(\varepsilon)$ is again given by (2.33).

As $F_0^{-1}(s+\varepsilon)-F_0^{-1}(s-\varepsilon)\downarrow 0$ for $\varepsilon\downarrow 0$, except at the discontinuity points of F_0^{-1}, we conclude that $b_1(\varepsilon)\leqslant b_+(\varepsilon)-b_-(\varepsilon)\downarrow 0$ iff the distribution function of M and F_0^{-1} do not have common discontinuity points, and then T is continuous at F_0. Since $b_1(\varepsilon)$ is finite for $\varepsilon<\alpha$, we must have $\varepsilon^*\geqslant\alpha$.

In particular, the α-trimmed mean with $0<\alpha<\frac{1}{2}$ is everywhere continuous. The α-Winsorized mean is continuous at F_0 if $F_0^{-1}(\alpha)$ and $F_0^{-1}(1-\alpha)$ are uniquely determined (i.e. if F_0^{-1} does not have jumps there).

The generalization to signed measures is immediate, as far as sufficiency is concerned: if $M=M^+-M^-$, then continuity of $T^+(F)=\int F^{-1}(s)M^+(ds)$ and $T^-(F)=\int F^{-1}(s)M^-(ds)$ implies continuity of $T(F)=\int F^{-1}(s)M(ds)$; if both T^+ and T^- have breakdown points $\geqslant\alpha$, then T also has.

The necessity part is trickier, but the arguments given above carry through if there are neighborhoods of the endpoints α and $1-\alpha$ of the support, respectively, where the measure M is of one sign only. We conjecture that $\varepsilon^*=\alpha$ holds generally, but it has not even been proved that $\alpha=0$ implies discontinuity of T in the signed case.

We summarize the results in a theorem.

THEOREM 3.1   Let $M=M^+-M^-$ be a finite signed measure on $(0,1)$ and let $T(F)=\int F^{-1}(s)M(ds)$. Let α be the largest real number such that $[\alpha,1-\alpha]$ contains the support of M^+ and M^-. If $\alpha>0$, then T is weakly continuous at F_0, provided M does not put any pointmass on a discontinuity point of F_0^{-1}. The breakdown point satisfies $\varepsilon^*\geqslant\alpha$. If M is positive, we have $\varepsilon^*=\alpha$, and $\alpha=0$ implies that T is discontinuous.

Since weak continuity of T at F implies consistency, $T(F_n)\rightarrow T(F)$, the above theorem also gives a simple sufficient condition for consistency. Of course, it does not cover the case $\alpha=0$.

The asymptotic properties of L-estimates are in fact rather tricky to establish. In the case $\alpha=0$ (which is only of limited interest to us, because of its lack of robustness) some awkward smoothness conditions on the tails of F and M seem to be needed [cf. Chernoff et al. (1967)]. Even if $\alpha>0$, there is no blanket theorem covering all the more interesting cases simultaneously. But if $\sqrt{n}\,(T(F_n)-T(F))$ is asymptotically normal, then $\int IC(x;F,T)^2F(dx)$ always seems to give the correct asymptotic variance. For our purposes the most useful version is the following.

THEOREM 3.2   Let M be an absolutely continuous signed measure with density m, whose support is contained in $[\alpha,1-\alpha]$, $\alpha>0$. Let $T(F)=\int F^{-1}(s)m(s)\,ds$. The $\sqrt{n}\,(T(F_n)-T(F))$ is asymptotically normal with mean 0 and variance $\int IC(x; F, T)^2 F(dx)$, provided both (1) and (2) hold:

(1) m is of bounded total variation (so all its discontinuities are jumps).

(2) No discontinuity of m coincides with a discontinuity of F^{-1}.

Proof See, for instance, Huber (1969). Condition (2) is necessary; without it not even the influence function would be well defined [see the remark at the end of Example 3.3, and Stigler (1969)]. ∎

3.4 ESTIMATES DERIVED FROM RANK TESTS (R-ESTIMATES)

Consider a two-sample rank test for shift: let $x_1, \ldots, x_m$ and $y_1, \ldots, y_n$ be two independent samples from the distributions $F(x)$ and $G(x) = F(x - \Delta)$, respectively. Merge the two samples into one of size $m + n$ and let R_i be the rank of x_i in the combined sample. Let $a_i = a(i)$, $1 \leqslant i \leqslant m + n$, be some given scores; then base a test of $\Delta = 0$ against $\Delta > 0$ on the test statistic

$$S_{m,n} = \frac{1}{m} \sum_{i=1}^{m} a(R_i). \qquad (4.1)$$

Usually, we assume that the scores a_i are generated by some function J as follows:

$$a_i = J\left(\frac{i}{m + n + 1}\right). \qquad (4.2)$$

There are several other possibilities for deriving scores a_i from J, for example,

$$a_i = J\left(\frac{i - \frac{1}{2}}{m + n}\right), \qquad (4.3)$$

or

$$a_i = (m + n) \int_{(i-1)/(m+n)}^{i/(m+n)} J(s)\, ds, \qquad (4.4)$$

and in fact we prefer to work with this last version. Of course, for "nice" J and F, all these scores lead to asymptotically equivalent tests. In the case of the Wilcoxon test, $J(t) = t - \frac{1}{2}$, the above three variants even create exactly the same tests.

To simplify the presentation, from now on we assume that $m=n$. In terms of functionals (4.1) can then be written as

$$S(F,G) = \int J\left[\tfrac{1}{2}F(x) + \tfrac{1}{2}G(x)\right] F(dx),\tag{4.5}$$

or, if we substitute $F(x)=s$,

$$S(F,G) = \int J\left[\tfrac{1}{2}s + \tfrac{1}{2}G(F^{-1}(s))\right] ds.\tag{4.6}$$

If F is continuous and strictly monotone, the two formulas (4.5) and (4.6) are equivalent. For discontinuous distributions, for instance if we insert the empirical distributions F_n and G_n corresponding to the x- and y-samples, the exact equivalence is destroyed. Moreover, (4.5) is no longer well defined (its value depends on the arbitrary convention about the value of $H = \tfrac{1}{2}F + \tfrac{1}{2}G$ at its jump points).

If we standardize $H(x) = \tfrac{1}{2}H(x-0) + \tfrac{1}{2}H(x+0)$, then (4.5) combined with the scores (4.3) gives (4.1). In any case (4.6) with (4.4) gives (4.1); we assume that there are no ties between x- and y-values. To fix the ideas from now on we work with (4.6) and (4.4). We also assume once and for all that

$$\int J(s)\,ds = 0,\tag{4.7}$$

corresponding to

$$\sum a_i = 0.\tag{4.8}$$

Then the expected value of (4.1) under the null hypothesis is 0.

We can derive estimates of shift Δ_n and of location T_n from such rank tests:

(1) In the two sample case, adjust Δ_n such that $S_{n,n} \approx 0$ when computed from $(x_1, \ldots, x_n)$ and $(y_1 - \Delta_n, \ldots, y_n - \Delta_n)$.

(2) In the one-sample case, adjust T_n such that $S_{n,n} \approx 0$ when computed from $(x_1, \ldots, x_n)$ and $(2T_n - x_1, \ldots, 2T_n - x_n)$. In this case a mirror image of the first sample serves as a stand-in for the missing second sample.

In other words we shift the second sample until the test is least able to detect a difference in location. Note that it may not be possible to achieve an exact zero, $S_{n,n}$ being a discontinuous function.

Example 4.1　The Wilcoxon test, $J(t)=t-\frac{1}{2}$, leads to the Hodges-Lehmann estimates $\Delta_n=\mathrm{med}\{y_i-x_j\}$ and $T_n=\mathrm{med}\{\frac{1}{2}(x_i+x_j)\}$. Note that our recipe in the second case leads to the median of the set of *all* n^2 pairs; the more customary versions use only the pairs $i<j$ or $i\leqslant j$, but asymptotically all three versions are equivalent.

Thus our location estimate T_n derives from a functional $T(F)$, defined by the implicit equation

$$\int J\{\tfrac{1}{2}[s+1-F(2T(F)-F^{-1}(s))]\}\,ds=0. \tag{4.9}$$

Influence Function of R-Estimates

We now derive the influence function of $T(F)$. To shorten the notation we introduce the distribution function of the pooled population:

$$K(x)=\tfrac{1}{2}[F(x)+1-F(2T(F)-x)]. \tag{4.10}$$

Assume that F has a strictly positive density f.

We insert $F_t=(1-t)F+tG$ for F in (4.9) and take the derivative $\partial/\partial t$ (denoted by a dot $\dot{\ }$) at $t=0$. This gives

$$\int J'(K(F^{-1}(s)))\left[\dot{F}(2T-F^{-1}(s))+\frac{f(2T-F^{-1}(s))}{f(F^{-1}(s))}\dot{F}(F^{-1}(s))\right.$$

$$\left.+2f(2T-F^{-1}(s))\dot{T}\right]ds=0. \tag{4.11}$$

We separate this expression in a sum of three integrals and substitute $x=2T-F^{-1}(s)$ in the first [thus $s=F(2T-x)$], but $x=F^{-1}(s)$ in the second and third integrals. This gives

$$\dot{T}\int J'(K(x))f(2T-x)f(x)\,dx$$

$$+\int\tfrac{1}{2}[J'(K(x))+J'(1-K(x))]f(2T-x)\dot{F}(x)\,dx=0. \tag{4.12}$$

Let us now assume that the scores-generating function is symmetric in the sense that

$$J(1-t) = -J(t), \qquad 0 < t < 1 \tag{4.13}$$

(asymmetric functions do not make much sense in the one-sample problem); then we can simplify (4.12) by introducing the function $U(x)$, being an indefinite integral of

$$U'(x) = J'\{\tfrac{1}{2}[F(x) + 1 - F(2T(F) - x)]\} f(2T(F) - x). \tag{4.14}$$

Then (4.12) turns into

$$\dot{T} \int U'(x) f(x)\, dx + \int U'(x) \dot{F}(x)\, dx = 0. \tag{4.15}$$

Integration by parts of the second integral yields

$$\int U'(x) \dot{F}(x)\, dx = - \int U(x) \dot{F}(dx).$$

As $\dot{F} = G - F$ any additive constant in U cancels out on the right-hand side. With $G = \delta_x$ we now obtain the influence function from (4.15) by solving for $\dot{T}$:

$$IC(x; F, T) = \frac{U(x) - \int U(x) f(x)\, dx}{\int U'(x) f(x)\, dx}. \tag{4.16}$$

For *symmetric* F this can be simplified considerably, since then $U(x) = J(F(x))$:

$$IC(x; F, T) = \frac{J(F(x))}{\int J'(F(x)) f(x)^2\, dx}. \tag{4.17}$$

Example 4.2 The influence function of the Hodges-Lehmann estimate $(J(t) = t - \tfrac{1}{2})$ is

$$IC(x; F, T) = \frac{\tfrac{1}{2} - F(2T(F) - x)}{\int f(2T(F) - x) f(x)\, dx}, \tag{4.18}$$

with $T(F)$ defined by

$$\int F(2T(F) - x)F(dx) = \tfrac{1}{2}. \tag{4.19}$$

For symmetric F this simplifies to

$$IC(x; F, T) = \frac{F(x) - \frac{1}{2}}{\int f(x)^2 \, dx}, \tag{4.20}$$

and the asymptotic variance of $\sqrt{n}\,[T(F_n) - T(F)]$ is indeed known to be

$$A(F, T) = \int IC^2 \, dF = \frac{1}{12\left[\int f(x)^2 \, dx\right]^2}. \tag{4.21}$$

[Formula (4.18) suggests that the Hodges-Lehmann estimate will be quite poor for certain asymmetric densities, since the denominator of the influence function might become very small.]

Example 4.3 The *normal scores estimate* is defined by $J(t) = \Phi^{-1}(t)$. For symmetric F its influence function is

$$IC(x; F, T) = \frac{\Phi^{-1}(F(x))}{\int \dfrac{f(x)^2}{\varphi\{\Phi^{-1}[F(x)]\}}\, dx}, \tag{4.22}$$

where $\varphi = \Phi'$ is the standard normal density. In particular, for $F = \Phi$, we obtain

$$IC(x; \Phi, T) = x. \tag{4.23}$$

Quantitative and Qualitative Robustness of R-Estimates

We now calculate the maximum bias (see Section 1.4) for R-estimates. We assume that the scores function J is monotone increasing and symmetric, $J(1 - t) = -J(t)$. In order that (4.6) be well defined, we must require

$$\int |J(s)| \, ds < \infty. \tag{4.24}$$

The function

$$\lambda(t; F) = \int J\{\tfrac{1}{2}[s + 1 - F(2t - F^{-1}(s))]\} \, ds \qquad (4.25)$$

then is monotone decreasing in t, and it increases if F is made stochastically larger. Thus among all F satisfying $d_L(F_0, F) \leqslant \varepsilon$ [or also $d_P(F_0, F) \leqslant \varepsilon$], $\lambda(t, F)$ is largest at the (improper) distribution F_1 of (2.35). Thus we have to calculate $\lambda(t; F_1)$.

We note first that

$$F_1^{-1}(s) = F_0^{-1}(s + \varepsilon) + \varepsilon, \qquad \text{for } 0 \leqslant s \leqslant 1 - \varepsilon$$

$$= \infty, \qquad \text{for } s > 1 - \varepsilon.$$

Thus provided the two side conditions

$$0 \leqslant s \leqslant 1 - \varepsilon$$

and

$$2t - F_1^{-1}(s) \geqslant x_0 + \varepsilon,$$

where

$$F_0(x_0) = \varepsilon,$$

are satisfied, we have

$$F_1[2t - F_1^{-1}(s)] = F_0[2t - 2\varepsilon - F_0^{-1}(s + \varepsilon)] - \varepsilon.$$

The second side condition can be written as

$$s \leqslant F_0(2t - 2\varepsilon - x_0) - \varepsilon.$$

Putting things together we obtain

$$\lambda(t; F_1) = \int_0^{s_0} J\left(\tfrac{1}{2}[s + \varepsilon + 1 - F_0(2(t - \varepsilon) - F_0^{-1}(s + \varepsilon))]\right) ds$$

$$+ \int_{s_0}^1 J\left[\tfrac{1}{2}(s + 1)\right] ds, \qquad (4.26)$$

with

$$s_0 = \left[F_0(2(t - \varepsilon) - x_0) - \varepsilon\right]^+.$$

We then have

$$b_+(\varepsilon) = \inf\{t \mid \lambda(t; F_1) < 0\},$$

and symmetrically, we also calculate $b_-(\varepsilon)$; if F_0 is symmetric, we have of course

$$b_1(\varepsilon) = b_+(\varepsilon) = -b_-(\varepsilon).$$

With regard to breakdown we note that $b_+(\varepsilon) < \infty$ iff

$$\lim_{t \to \infty} \lambda(t; F_1) < 0.$$

Since

$$\lim_{t \to \infty} \lambda(t; F_1) = \int_0^{1-\varepsilon} J\left[\tfrac{1}{2}(s+\varepsilon)\right] ds + \int_\varepsilon^1 J\left[\tfrac{1}{2}(s+1)\right] ds$$

$$= 2\int_{\varepsilon/2}^{1/2} J(s)\, ds + \int_{1-\varepsilon/2}^1 J(s)\, ds$$

(using symmetry of J):

$$= 2\left[\int_{1-\varepsilon/2}^1 J(s)\, ds - \int_{1/2}^{1-\varepsilon/2} J(s)\, ds\right],$$

the breakdown point ε^* is that value ε for which

$$\int_{1/2}^{1-\varepsilon/2} J(s)\, ds = \int_{1-\varepsilon/2}^1 J(s)\, ds. \qquad (4.27)$$

Example 4.4 For the Hodges-Lehmann estimates, $J(t) = t - \tfrac{1}{2}$, we obtain as breakdown point

$$\varepsilon^* = 1 - \frac{1}{\sqrt{2}} \approx 0.293.$$

Example 4.5 For the normal scores estimate, $J(t) = \Phi^{-1}(t)$, we obtain as breakdown point

$$\varepsilon^* = 2\Phi(-\sqrt{\ln 4}\,) \approx 0.239.$$

When $\varepsilon \downarrow 0$ the integrand in (4.26) decreases and converges to the integrand corresponding to F_0 for almost all s and t. It follows from the monotone convergence theorem that $\lambda(t; F_1) \downarrow \lambda(t; F_0)$ at the continuity points of $\lambda(\cdot; F_0)$. Hence if $\lambda(t; F_0)$ has a unique zero, that is, if $T(F_0)$ is uniquely defined, T is continuous at F_0. If $T(F_0)$ is not unique, then T of course cannot be continuous at F_0. A sufficient condition for uniqueness is, for instance, that the derivative of $\lambda(t; F_0)$ with regard to t exists and is not equal to 0 at $T = T(F_0)$; this derivative occurred already (with the opposite sign) as the denominator of (4.16) and (4.17).

We summarize the results in a theorem.

THEOREM 4.1 Assume that the scores generating function J is monotone increasing, integrable, and symmetric: $J(1-t) = -J(t)$. If the R-estimate $T(F_0)$ is uniquely defined by (4.9), then T is weakly continuous at F_0. The breakdown point of T is given by (4.27).

3.5 ASYMPTOTICALLY EFFICIENT M-, L-, AND R-ESTIMATES

The main purpose of this section is to develop some heuristic guidelines for the selection of the functions ψ, m, and J characterizing M-, L-, and R-estimates, respectively. The arguments, as they stand, are rigorous for Fréchet differentiable functionals only.

Let $(F_\theta)_{\theta \in \Theta}$ be a parametric family of distributions, and let the functional T be a Fisher consistent estimate of θ, that is,

$$T(F_\theta) = \theta, \qquad \text{for all } \theta. \tag{5.1}$$

Assume that T is Fréchet differentiable at F. We intend to show that the corresponding estimate is asymptotically efficient at F_θ iff its influence function satisfies

$$IC(x; F_\theta, T) = \frac{1}{I(F_\theta)} \frac{\partial}{\partial \theta} (\log f_\theta). \tag{5.2}$$

Here, f_θ is the density of F_θ, and

$$I(F_\theta) = \int \left(\frac{\partial}{\partial \theta} \log f_\theta \right)^2 dF_\theta \tag{5.3}$$

is the Fisher information.

Assume that $d_L(F_\theta, F_{\theta+\delta}) = O(\delta)$, that

$$\frac{f_{\theta+\delta} - f_\theta}{\delta f_\theta} \to \frac{\partial}{\partial \theta} \log f_\theta \qquad (5.4)$$

converges in the $L_2(F_\theta)$-sense, and that

$$0 < I(F_\theta) < \infty. \qquad (5.5)$$

Then by definition of the Fréchet derivative,

$$T(F_{\theta+\delta}) - T(F_\theta) - \int IC(x; F_\theta, T)(f_{\theta+\delta} - f_\theta)\, dx = o(d_L(F_\theta, F_{\theta+\delta})) = o(\delta). \qquad (5.6)$$

We divide this by δ and let $\delta \to 0$. In view of (5.1) and (5.4) we obtain

$$\int IC(x; F_\theta, T) \frac{\partial}{\partial \theta} (\log f_\theta) f_\theta\, dx = 1. \qquad (5.7)$$

The Schwarz inequality applied to (5.7) gives: first, that the asymptotic variance $A(F_\theta, T)$ of $\sqrt{n}\, [T(F_n) - T(F_\theta)]$ satisfies

$$A(F_\theta, T) = \int IC(x; F_\theta, T)^2\, dF_\theta \geqslant \frac{1}{I(F_\theta)}; \qquad (5.8)$$

and second, that we can have equality in (5.8) (i.e., asymptotic efficiency) only if $IC(x; F_\theta, T)$ is proportional to $(\partial/\partial\theta)\log f_\theta$. The factor of proportionality is easy to determine, and this gives the result announced in (5.2).

Remark   It is possible to establish a variant of (5.2), not even assuming Gâteaux differentiability of T. Assume (5.4), and that the sequence T_n is efficient at F_θ, or, more precisely, that the limit of an expression similar to (1.4.9) satisfies

$$\lim_{\varepsilon \to 0} \lim_{n} \sup_{|\delta| < \varepsilon} Q_t(F_{\theta+\delta}, T_n)^2 \leqslant \frac{1}{I(F_\theta)}. \qquad (5.9)$$

Then it follows that $\sqrt{n}\,(T_n - \theta)$ is asymptotically normal with mean 0 and variance $1/I(F_\theta)$, and that, in fact, we must have asymptotic equivalence

$$\sqrt{n}\,(T_n - \theta) \sim \frac{1}{I(F_\theta)} \sum \frac{\partial}{\partial \theta} \log f_\theta(x_i). \qquad (5.10)$$

This is, for all practical purposes, the same as (5.2). For details see Hájek (1972), and earlier work by LeCam (1953) and Huber (1966).

Let us now check whether it is possible to achieve (5.2) with M-, L-, and R-estimates, at least in the case of a location parameter, $f_\theta(x) = f_0(x - \theta)$.

(1) For M-estimates it suffices to choose

$$\psi(x) = -c\frac{f_0'(x)}{f_0(x)}, \qquad c \neq 0; \tag{5.11}$$

compare (2.14).

(2) For L-estimates we must take $h(x) = x$ (otherwise we do not have translation invariance and thus lose consistency). Then the proper choice, suggested by (3.13), is

$$m(F_0(x)) = \frac{-1}{I(F_0)}(\log f_0(x))'', \tag{5.12}$$

and it is easy to check that $\int m(s)\,ds = 1$ (translation invariance). If f_0 is not twice differentiable, we have to replace (5.12) by a somewhat more complicated integrated version for M itself.

(3) For R-estimates we assume that F_0 is symmetric. Then (4.17) suggests the choice

$$J(F_0(x)) = -c\frac{f_0'(x)}{f_0(x)}, \qquad c \neq 0, \tag{5.13}$$

and this indeed gives (5.2). For asymmetric F_0 we cannot achieve full efficiency with R-estimates.

Of course, we must check in each individual case whether these estimates are indeed efficient (the stringent regularity conditions—Fréchet differentiability—that we used to derive efficiency will rarely be satisfied).

Example 5.1 *Normal Distribution* $f_0(x) = (1/\sqrt{2\pi})e^{-x^2/2}$

M:	$\psi(x) = x$	sample mean, nonrobust,
L:	$m(t) = 1$	sample mean, nonrobust,
R:	$J(t) = \Phi^{-1}(t)$	normal scores estimate, robust.

Example 5.2   *Logistic Distribution* $F_0(x)=1/(1+e^{-x})$

M:	$\psi(x)=\tanh(x/2)$	robust,
L:	$m(t)=6t(1-t)$	nonrobust,
R:	$J(t)=t-\frac{1}{2}$	Hodges-Lehmann, robust.

Example 5.3   *Cauchy Distribution* $f_0(x)=1/[\pi(1+x^2)]$

M:	$\psi(x)=2x/(1+x^2)$	robust,
L:	$m(t)=2\cos(2\pi t)[\cos(2\pi t)-1]$	nonrobust,
R:	$J(t)=-\sin(2\pi t)$	robust(?).

Example 5.4   *"Least Informative" Distribution*   (see Example 4.5.2)

$$f_0(x)=Ce^{-x^2/2}, \qquad |x|\leq c,$$
$$=Ce^{-c|x|+c^2/2}, \qquad |x|>c.$$

M:	$\psi(x)=\max[-c,\min(c,x)],$	Huber-estimate, robust.
L:	$m(t)=\dfrac{1}{1-2\alpha},$	for $\alpha<t<1-\alpha$, $\alpha=F_0(-c)$,
	$=0,$	otherwise,
		α-trimmed mean, robust,
R:	the corresponding estimate has occasionally been mentioned in the literature, but does not have a simple description; robust.	

Some of these estimates deserve a closer look:

(1) The efficient R-estimate for the normal distribution, the normal scores estimate, has an unbounded influence curve and hence infinite gross error sensitivity $\gamma^*=\infty$ (Section 1.5). Nevertheless it is robust! I would hesitate, though, to recommend it for practical use; its quantitative robustness indicators $b(\varepsilon)$ and $v(\varepsilon)$ increase steeply when we depart from the normal model, and the estimate very soon falls behind, for example, the Hodges-Lehmann estimate, (see Exhibit 6.6.2).

(2) The efficient L-estimate for the logistic is not robust, and $b_1(\varepsilon)=\infty$ for all $\varepsilon>0$, even though its "gross error sensitivity" γ^* at F_0

(Section 1.5) is finite. But note that its influence function for general (not necessarily logistic) F satisfies

$$\frac{d}{dx} IC(x; F, T) = 6F(x)[1 - F(x)].$$

Thus if F has Cauchy-like tails, the influence function becomes unbounded.

The lesson to be learned from these two examples is that it is not enough to look at the influence function at the model distribution only; we must also take into account its behavior in a neighborhood of the model. In the case of the normal scores estimate, a longer tailed F deflates the tails of the influence curve; in the case of the logistic L-estimate, the opposite happens. M-estimates are more straightforward to handle, since for them the shape of the influence function is fixed by ψ.

It is somewhat tricky to construct L- and R-estimates with prescribed robustness properties. For M-estimates the task is more straightforward. If we want to make a robust estimate that has good efficiency at the model F_0, we should choose a ψ that is bounded, but otherwise closely proportional to $-(\log f_0)'$. If we feel that very far-out outliers should be totally discarded, we should choose a ψ that goes to zero (or is zero) for large absolute 0. This finds its theoretical justification also in the remark that, for heavier-than-exponential tails, the influence curve of the efficient estimate decreases to zero (compare Examples 5.3 and 5.4). For L-estimates such an effect would be impossible to achieve over an entire range of distributions. With R-estimates we could do it, but not particularly well, because a change of the influence function in the extreme x-range selectively affects long-tailed distributions, while changes in the extreme t-range $[t = F(x)]$ affect all distributions equally.

In one-parameter location problems, L-estimates, in particular trimmed means, are very attractive because they are simple to calculate. However, unless we use relatively inefficient high trimming rates (i.e., essentially the sample median), the α-trimmed mean has very poor breakdown properties. The situation is particularly bad for small sample sizes. For instance, for sample sizes below 20 the 10% trimmed mean cannot cope with more than one outlier!

CHAPTER 4

Asymptotic Minimax Theory for Estimating a Location Parameter

4.1 GENERAL REMARKS

Qualitative robustness is of little help in the actual selection of a robust procedure suited for a particular application. In order to make a rational choice, we must introduce quantitative aspects as well.

Anscombe's (1960) comparison of the situation with an insurance problem is very helpful. Typically a so-called classical procedure is *the* optical procedure for some ideal (usually normal) model. If it happens to be nonrobust and we want to insure against accidents caused by deviations from the model, we clearly will have to pay for it by sacrificing some efficiency at the model. The questions are, of course, how much efficiency we are willing to sacrifice, and against how bad a deviation we would like to insure.

One possible approach is to fix a certain neighborhood of the model and to safeguard within that neighborhood (Huber 1964). In the simple location case, this leads to quite manageable minimax problems (even though the space of pure strategies for Nature is not dominated), both for asymptotic performance criteria (asymptotic bias or variance, treated in this chapter) and for finite sample ones (Chapter 10). If we take asymptotic variance as our performance criterion, then the least favorable situation F_0 (the minimax strategy for Nature) can be characterized intrinsically; it minimizes Fisher information in the chosen neighborhood, and the minimax strategy for the Statistician is efficient for F_0. Typically, if the neighborhood of the model is chosen not too large, the least favorable F_0 is a quite realistic distribution (which is closer to the error distributions observed in actual samples than the normal distribution), and so we even escape the perennial criticism directed against minimax methods, namely, that they safeguard against unlikely contingencies.

Unfortunately, this approach does not carry beyond problems possessing a high degree of symmetry (e.g., translation or scale invariance). Still it

73

suffices to deal successfully with a very large part of traditional statistics; in particular, the results carry over straightforwardly to regression.

Another approach [proposed by Hampel (1968)] remains even closer to Anscombe's idea; it minimizes the asymptotic variance at the model (i.e., it minimizes the efficiency loss), subject to a bound on the gross error sensitivity (also at the model). This approach has the conceptual flaw that it allows only infinitesimal deviations from the model, but, precisely because of this, it works for arbitrary one-parameter families of distributions; it is discussed in Chapter 11.

4.2 MINIMAX BIAS

Assume that the true underlying shape F lies in some neighborhood $\mathcal{P}_\varepsilon$ of the assumed model distribution F_0, that the observations are independent with common distribution $F(x-\theta)$, and that the location parameter θ is to be estimated. In this section we plan to optimize the robustness properties of such a location estimate by minimizing its maximum asymptotic bias $b(\varepsilon)$ for distributions $F \in \mathcal{P}_\varepsilon$. For the reasons mentioned in Section 1.4, we begin with minimizing the maximum bias $b_1(\varepsilon)$ of the functional T underlying the estimate; it is then a trivial matter to verify that $b(\varepsilon) = b_1(\varepsilon)$; compare Theorems 1.4.1 and 1.4.2.

To fix the idea, consider the case of ε-contaminated normal distributions

$$\mathcal{P}_\varepsilon = \{ F \mid F = (1-\varepsilon)\Phi + \varepsilon H, \, H \in \mathfrak{M} \}. \tag{2.1}$$

We show that the median minimizes $b_1(\varepsilon)$.

Clearly the maximum absolute bias $b_1(\varepsilon)$ of the median is attained whenever the total contaminating mass sits on one side, say on the right, and then its value is given by the solution x_0 of

$$(1-\varepsilon)\Phi(x_0) = \tfrac{1}{2},$$

or

$$b_1(\varepsilon) = x_0 = \Phi^{-1}\left(\frac{1}{2(1-\varepsilon)}\right) \tag{2.2}$$

We now construct two ε-contaminated normal distributions F_+ and F_-, which are symmetric about x_0 and $-x_0$, respectively, and which are translates of each other. F_+ is given by its density (cf. Exhibit 4.2.1)

$$f_+(x) = (1-\varepsilon)\varphi(x), \qquad \text{for } x \leqslant x_0,$$

$$= (1-\varepsilon)\varphi(x-2x_0), \qquad \text{for } x > x_0, \tag{2.3}$$

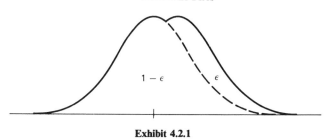

Exhibit 4.2.1

where $\varphi = \Phi'$ is the standard normal density, and

$$F_-(x) = F_+(x + 2x_0). \qquad (2.4)$$

Thus

$$T(F_+) - T(F_-) = 2x_0 \qquad (2.5)$$

for any translation invariant functional, and it is evident that none can have an absolute bias smaller than x_0 at F_+ and F_- simultaneously.

This shows that the median achieves the smallest maximum bias among all translation invariant functionals. It is trivial to verify that, for the median, $b(\varepsilon) = b_1(\varepsilon)$, so we have proved that the sample median solves the minimax problem of minimizing the maximum asymptotic bias.

Evidently, we have not used any particular property of the normal distribution, except symmetry and unimodality, and the same kind of argument also carries through for other neighborhoods. For example, with a Lévy neighborhood

$$\mathcal{P}_{\varepsilon, \delta} = \{ F \mid \forall x \ \Phi(x - \varepsilon) - \delta \leqslant F(x) \leqslant \Phi(x + \varepsilon) + \delta \}, \qquad (2.6)$$

the expression (2.2) for b_1 is replaced by

$$b_1(\varepsilon, \delta) = \Phi^{-1}\left(\tfrac{1}{2} + \delta\right) + \varepsilon, \qquad (2.7)$$

but everything else goes through without change.

Thus minimizing the maximum bias leads to a rather uneventful theory; for symmetric unimodal distributions, the solution invariably is the sample median.

The sample median thus is the estimate of choice for extremely large samples, where the standard deviation of the estimate (which is of the order $1/\sqrt{n}$) is comparable to or smaller than the bias $b(\varepsilon)$. Exhibit 4.2.2 evaluates (2.2) and gives the values of n for which $b(\varepsilon) = 1/\sqrt{n}$. It appears

ε	$b(\varepsilon)$	$n = b(\varepsilon)^{-2}$
0.25	0.4307	5
0.1	0.1396	50
0.05	0.0660	230
0.01	0.0126	6300

Exhibit 4.2.2

from this table that, for the customary sample sizes and not too large ε (i.e., $\varepsilon \leqslant 0.1$), the statistical variability of the estimate will be more important than its bias.

4.3   MINIMAX VARIANCE: PRELIMINARIES

Minimizing the maximal variance $v(\varepsilon)$ leads to a deeper theory. We begin by minimizing the more tractable

$$v_1(\varepsilon) = \sup_{F \in \mathscr{P}_\varepsilon} A(F, T) \qquad (3.1)$$

(cf. Section 1.4), and, since ε will be kept fixed, we suppress it in the notation.

We assume that the observations are independent with common distribution function $F(x - \theta)$. The location parameter θ is to be estimated, while the shape F may lie anywhere in some given set $\mathscr{P} = \mathscr{P}_\varepsilon$ of distribution functions. There are some difficulties of a topological nature; for certain existence proofs we would like $\mathscr{P}$ to be compact, but the more interesting neighborhoods $\mathscr{P}_\varepsilon$ are not tight, and thus their closure is not compact in the weak topology. As a way out we propose to take an even weaker topology, the vague topology (see below); then we can enforce compactness, but at the cost of including substochastic measures in $\mathscr{P}$ (or, equivalently, probability measures that put nonzero mass at $\pm \infty$). These measures may be thought to formalize the possibility of infinitely bad outliers. From now on we assume that $\mathscr{P}$ is vaguely closed and hence compact.

The vague topology in the space $\mathfrak{M}_+$ of substochastic measures on Ω is the weakest topology making the maps

$$F \to \int \psi \, dF$$

continuous for all continuous ψ having a compact support. Note that we are now working on the real line; thus $\Omega = \mathbb{R}$ is not only Polish, but also locally compact. Then $\mathfrak{M}_+$ is compact [see, e.g., Bourbaki (1952), Ch. III].

Let F_0 be the distribution having the smallest Fisher information

$$I(f) = \int \left(\frac{f'}{f}\right)^2 f \, dx \qquad (3.2)$$

among the members of $\mathscr{P}$. Under quite general conditions there is one and only one such F_0, as we see below.

For any sequence (T_n) of estimates, the asymptotic variance of $\sqrt{n} \, T_n$ at F_0 is at best $1/I(F_0)$; see Section 3.5. If we can find a sequence (T_n) such that its asymptotic variance does not exceed $1/I(F_0)$ for any $F \in \mathscr{P}$, we have clearly solved the minimax problem.

In particular, this sequence (T_n) must be asymptotically efficient for F_0, which gives a hint where to look for asymptotic minimax estimates.

4.4 DISTRIBUTIONS MINIMIZING FISHER INFORMATION

First of all, we extend the definition of Fisher information so that it is infinite whenever the classical expression (3.2) does not make sense. More precisely, we define it as follows.

DEFINITION 4.1 The Fisher information for location of a distribution F on the real line is

$$I(F) = \sup_{\psi} \frac{\left(\int \psi' \, dF\right)^2}{\int \psi^2 \, dF} \qquad (4.1)$$

where the supremum is taken over the set $\mathcal{C}_K^1$ of all continuously differentiable functions with compact support, satisfying $\int \psi^2 \, dF > 0$.

THEOREM 4.2 The following two assertions are equivalent:

(1) $I(F) < \infty$.

(2) F has an absolutely continuous density f, and $\int (f'/f)^2 f \, dx < \infty$.

In either case, we have $I(F) = \int (f'/f)^2 f \, dx$.

Proof If $\int (f'/f)^2 f \, dx < \infty$, then integration by parts and the Schwarz inequality give

$$\left(\int \psi' f \, dx \right)^2 = \left(\int \psi \frac{f'}{f} f \, dx \right)^2 \leqslant \int \psi^2 f \, dx \int \left(\frac{f'}{f} \right)^2 f \, dx;$$

hence

$$I(F) \leqslant \int \left(\frac{f'}{f} \right)^2 f \, dx < \infty.$$

Conversely, assume that $I(F) < \infty$, or, which is the same, the linear functional A, defined by

$$A\psi = - \int \psi' \, dF \tag{4.2}$$

on the dense subset $\mathcal{C}_K^1$ of the Hilbert space $L_2(F)$ of square F-integrable functions, is bounded:

$$\| A \|^2 = \sup \frac{|A\psi|^2}{\|\psi\|^2} = I(F) < \infty. \tag{4.3}$$

Hence A can be extended by continuity to the whole Hilbert space $L_2(F)$, and moreover, by Riesz' theorem, there is a $g \in L_2(F)$ such that

$$A\psi = \int \psi g \, dF \tag{4.4}$$

for all $\psi \in L_2(F)$. Note that

$$A1 = \int g \, dF = 0 \tag{4.5}$$

[this follows easily from the continuity of A and (4.2), if we approximate 1 by smooth functions with compact support].

We do not know, at this stage of the proof, whether F has an absolutely continuous density f, but if it has, then integration by parts of (4.2) gives

$$A\psi = - \int \psi' f \, dx = \int \psi \frac{f'}{f} f \, dx;$$

hence $g = f'/f$. So we *define* a function f by

$$f(x) = \int_{y<x} g(y) F(dy), \qquad (4.6)$$

and we have to check that this is indeed a version of the density of F.

The Schwarz inequality applied to (4.6) yields that f is bounded,

$$|f(x)|^2 \leqslant F(x) \int g^2 \, dF,$$

and tends to 0 for $x \to -\infty$ [and symmetrically also for $x \to +\infty$; here we use (4.5)]. If $\psi \in \mathcal{C}_K^1$, then Fubini's theorem gives

$$-\int \psi'(x) f(x) \, dx = -\int\int_{y<x} \psi'(x) g(y) F(dy) \, dx = \int \psi(y) g(y) F(dy) = A\psi.$$

A comparison with the definition (4.2) of A now shows that $f(x) \, dx$ and $F(dx)$ define the same linear functional on the set $\{\psi' | \psi \in \mathcal{C}_K^1\}$, which is dense in $L_2(F)$. It follows that they define the same measure, and so f is a version of the density of F.

Evidently, we then have

$$I(F) = \|A\|^2 = \int g^2 \, dF = \int \left(\frac{f'}{f}\right)^2 f \, dx. \qquad \blacksquare$$

[This theorem was first proved by Huber (1964); the elegant proof given above is based on an oral suggestion by T. Liggett.]

If the set $\mathcal{P}$ is endowed with the vague topology, then Fisher information (4.1) is lower-semicontinuous as a function of F (it is the pointwise supremum of a set of vaguely continuous functions).

It follows that $I(F)$ attains its infimum on any vaguely compact set $\mathcal{P}$, so we have proved the following proposition.

PROPOSITION 4.3   (EXISTENCE)   If $\mathcal{P}$ is vaguely compact, then there is an $F_0 \in \mathcal{P}$ minimizing $I(F)$.

We note furthermore that $I(F)$ is a convex function of F. This follows at once from the remark that $\int \psi' \, dF$ and $\int \psi^2 \, dF$ are linear functions of F, and from the following lemma.

LEMMA 4.4   Let $u(t)$, $v(t)$ be linear functions of t such that $v(t) > 0$ for $0 < t < 1$. Then $w(t) = u(t)^2/v(t)$ is convex for $0 < t < 1$.

Proof   The second derivative of w is

$$w''(t) = \frac{2[u'v(t) - u(t)v']^2}{v(t)^3} \geqslant 0$$

for $0 < t < 1$. $\blacksquare$

We are now ready to prove also the uniqueness of F_0.

PROPOSITION 4.5 (UNIQUENESS)   Assume that

(1) $\mathcal{P}$ is convex.
(2) $F_0 \in \mathcal{P}$ minimizes $I(F)$ in $\mathcal{P}$, and $0 < I(F_0) < \infty$.
(3)   The set where the density f_0 of F_0 is strictly positive is convex.
Then F_0 is the unique member of $\mathcal{P}$ minimizing $I(F)$.

Proof   Assume that F_1 also minimizes $I(F)$. Then by convexity $I(F_t)$ must be constant on the segment $0 \leqslant t \leqslant 1$, where $F_t = (1-t)F_0 + tF_1$. Without loss of generality we may assume that F_0 is absolutely continuous with respect to F_1 (if not, replace F_1 by F_{t_0} for some fixed $0 < t_0 < 1$).

Evidently, the integrand in

$$I(F_t) = \int \frac{(f_t')^2}{f_t} dx \tag{4.7}$$

is a convex function of t. If we may differentiate twice under the integral sign, we obtain

$$0 = \frac{d^2}{dt^2} I(F_t) = \int 2\left(\frac{f_1'}{f_1} - \frac{f_0'}{f_0}\right)^2 \frac{f_0^2 f_1^2}{f_t^3} dx. \tag{4.8}$$

This is indeed permissible; if

$$Q(t) = \int q_t(x) dx$$

where $q_t(x)$ is any function convex in t, then the integrand in

$$\frac{Q(t+h) - Q(t)}{h} = \int \frac{q_{t+h} - q_t}{h} dx$$

is monotone in h. Hence

$$Q'(t) = \int q_t' dx$$

by the monotone convergence theorem. Moreover, the integrand in

$$\frac{Q'(t+h) - Q'(t)}{h} = \int \frac{q_{t+h}' - q_t'}{h} dx$$

is positive; hence, by Fatou's lemma,

$$Q''(t) \geqslant \int q_t'' \, dx \geqslant 0,$$

and (4.8) follows.

Thus we must have

$$\frac{f_1'}{f_1} = \frac{f_0'}{f_0} \qquad \text{a.e.} \tag{4.9}$$

If we integrate this relation, we obtain

$$f_1 = cf_0 \tag{4.10}$$

for some constant c (here we have used assumption (3) of Proposition 4.5: the set where f_0 and f_1 are different from 0 is convex and hence, in particular, connected). Since

$$I(F_1) = \int \left(\frac{f_1'}{f_1} \right)^2 f_1 \, dx = \int \left(\frac{f_0'}{f_0} \right)^2 cf_0 \, dx = cI(F_0),$$

it follows that $c = 1$. ∎

NOTE 1 We have not assumed that our measures have total mass 1 [note in particular the argument showing that $c = 1$ in (4.10)]. In principle the minimizing F_0 could be substochastic. However, we do not know of any realistic set $\mathscr{P}$ where this occurs, that is, where the least informative F_0 would put pointmasses at $\pm \infty$, and there is a good intuitive reason for this. For a "realistic" $\mathscr{P}$, any masses at $\pm \infty$ are not genuinely at infinity, but must have arisen as a limit of contamination that has escaped to infinity, and it is intuitively clear that, by shifting these masses again to finite values, the task of the statistician can be made harder, since they would no longer be immediately recognizable as outliers.

NOTE 2 Proposition 4.5 is wrong without some form of assumption (3); this was overlooked in Huber (1964). For example, let F_0 and F_1 be defined by their densities

$$f_0(x) = Cx^2(1 + x)^2, \quad \text{for } -1 \leqslant x \leqslant 0$$
$$= 0, \qquad\qquad \text{otherwise,}$$
$$f_1(x) = Cx^2(1 - x)^2, \quad \text{for } 0 \leqslant x \leqslant 1$$
$$= 0, \qquad\qquad \text{otherwise,} \tag{4.11}$$

and let $\mathscr{P} = \{F_t | t \in [0, 1]\}$. Then $I(F)$ is finite and constant on $\mathscr{P}$.

There are several other equivalent expressions for Fisher information if $f(x; \theta)$ is sufficiently smooth. For the sake of reference, we list a few (we denote differentiation with respect to θ by a prime):

$$I(F; \theta) = \int [(\log f)']^2 f \, dx$$

$$= - \int (\log f)'' f \, dx$$

$$= -4 \int \frac{(\sqrt{f})''}{\sqrt{f}} f \, dx. \tag{4.12}$$

4.5   DETERMINATION OF F_0 BY VARIATIONAL METHODS

Assume that $\mathscr{P}$ is convex. Because of convexity of $I(\cdot)$, $F_0 \in \mathscr{P}_0$ minimizes Fisher information iff $(d/dt)I(F_t) \geq 0$ at $t=0$ for every $F_1 \in \mathscr{P}_1$, where $\mathscr{P}_1$ is the set of all $F \in \mathscr{P}$ with $I(F) < \infty$. A straightforward differentiation of (4.7) under the integral sign, justified by the monotone convergence theorem, gives

$$\left[\frac{d}{dt} I(F_t) \right]_{t=0} = \int \left[2\frac{f_0'}{f_0}(f_1' - f_0') - \left(\frac{f_0'}{f_0} \right)^2 (f_1 - f_0) \right] dx \geq 0. \tag{5.1}$$

If we introduce $\psi(x) = -f_0'(x)/f_0(x)$, and if ψ has a derivative ψ' so that integration by parts is possible, (5.1) can be rewritten in the more convenient form

$$\int (2\psi' - \psi^2)(f_1 - f_0) \, dx \geq 0, \tag{5.2}$$

or also as

$$-4 \int \frac{(\sqrt{f_0})''}{\sqrt{f_0}} (f_1 - f_0) \, dx \geq 0, \tag{5.3}$$

for all $F_1 \in \mathscr{P}_1$.

Among the following examples the first one highlights an amusing connection between least informative distributions and the ground-state solution in quantum mechanics; the second one is of central importance to robust estimation.

Example 5.1   Let $\mathcal{P}$ be the set of all probability distributions F such that

$$\int V(x)F(dx) \leqslant 0, \tag{5.4}$$

where V is some given function. For the F_0 minimizing Fisher information in $\mathcal{P}$, we have equality in (5.3) and (5.4). If we combine (5.3), (5.4), and

$$\int F(dx) = 1 \tag{5.5}$$

with the aid of Lagrange multipliers α and β, we obtain the differential equation

$$4\frac{\sqrt{f_0}''}{\sqrt{f_0}} - \alpha V + \beta = 0, \tag{5.6}$$

or, with $u = \sqrt{f_0}$,

$$4u'' - (\alpha V - \beta)u = 0. \tag{5.7}$$

This is, essentially, the Schrödinger equation for an electron moving in the potential V.

If f_0 is a solution of (5.6) satisfying the side conditions (5.4) and (5.5), then (5.3) holds provided $\alpha > 0$. If we multiply (5.6) by f_0 and integrate over x, we obtain $I(F_0) = \beta$; hence (using the quantum mechanical jargon) we are interested in the ground-state solution corresponding to the lowest eigenvalue β.

In the particular case $V(x) = x^2 - 1$, the well-known solution for the ground-state of the harmonic oscillator yields the result, which is also well-known, that, among all distributions with variance $\leqslant 1$, the standard normal has the smallest Fisher information for location.

From the point of view of robust estimation, a "box" potential is more interesting:

$$V(x) = -a < 0, \qquad \text{for } |x| \leqslant 1$$

$$= b > 0, \qquad \text{for } |x| > 1. \tag{5.8}$$

It is easy to see that the solution of (5.6) then is of the general form

$$f_0(x) = \frac{C}{\cos^2(\omega/2)} \cos^2\left(\frac{\omega x}{2}\right), \qquad \text{for } |x| \leq 1$$

$$= C e^\lambda e^{-\lambda|x|}, \qquad \text{for } |x| > 1, \tag{5.9}$$

for some constants ω and λ. In order that f_0 be strictly positive, we should have $0 < \omega < \pi$. We have already arranged the integration constants so that f_0 is continuous; if $\psi = -(\log f_0)'$ is also to be continuous, we must have

$$\lambda = \omega \tan \frac{\omega}{2}, \tag{5.10}$$

and C must be determined such that $\int f_0 \, dx = 1$, that is,

$$C = \frac{\cos^2(\omega/2)}{1 + 2/[\omega \tan(\omega/2)]}. \tag{5.11}$$

Note that then

$$-4 \frac{\sqrt{f_0}''}{\sqrt{f_0}} = \omega^2, \qquad \text{for } |x| < 1$$

$$= -\lambda^2, \qquad \text{for } |x| > 1; \tag{5.12}$$

hence

$$I(F_0) = -4 \int \frac{\sqrt{f_0}''}{\sqrt{f_0}} f_0 \, dx = \frac{\omega^2}{1 + 2/[\omega \tan(\omega/2)]}. \tag{5.13}$$

It is now straightforward to check that (5.3) is satisfied, that is, that this F_0 minimizes Fisher information among all probability distributions F satisfying

$$F\{(-1, 1)\} \geq F_0\{(-1, 1)\} = 1 - \frac{2C}{\lambda}. \tag{5.14}$$

Example 5.2 Let G be a fixed probability distribution having a twice differentiable density g, such that $-\log g(x)$ is convex on the convex support of G. Let $\varepsilon > 0$ be given, and let $\mathscr{P}$ be the set of all probability distributions arising from G through ε-contamination:

$$\mathcal{P} = \{F \,|\, F = (1-\varepsilon)G + \varepsilon H, \; H \in \mathfrak{M}\}. \tag{5.15}$$

Here $\mathfrak{M}$ is, as usual, the set of all probability measures on the real line, but we can also take $\mathfrak{M}$ to be the set of all substochastic measures, in order to make $\mathcal{P}$ vaguely compact.

In view of (5.3) it is plausible that the density f_0 of the least informative distribution behaves as follows. There is a central part where f_0 touches the boundary, $f_0(x) = (1-\varepsilon)g(x)$; in the tails $(\sqrt{f_0}\,)''/\sqrt{f_0}$ is constant, that is, f_0 is exponential, $f_0(x) = Ce^{-\lambda|x|}$. This is indeed so, and we now give the solution f_0 explicitly.

Let $x_0 < x_1$ be the endpoints of the interval where $|g'/g| \leqslant k$, and where k is related to ε through

$$\int_{x_0}^{x_1} g(x)\,dx + \frac{g(x_0) + g(x_1)}{k} = \frac{1}{1-\varepsilon}. \tag{5.16}$$

Either x_0 or x_1 may be infinite. Then put

$$f_0(x) = (1-\varepsilon)g(x_0)e^{k(x-x_0)}, \qquad \text{for } x \leqslant x_0$$

$$= (1-\varepsilon)g(x), \qquad\qquad \text{for } x_0 < x < x_1 \tag{5.17}$$

$$= (1-\varepsilon)g(x_1)e^{-k(x-x_1)}, \qquad \text{for } x \geqslant x_1.$$

Condition (5.16) ensures that f_0 integrates to 1; hence the contamination distribution $H_0 = [F_0 - (1-\varepsilon)G]/\varepsilon$ also has total mass 1, and it remains to be checked that its density h_0 is nonnegative. But this follows at once from the remark that the convex function $-\log g(x)$ lies above its tangents at the points x_0 and x_1, that is

$$g(x) \leqslant g(x_0)e^{k(x-x_0)} \qquad \text{and} \qquad g(x) \leqslant g(x_1)e^{-k(x-x_1)}.$$

Clearly, both f_0 and its derivative are continuous; we have

$$\psi(x) = -[\log f_0(x)]' = -k, \qquad \text{for } x \leqslant x_0$$

$$= \frac{-g'(x)}{g(x)}, \qquad\qquad \text{for } x_0 < x < x_1$$

$$= k, \qquad\qquad\qquad \text{for } x \geqslant x_1. \tag{5.18}$$

We now check that (5.2) holds. As $\psi'(x) \geq 0$ and as

$$k^2 + 2\psi' - \psi^2 \geq 0, \qquad \text{for } x_0 \leq x \leq x_1$$

$$= 0, \qquad \text{otherwise,}$$

it follows that

$$\int (2\psi' - \psi^2)(f_1 - f_0)\,dx = \int_{x_0}^{x_1}(k^2 + 2\psi' - \psi^2)(f_1 - f_0)\,dx - k^2\int(f_1 - f_0)\,dx$$

$$\geq 0, \tag{5.19}$$

since $f_1 \geq f_0$ in the interval $x_0 < x < x_1$, and since $\int(f_1 - f_0)\,dx \leq 0$ (we may allow F_1 to be substochastic!).

Because of their importance we state the results for the case where $G = \Phi$ is the standard normal cumulative separately. In this case Fisher information is minimized by

$$f_0(x) = \frac{1-\varepsilon}{\sqrt{2\pi}}e^{-x^2/2}, \qquad \text{for } |x| \leq k$$

$$= \frac{1-\varepsilon}{\sqrt{2\pi}}e^{k^2/2 - k|x|}, \qquad \text{for } |x| > k, \tag{5.20}$$

with k and ε connected through

$$\frac{2\varphi(k)}{k} - 2\Phi(-k) = \frac{\varepsilon}{1-\varepsilon} \tag{5.21}$$

($\varphi = \Phi'$ being the standard normal density). In this case

$$\psi(x) = -[\log f_0(x)]' = \max[-k, \min(k, x)]. \tag{5.22}$$

Compare Exhibit 4.5.1 for some numerical results.

Example 5.3 Let $\mathscr{P}$ be the set of all distributions differing at most ε in Kolmogorov distance from the standard normal cumulative

$$\sup|F(x) - \Phi(x)| \leq \varepsilon. \tag{5.23}$$

It is easy to guess that the solution F_0 is symmetric and that there will be two (possibly coinciding) constants $0 < x_0 \leq x_1$ such that $F_0(x) = \Phi(x) - \varepsilon$

ε	k	$F_0(-k)$	$1/I(F_0)$
0	∞	0	1.000
0.001	2.630	0.005	1.010
0.002	2.435	0.008	1.017
0.005	2.160	0.018	1.037
0.01	1.945	0.031	1.065
0.02	1.717	0.052	1.116
0.05	1.399	0.102	1.256
0.10	1.140	0.164	1.490
0.15	0.980	0.214	1.748
0.20	0.862	0.256	2.046
0.25	0.766	0.291	2.397
0.3	0.685	0.323	2.822
0.4	0.550	0.375	3.996
0.5	0.436	0.416	5.928
0.65	0.291	0.460	12.48
0.80	0.162	0.487	39.0
1	0	0.5	∞

Exhibit 4.5.1 The ε-contaminated normal distributions least informative for location.

for $x_0 \leqslant x \leqslant x_1$, with strict inequality $|F_0(x) - \Phi(x)| < \varepsilon$ for all other positive x. In view of (5.3) we expect that $\sqrt{f_0}'' / \sqrt{f_0}$ is constant in the intervals $(0, x_0)$ and (x_1, ∞); hence we try a solution of the form

$$f_0(x) = f_0(-x) = \frac{\varphi(x_0)}{\cos^2(\omega x_0/2)} \cos^2\left(\frac{\omega x}{2}\right), \qquad \text{for } 0 \leqslant x \leqslant x_0$$

$$\varphi(x), \qquad \text{for } x_0 \leqslant x \leqslant x_1$$

$$= \varphi(x_1)e^{-\lambda(x-x_1)}, \qquad \text{for } x \geqslant x_1. \qquad (5.24)$$

We now distinguish two cases.

Case A *Small Values of* ε, $x_0 < x_1$ In order that

$$\psi(x) = -[\log f_0(x)]' = \omega \tan\left(\frac{\omega x}{2}\right), \qquad \text{for } 0 \leqslant x \leqslant x_0$$

$$= x, \qquad \text{for } x_0 \leqslant x \leqslant x_1$$

$$= \lambda, \qquad \text{for } x \geqslant x_1 \qquad (5.25)$$

be continuous, we must require

$$\omega \tan\left(\frac{\omega x_0}{2}\right) = x_0, \tag{5.26}$$

$$\lambda = x_1. \tag{5.27}$$

In order that $F_0(x) = \Phi(x) - \varepsilon$ for $x_0 \leqslant x \leqslant x_1$, and that its total mass be 1, we must have

$$\int_0^{x_0} f_0(x)\, dx = \int_0^{x_0} \varphi(x)\, dx - \varepsilon \tag{5.28}$$

and

$$\int_{x_1}^{\infty} f_0(x)\, dx = \int_{x_1}^{\infty} \varphi(x)\, dx + \varepsilon. \tag{5.29}$$

For a given ε, (5.26) to (5.29) determine the four quantities x_0, x_1, ω, and λ. For the actual calculation it is advantageous to use

$$u = \omega x_0 \tag{5.30}$$

as the independent variable, $0 < u < \pi$, and to express everything in terms of u instead of ε. Then from (5.26), (5.30), and (5.28), we obtain, respectively,

$$x_0 = \left(u \tan\frac{u}{2}\right)^{1/2}, \tag{5.31}$$

$$\omega = \frac{u}{x_0}, \tag{5.32}$$

$$\varepsilon = \Phi(x_0) - \tfrac{1}{2} - x_0 \varphi(x_0) \frac{1 + \sin u / u}{1 + \cos u}, \tag{5.33}$$

and finally, x_1 has to be determined from (5.29), that is, from

$$\varepsilon = \frac{\varphi(x_1)}{x_1} - \Phi(-x_1). \tag{5.34}$$

It turns out that $x_0 < x_1$ so long as $\varepsilon < \varepsilon_0 \cong 0.0303$. It remains to check (5.23) and (5.3). The first one follows easily from $f_0(x_0) = \varphi(x_0)$, $f_0(x_1) = \varphi(x_1)$, and from the remark that

$$-\left[\log f_0(x)\right]' = \psi(x) \leqslant -\left[\log \varphi(x)\right]', \qquad \text{for } x \geqslant 0.$$

If we integrate this relation, we obtain that $f_0(x) \leqslant \varphi(x)$ for $0 \leqslant x \leqslant x_0$ and $f_0(x) \geqslant \varphi(x)$ for $x \geqslant x_1$. In conjunction with $F_0(x) = \Phi(x) - \varepsilon$ for $x_0 \leqslant x \leqslant x_1$, this establishes (5.23). In order to check (5.3), we first note that it suffices to consider symmetric distributions for F_1 (since $I(F)$ is convex, the symmetrized distribution $\tilde{F}(x) = \frac{1}{2}[F(x) + 1 - F(-x)]$ has a smaller Fisher information than F). We have

$$-4 \frac{\sqrt{f_0}''}{\sqrt{f_0}} = \omega^2, \qquad \text{for } 0 \leqslant x < x_0$$

$$= 2 - x^2, \qquad \text{for } x_0 < x < x_1$$

$$= -x_1^2, \qquad \text{for } x > x_1.$$

Thus with $G = F_1 - F_0$ the left-hand side of (5.3) becomes twice

$$\int_0^{x_0} \omega^2 \, dG + \int_{x_0}^{x_1} (2 - x^2) \, dG - \int_{x_1}^{\infty} x_1^2 \, dG$$

$$= (\omega^2 + x_0^2 - 2) G(x_0) + 2G(x_1) - x_1^2 G(\infty) + \int_{x_0}^{x_1} xG(x) \, dx.$$

We note that

$$\omega^2 + x_0^2 - 2 = \frac{u}{\tan(\frac{1}{2}u)} + u\tan(\frac{1}{2}u) - 2 = 2\left(\frac{u}{\sin u} - 1\right) \geqslant 0,$$

that $G(x) \geqslant 0$ for $x_0 \leqslant x \leqslant x_1$, and that $G(\infty) \leqslant 0$. Hence all terms are positive and (5.3) is verified.

Case B   *Large Values of ε, $x_0 = x_1$*   In this case (5.24) simplifies to

$$f_0(x) = f_0(-x) = \frac{\varphi(x_0)}{\cos^2(\omega x_0/2)} \cos^2\left(\frac{\omega x}{2}\right), \qquad \text{for } 0 \leqslant x \leqslant x_0$$

$$= \varphi(x_0) e^{-\lambda(x - x_0)}, \qquad \text{for } x > x_0. \qquad (5.35)$$

Apart from a change of scale, this is the distribution already encountered in (5.9).

In order that

$$\psi(x) = -[\log f_0(x)]' = \omega \tan\left(\frac{\omega x}{2}\right), \qquad \text{for } 0 \leqslant x \leqslant x_0$$

$$= \lambda, \qquad \text{for } x > x_0 \qquad (5.36)$$

be continuous, we must require that

$$\lambda x_0 = \omega x_0 \tan \frac{\omega x_0}{2}, \qquad (5.37)$$

and f_0 integrates to 1 if

$$x_0 \varphi(x_0) = \cos^2 \left(\frac{u}{2} \right) \frac{1}{1 + 2/[u \tan(u/2)]}, \qquad (5.38)$$

with $u = \omega x_0$; compare (5.11).

It is again convenient to use $u = \omega x_0$ instead of ε as the independent variable. We first determine $x_0 \geqslant 1$ from (5.38) (there is also a solution <1), and then λ from (5.37). From (5.29) we get

$$\varepsilon = \frac{\varphi(x_0)}{\lambda} - \Phi(-x_0). \qquad (5.39)$$

This solution holds for $\varepsilon \geqslant \varepsilon_0 \cong 0.0303$. It is somewhat tricky to prove that F_0 satisfies (5.23); see Sacks and Ylvisaker (1972). Exhibit 4.5.2 gives some numerical results.

ε	x_0	ω	$\lambda(=x_1)$	$1/I(F_0)$
0	0	1.4142	∞	1.
0.001	0.6533	1.3658	2.4364	1.019
0.002	0.7534	1.3507	2.2317	1.034
0.005	0.9118	1.3234	1.9483	1.075
0.01	1.0564	1.2953	1.7241	1.136
0.02	1.2288	1.2587	1.4921	1.256
0.03033	1.3496	1.2316	1.3496	1.383
0.05	1.3216	1.1788	1.1637	1.656
0.10	1.3528	1.0240	0.8496	2.613
0.15	1.4335	0.8738	0.6322	4.200
0.20	1.5363	0.7363	0.4674	6.981
0.25	1.6568	0.6108	0.3384	12.24
0.3	1.7974	0.4950	0.2360	23.33
0.4	2.1842	0.2803	0.0886	144.2

Exhibit 4.5.2 Least informative distributions for $\sup |F(x) - \Phi(x)| \leqslant \varepsilon$ [cf. Example 4.5.3; (5.25), (5.35)].

We have now determined a small collection of least informative situations and we should take some time out to reflect how realistic or unrealistic they are.

First, it may surprise us that the least informative F_0 do *not* have excessively long tails. On the contrary we might perhaps argue that they have unrealistically *short* tails, since they do not provide for the extreme outliers we sometimes encounter.

Second, we should compare them with actual, supposedly normal, distributions. For that we need very large samples, and these seem to be quite rare; some impressive examples have been collected by Romanowski and Green (1965). Their largest sample ($n = 8688$), when plotted on normal probability paper (Exhibit 4.5.3), is seen to behave very much like a least informative 2%-contaminated normal distribution [it lies between the slightly different curves for the least favorable F_0 for location and the least favorable one for scale (5.6.15)]. For their smaller samples the conclusions are less clear-cut because of the higher random variability, but there also the sample distribution functions are close to some least informative ε-contaminated F_0 (with ε in the range 0.01 to 0.1).

Thus it makes very good sense to use minimax procedures safeguarding against ε-contamination, for an ε in the range just mentioned.

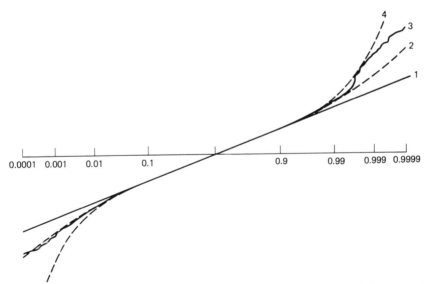

Exhibit 4.5.3   1: Normal cumulative. 2: Least favorable for location ($\varepsilon = 0.02$). 3: Empirical cumulative. 4: Least favorable for scale ($\varepsilon = 0.02$). $n = 8688$. Data from Romanowski and Green (1965).

92

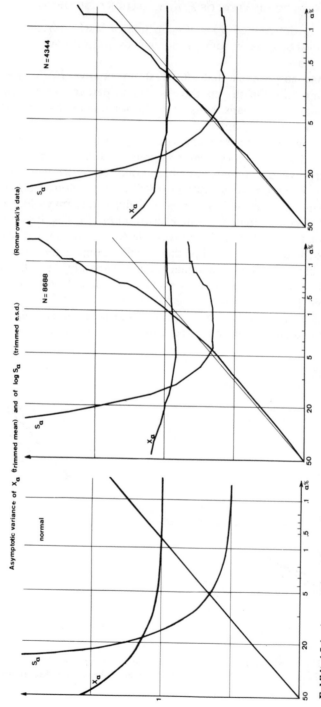

Asymptotic variance of x_α (trimmed mean) and of $\log S_\alpha$ (trimmed e.s.d.)

(Romanowski's data)

Exhibit 4.5.4 Asymptotic variance of x_α (trimmed mean) and of $\log S_\alpha$ (trimmed e.s.d.). Data from Romanowski and Green (1965).

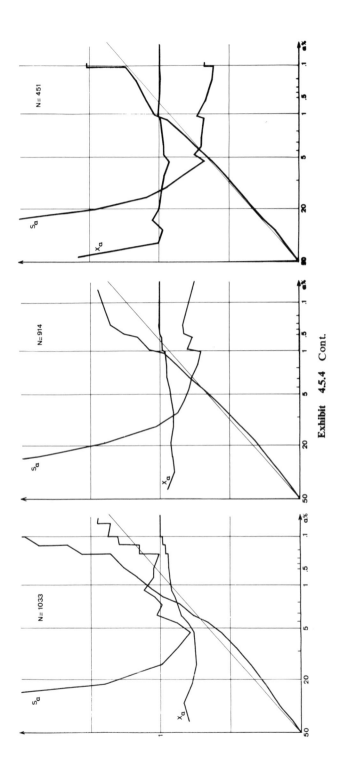

Exhibit 4.5.4 Cont.

Exhibit 4.5.4 plots, on normal probability paper, the symmetrized empirical distributions of several large samples taken from Romanowski and Green (1965). Also shown are the asymptotic variances of the α-trimmed mean and of the logarithm of the α-trimmed standard deviation (this corresponds to sampling with replacement from the symmetrized empirical distributions).

These are all good data sets, so the classical estimates do not fare badly, but note that moderate trimming would never do much harm, but sometimes considerable good.

4.6 ASYMPTOTICALLY MINIMAX M-ESTIMATES

Assume that F_0 has minimal Fisher information for location in the convex set $\mathscr{P}$ of distribution functions. We now show that the asymptotically efficient M-estimate of location for F_0 in fact possesses certain minimax properties in $\mathscr{P}$.

According to (3.5.11) we must choose

$$\psi(x) = \frac{-cf_0'(x)}{f_0(x)} \tag{6.1}$$

in order to achieve asymptotic efficiency at F_0 (the value of the constant $c \neq 0$ is irrelevant). We do not worry about regularity conditions for the moment, but we note that, in all examples of Section 4.5, the function (6.1) is monotone, so the theory of Section 3.2 is applicable, and the M-estimate, defined by

$$\int \psi(x - T(F))F(dx) = 0, \tag{6.2}$$

is asymptotically normal

$$\mathscr{L}\left\{ \sqrt{n}\,[T(F_n) - T(F)]\right\} \to \mathfrak{N}(0, A(F, T)) \tag{6.3}$$

with asymptotic variance

$$A(F, T) = \frac{\int \psi(x - T(F))^2 F(dx)}{[\lambda'(T(F))]^2} = \frac{\int \psi(x - T(F))^2 F(dx)}{\left[\int \psi'(x - T(F))F(dx)\right]^2}. \tag{6.4}$$

In particular,

$$A(F_0, T) = \frac{1}{I(F_0)}.$$ (6.5)

Without loss of generality we may assume $T(F_0) = 0$.

But we now run into an awkward technical difficulty, caused by the variable term $T(F)$ in the expression (6.4) for the asymptotic variance. If $\mathcal{P}$ consists of symmetric distributions only, then

$$T(F) = 0, \qquad \text{for all } F \in \mathcal{P},$$ (6.6)

and the difficulty disappears.

Traditionally and conveniently, most of the robustness literature therefore adopts the assumption of symmetry. However, it should be pointed out that a restriction to exactly symmetric distributions:

(1) Violates the very spirit of robustness.

(2) Is out of the question if the model distribution itself already is asymmetric.

We therefore adopt a slightly different approach. We replace $\mathcal{P}$ by the convex subset

$$\mathcal{P}_0 = \{ F \in \mathcal{P} \mid T(F) = 0 \}.$$ (6.7)

This enforces (6.6) and eliminates the explicit dependence of (6.4) on $T(F)$. Moreover, it leads to a "cleaner" problem; we do not have to worry about the asymptotic bias of $T(F_n)$ while investigating its asymptotic variance on $\mathcal{P}_0$. Clearly, the behavior of $T(F)$ and $A(F, T)$ on $\mathcal{P} \setminus \mathcal{P}_0$ still must be checked separately (see Section 4.9).

According to Lemma 4.4, $1/A(F, T)$ is a convex function of $F \in \mathcal{P}_0$. Let $F_t = (1 - t)F_0 + tF_1$ with $F_1 \in \mathcal{P}_0 \cap \mathcal{P}_1$, where $\mathcal{P}_1$ is the subset of $\mathcal{P}$ consisting of distributions with finite Fisher information (cf. Section 4.5). Then an explicit calculation and a comparison with (5.1) and (5.2) gives

$$\left[\frac{d}{dt} \frac{1}{A(F, T)} \right]_{t=0} = \int \left[2\psi'(x) - \psi(x)^2 \right] \left[F_1(dx) - F_0(dx) \right]$$

$$= \left[\frac{d}{dt} I(F_t) \right]_{t=0} \geq 0.$$ (6.8)

It follows from the convexity of $1/A(F, T)$ that

$$A(F, T) \leq A(F_0, T), \qquad \text{for all } F \in \mathcal{P}_0 \cap \mathcal{P}_1$$ (6.9)

In other words the maximum likelihood estimate for location based on the least informative F_0 minimizes the maximum asymptotic variance for alternatives in $\mathscr{P}_0 \cap \mathscr{P}_1$. If $\mathscr{P}_1$ is dense in $\mathscr{P}$, the estimate usually is minimax for the whole of $\mathscr{P}_0$, but each case seems to need a separate investigation.

For instance, take the case of Example 5.2, assuming that $(-\log g)''$ is continuous. We rely heavily on the asymptotic normality proof given in Section 3.2.

First, it is evident that

$$\int \psi(x)^2 F(dx) \leqslant \int \psi(x)^2 F_0(dx), \qquad \text{for all } F \in \mathscr{P}_0, \qquad (6.10)$$

since F_0 puts all contamination on the maximum of ψ^2.

Some difficulties arise with

$$\lambda(t, F) = \int \psi(x-t) F(dx),$$

since it may fail to have a derivative. To see what is going on, put $u_i = (-\log g)''(x_i)$, $i = 0, 1$, with x_i as in Example 5.2. If F puts pointmasses ε_i at x_i, then a straightforward calculation shows that $\lambda(\cdot, F)$ still has (possibly different) one-sided derivatives at $t = 0$; in fact

$$\lambda'(+0; F) - \lambda'(-0; F) = \varepsilon_0 u_0 - \varepsilon_1 u_1. \qquad (6.11)$$

In any case we have

$$-\lambda'(\pm 0; F) \geqslant -\lambda'(0; F_0) > 0 \qquad (6.12)$$

for all $F \in \mathscr{P}_0$.

Theorem 3.2.4 remains valid; a closer look at the limiting distribution of $\sqrt{n}\, T(F_n)$ shows that it is no longer normal, but pieced together from the right half of a normal distribution whose variance (6.4) is determined by the right derivative of λ, and from the left half of a normal distribution whose variance is determined by the left derivative of λ.

But (6.10) and (6.12) together imply that, nevertheless,

$$A(F; T) \leqslant A(F_0; T)$$

even if $A(F; T)$ now may have different values on the left- and the right-hand sides of the median of the distribution of $\sqrt{n}\, T(F_n)$. Moreover, there is enough uniformity in the convergence of (3.2.29) to imply

$$v(\varepsilon) = v_1(\varepsilon) = A(F_0; T)$$

(see Section 1.4) when F varies over $\mathcal{P}_0$.

Remark An interesting limiting case.

Consider the general ε-contaminated case of Example 5.2, and let $\varepsilon \to 1$. Then $k \to 0$ and $f_0 \to 0$, so there is no proper limiting distribution. But the asymptotically efficient M-estimate for F_0 tends to a nontrivial limit, namely, apart from an additive constant, to the sample median. This may be seen as follows: ψ can be multiplied by a constant, without changing the estimate, and in particular

$$\lim_{\varepsilon \to 0} \frac{1}{k} \psi(x) = -1, \qquad \text{for } x < x^*$$

$$= 1, \qquad \text{for } x > x^*$$

where x^* is defined by $g'(x^*)/g(x^*) = 0$. Hence the limiting estimate is determined as the solution of

$$\sum_{i=1}^{n} \text{sign}(x_i - x^* - T_n) = 0,$$

and thus

$$T_n = \text{med}\{x_i\} - x^*.$$

Example 6.1 Because of its importance, we single out the minimax M-estimate of location for the ε-contaminated normal distribution. There, the least informative distribution is given by (5.20) and (5.21), and the estimate T_n is defined by

$$\sum \psi(x_i - T_n) = 0$$

with ψ given by (5.22).

4.7 ON THE MINIMAX PROPERTY FOR L- AND R-ESTIMATES

For L- and R-estimates $1/A(F; T)$ is no longer a convex function of F. Although (6.8) still holds [this is shown either by explicit calculation, or it can also be inferred on general grounds from the remark that $I(F) = \sup_T 1/A(F, T)$, with T ranging over either class of estimates] we can no longer conclude that the asymptotically efficient estimate for F_0 is asymptotically minimax, even if we restrict $\mathscr{P}$ to symmetric and smooth distributions. In fact Sacks and Ylvisaker (1972) have constructed counterexamples. However, in the important Example 5.2 (ε-contamination), the conclusion is true (Jaeckel 1971a). We assume throughout that all distributions are symmetric.

Consider first the case of L-estimates, where the efficient one (cf. Section 3.5) is characterized by the weight density

$$m(F_0(x)) = -\left(\frac{g'(x)}{g(x)}\right)' \frac{1}{I(F_0)} \geq 0, \qquad \text{for } |x| \leq x_1 = -x_0$$
$$= 0, \qquad\qquad\qquad \text{otherwise,}$$

with g as in Example 5.2.

The influence function is skew symmetric, and, for $x \geq 0$, it satisfies

$$IC(x; F, T) = \int_0^x m(F(y)) \, dy,$$

or, for $\frac{1}{2} \leq t < 1$,

$$IC(F^{-1}(t); F, T) = \int_{1/2}^t \frac{m(s)}{f(F^{-1}(s))} \, ds.$$

We have

$$F(x) \geq F_0(x), \qquad \text{for } 0 \leq x \leq x_1,$$

and

$$F^{-1}(t) \leq F_0^{-1}(t), \qquad \text{for } \frac{1}{2} \leq t \leq F_0(x_1).$$

Thus for $\frac{1}{2} \leq t \leq F_0(x_1)$,

$$IC(F^{-1}(t); F, T) = \int_{1/2}^t \frac{m(s)}{f(F^{-1}(s))} \, ds$$

$$\leq \int_{1/2}^t \frac{m(s)}{f_0(F^{-1}(s))} \, ds \leq \int_{1/2}^t \frac{m(s)}{f_0(F_0^{-1}(s))} \, ds$$

$$= IC(F_0^{-1}(t); F, T).$$

Since $IC(F^{-1}(t); F, T)$ is constant for $F_0(x_1) \leqslant t \leqslant 1$, and as

$$A(F, T) = 2 \int_{1/2}^{1} IC(F^{-1}(t); F, T)^2 \, dt,$$

it follows that $A(F, T) \leqslant A(F_0, T)$; hence the minimax property holds.

Now consider the R-estimate. The optimal scores function $J(t)$ is given by

$$J(F_0(x)) = -\frac{f_0'(x)}{f_0(x)}.$$

The value of the influence function at $x = F^{-1}(t)$ is

$$IC(F^{-1}(t); F, t) = \frac{J(t)}{\int J'(F(x)) f(x)^2 \, dx} = \frac{J(t)}{\int J'(s) f(F^{-1}(s)) \, ds}.$$

Since $J'(t) = 0$ outside of the interval $(F_0(x_0), F_0(x_1))$, and since in this interval

$$f(F^{-1}(t)) \geqslant f_0(F^{-1}(t)) \geqslant f_0(F_0^{-1}(t)),$$

we conclude that, for $t \geqslant \frac{1}{2}$,

$$IC(F^{-1}(t); F, T) \leqslant IC(F_0^{-1}(t); F_0, T);$$

hence, as above,

$$A(F, T) \leqslant A(F_0, T),$$

and the minimax property holds.

Example 7.1 In the ε-contaminated normal case, the least informative distribution F_0 is given by (5.20) and (5.21), and all of the following three estimates are asymptotically minimax:

(1) The M-estimate with ψ given by (5.22).
(2) The α-trimmed mean with $\alpha = F_0(-k) = (1 - \varepsilon)\Phi(-k) + \varepsilon/2$.
(3) The R-estimate defined through the scores generating function $J(t) = \psi(F_0^{-1}(t))$, that is,

$$
\begin{aligned}
J(t) &= -k, & \text{for } t \leqslant \alpha \\
&= \Phi^{-1}\left(\frac{t - \varepsilon/2}{1 - \varepsilon}\right), & \text{for } \alpha \leqslant t \leqslant 1 - \alpha \\
&= k, & \text{for } t \geqslant 1 - \alpha.
\end{aligned}
$$

4.8   DESCENDING *M*-ESTIMATES

We have already noted that the least informative distributions tend to have exponential tails, that is, they might be slimmer (!) than what we would expect in practice. So it might be worthwhile to increase the maximum risk slightly beyond its minimax value in order to gain a better performance at very long-tailed distributions.

This can be done as follows. Consider *M*-estimates, and minimize the maximal asymptotic variance subject to the side condition

$$\psi(x)=0, \qquad \text{for } |x|>c, \tag{8.1}$$

where c can be chosen arbitrarily.

For ε-contaminated normal distributions, the solution is of the following form:

$$
\begin{aligned}
\psi(x)=-\psi(-x)&=x, & \text{for } 0 \leqslant x \leqslant a \\
&= b \tanh[\tfrac{1}{2}b(c-x)], & \text{for } a \leqslant x \leqslant c \\
&= 0, & \text{for } x \geqslant c;
\end{aligned}
\tag{8.2}
$$

see Exhibit 4.8.1. The values of a and b, of course, depend on ε.

The above estimate is a maximum likelihood estimate based on a truncated sample, for an underlying density

$$
\begin{aligned}
f_0(x)=f_0(-x)&=(1-\varepsilon)\varphi(x), & \text{for } 0 \leqslant x \leqslant a \\
&= \frac{(1-\varepsilon)\varphi(a)}{\cosh^2\left[\tfrac{1}{2}b(c-a)\right]} \cosh^2[\tfrac{1}{2}b(c-x)], & \text{for } a \leqslant x \leqslant c \\
&= (1-\varepsilon)\varphi(x), & \text{for } x \geqslant c.
\end{aligned}
\tag{8.3}
$$

Note that this density is discontinuous at $\pm c$. In order that f_0 integrate to 1, we must have

$$2\int_a^c \left[f_0(x)-(1-\varepsilon)\varphi(x) \right] dx = \varepsilon; \tag{8.4}$$

this gives one relation between ε and a, b; the other one is continuity of ψ at a:

$$a=b\tanh\left[\tfrac{1}{2}b(c-a)\right]. \tag{8.5}$$

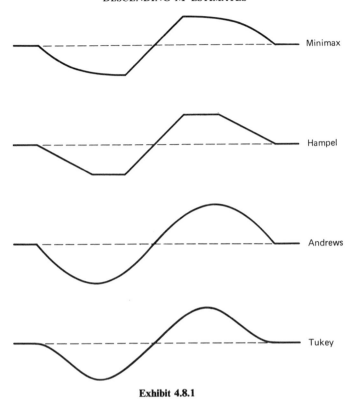

Minimax

Hampel

Andrews

Tukey

Exhibit 4.8.1

This solution can be found by essentially the same variational methods as used in Section 4.5; for a given F the best choice of ψ is

$$\psi_F(x) = -\frac{f'(x)}{f(x)}, \qquad \text{for } |x| < c$$
$$= 0, \qquad\qquad \text{otherwise,} \qquad (8.6)$$

and the corresponding asymptotic variance is $1/I_c(F)$, with

$$I_c(F) = \int \psi_F^2 \, dF; \qquad (8.7)$$

compare Section 6.3. Now minimize $I_c(F)$; the variational conditions imply that $-4\sqrt{f_0}\,''/\sqrt{f_0} = \text{const.}$ on the set where $f_0(x) > (1-\varepsilon)\varphi(x)$, and that $\psi_{F_0}(\pm c) = 0$. This yields (8.2) to (8.5), and it only remains to check that this indeed is a solution. For details see Collins (1976).

	c	b	a	$1/I_c(F_0)$
$\varepsilon=0.01$	2	2.747	1.539	1.727
	3	2.451	2.032	1.166
	4	2.123	2.055	1.082
	5	1.991	1.982	1.068
	∞	1.945	1.945	1.065
$\varepsilon=0.05$	2	1.714	1.105	2.640
	3	1.693	1.460	1.503
	4	1.550	1.488	1.314
	5	1.461	1.445	1.271
	∞	1.399	1.399	1.256
$\varepsilon=0.10$	2	1.307	0.838	4.129
	3	1.376	1.171	1.963
	4	1.289	1.220	1.621
	5	1.217	1.194	1.532
	∞	1.140	1.140	1.490
$\varepsilon=0.25$	2	0.692	0.356	21.741
	3	0.912	0.711	4.575
	4	0.905	0.810	3.089
	5	0.865	0.820	2.683
	∞	0.766	0.766	2.397

Exhibit 4.8.2   The minimax descending M-estimate [cf. (8.2) and (8.3)].

Exhibit 4.8.2 shows some of the quantitative aspects. The last column gives the maximal risk $1/I_c(F_0)$. Clearly, a choice $c \geqslant 5$ will increase it only by a negligible amount beyond its minimax value ($c=\infty$), but a choice $c \leqslant 3$ may have quite poor consequences. In other words it appears that descending ψ-functions are much more sensitive to wrong scaling than monotone ones.

The actual performance of such an estimate does not seem to depend very much on the exact shape of ψ. Other proposals for redescending M-estimates have been Hampel's piecewise linear function:

$$
\begin{aligned}
\psi(x)=-\psi(-x)&=x, && \text{for } 0 \leqslant x \leqslant a \\
&=a, && \text{for } a \leqslant x < b \\
&=\frac{c-x}{c-b}a, && \text{for } b \leqslant x < c \\
&=0, && \text{for } x \geqslant c;
\end{aligned}
\tag{8.8}
$$

Andrews' sine wave:

$$\psi(x) = \sin(x), \qquad \text{for } -\pi \leqslant x \leqslant \pi$$

$$= 0, \qquad \text{otherwise;} \tag{8.9}$$

and Tukey's biweight:

$$\psi(x) = x(1 - x^2)^2, \qquad \text{for } |x| \leqslant 1$$

$$= 0, \qquad \text{otherwise.} \tag{8.10}$$

Compare Andrews et al. (1972) and Exhibit 4.8.1.

When choosing a redescending ψ, we must take care that it does not descend too steeply; if it does contamination sitting on the slopes may play havoc with the denominator in the expression for the asymptotic variance

$$A(F, T) = \frac{\int \psi^2 \, dF}{\left(\int \psi' \, dF \right)^2}.$$

This effect is particularly harmful when a large negative value $\psi'(x)$ combines with a large positive value $\psi(x)^2$, and there is a cluster of outliers near x. (Some people have used quite dangerous Hampel estimates in their computer programs, with slopes between b and c that are much too steep.)

A Word of Caution

It seems to me that, in some discussions, the importance of using redescending ψ-functions has been exaggerated beyond all proportion. They are certainly beneficial if there are extreme outliers, but the improvement is relatively minor (a few percent of the asymptotic variance) and is counterbalanced by an increase of the minimax risk. If we are really interested in these few percentage points of potential improvement, then a removal of the "impossible" data points through a careful data screening based on physical expertise might be more effective and less risky than the routine use of a poorly tuned redescending ψ. Note, in particular, the increased sensitivity to a wrong scale. Unless we are careful we may even get trapped in a local minimum of $\Sigma \rho(x_i - T_n)$. The situation gets particularly acute in multiparameter regression.

4.9   QUESTIONS OF ASYMMETRIC CONTAMINATION

In the preceding sections we have determined estimates minimizing the maximal asymptotic variance over some subset of $\mathcal{P} = \mathcal{P}_\varepsilon$; only symmetric F, or, slightly more generally, only those $F \in \mathcal{P}$ were admitted whose bias for the selected estimate was zero, $T(F) = 0$. We now check the behavior of these estimates over the rest of $\mathcal{P}_\varepsilon$.

We have to answer two questions:

(1)   How large is the maximal asymptotic bias $b(\varepsilon)$ and how does it compare to the bias of the median (which is minimax, Section 4.2)?

(2)   How large is the maximal asymptotic variance $v_a(\varepsilon)$ when F ranges over *all* of $\mathcal{P}_\varepsilon$, and how does it compare to the restricted maximal asymptotic variance $v_s(\varepsilon)$, where F ranges only over the symmetric $F \in \mathcal{P}_\varepsilon$?

The discussion of breakdown properties (Sections 3.2 to 3.4) suggests that L-estimates are more sensitive to asymmetries than either M- or R-estimates. We therefore restrict ourselves to α-trimmed means and ε-contaminated normal distributions.

For small ε we have [see (1.5.8)]

$$b(\varepsilon) \cong \varepsilon \sup_x |IC(x; \Phi, T)|. \tag{9.1}$$

We thus tabulate $b(\varepsilon)/\varepsilon$ in order to obtain more nearly constant numbers; see Exhibit 4.9.1. The bottom row ($\alpha = 0.5$) corresponds to the median.

					ε						
α	0	0.01	0.02	0.05	0.1	0.15	0.2	0.25	0.3	0.4	0.5
0.01	2.37	2.71	∞	∞	∞	∞	∞	∞	∞	∞	∞
0.02	2.14	2.26	2.51	∞	∞	∞	∞	∞	∞	∞	∞
0.05	1.83	1.88	1.94	2.27	∞	∞	∞	∞	∞	∞	∞
0.1	1.60	1.63	1.66	1.78	2.13	∞	∞	∞	∞	∞	∞
0.15	1.48	1.50	1.53	1.60	1.77	2.10	∞	∞	∞	∞	∞
0.2	1.40	1.42	1.44	1.50	1.63	1.80	2.12	∞	∞	∞	∞
0.25	1.35	1.37	1.38	1.44	1.54	1.67	1.85	2.18	∞	∞	∞
0.3	1.31	1.33	1.34	1.39	1.48	1.59	1.73	1.93	2.29	∞	∞
0.4	1.26	1.28	1.29	1.33	1.41	1.51	1.62	1.76	1.95	2.73	∞
0.5	1.25	1.26	1.28	1.32	1.39	1.48	1.59	1.72	1.89	2.42	∞

Exhibit 4.9.1   Maximal bias of α-trimmed means for ε-contaminated normal distributions (tabulated: $b(\varepsilon)/\varepsilon$).

α	0	0.01	0.02	0.05	0.1	0.15	0.2	0.25	0.3	0.4	0.5
0.01	1.004	1.08	∞	∞	∞	∞	∞	∞	∞	∞	∞
	1	∞	∞	∞	∞	∞	∞	∞	∞	∞	∞
0.02	1.009	1.07	1.14	∞	∞	∞	∞	∞	∞	∞	∞
	1	1.0065	∞	∞	∞	∞	∞	∞	∞	∞	∞
0.05	1.027	1.07	1.12	1.30	∞	∞	∞	∞	∞	∞	∞
	1	1.0017	1.0084	∞	∞	∞	∞	∞	∞	∞	∞
0.1	1.061	1.09	1.13	1.26	1.54	2.03	∞	∞	∞	∞	∞
	1	1.0007	1.0031	1.027	∞	∞	∞	∞	∞	∞	∞
0.15	1.100	1.13	1.16	1.27	1.49	1.80	2.25	3.07	∞	∞	∞
	1	1.0004	1.0019	1.014	1.08	∞	∞	∞	∞	∞	∞
0.2	1.144	1.17	1.20	1.30	1.50	1.75	2.08	2.56	3.28	∞	∞
	1	1.0003	1.0013	1.010	1.05	1.18	∞	∞	∞	∞	∞
0.25	1.195	1.22	1.25	1.35	1.53	1.76	2.05	2.42	2.93	4.81	∞
	1	1.0003	1.0010	1.007	1.04	1.11	1.29	∞	∞	∞	∞
0.3	1.252	1.28	1.31	1.40	1.58	1.79	2.06	2.40	2.83	4.20	7.26
	1	1.0002	1.0009	1.006	1.03	1.08	1.18	1.45	∞	∞	∞
0.4	1.393	1.42	1.45	1.55	1.73	1.94	2.20	2.51	2.90	4.01	5.94
	1	1.0002	1.0007	1.005	1.02	1.06	1.12	1.24	1.47	∞	∞
0.5	1.571	1.60	1.64	1.74	1.94	2.17	2.45	2.79	3.21	4.36	6.28
	1	1.0002	1.0007	1.004	1.02	1.05	1.11	1.20	1.38	2.55	∞
Minimax bound for v_s	1.000	1.065	1.116	1.256	1.490	1.748	2.046	2.397	2.822	3.996	5.928

The header ε spans the ε-value columns.

Exhibit 4.9.2 Maximal symmetric and asymmetric variance of α-trimmed means for ε-contaminated normal distributions [tabulated: $v_s(\varepsilon)$, $v_a(\varepsilon)/v_s(\varepsilon)$].

Exhibit 4.9.2 is concerned with asymptotic variances; it tabulates $v_s(\varepsilon)$ and $v_a(\varepsilon)/v_s(\varepsilon)$ (cf. the second question above).

For the α-trimmed mean, asymptotic bias and variance apparently are maximized if the entire contaminating mass ε is put at $+\infty$. This is trivially true for the bias, and highly plausible (but not yet proved) for the variance. Calculating Exhibits 4.9.1 and 4.9.2 is, by the way, an instructive exercise in the use of several formulas derived in Section 3.3.

The following features deserve some comments. Note that $b(\varepsilon)/\varepsilon$ increases only very slowly with ε and in fact stays bounded right up to the breakdown point.

For small ε the excess of v_a beyond v_s is negligible (it is of the order ε^2). This gives an *a posteriori* justification for restricting attention to symmetric distributions when minimizing asymptotic variances. For larger ε, however, the discrepancies can become sizeable. Take $\varepsilon = 0.2$, and the 25%-trimmed mean, which is very nearly minimax there for symmetric contamination; then $v_a/v_s \cong 1.29$.

Exhibits 4.9.1 and 4.9.2 also illustrate the two breakdown points ε^* and ε^{**}, defined in Section 1.4: $b(\varepsilon) = \infty$ for $\varepsilon > \varepsilon^* = \alpha$, and $v_s(\varepsilon) = \infty$ for $\varepsilon \geqslant \varepsilon^{**} = 2\alpha$.

CHAPTER 5

Scale Estimates

5.1 GENERAL REMARKS

By *scale estimate* we denote any positive statistic S_n that is equivariant under scale transformations:

$$S_n(ax_1,\ldots,ax_n) = aS_n(x_1,\ldots,x_n), \qquad \text{for} \quad a > 0. \qquad (1.1)$$

Many scale estimates are also invariant under changes of sign and shifts:

$$S_n(-x_1,\ldots,-x_n) = S_n(x_1,\ldots,x_n), \qquad (1.2)$$

$$S_n(x_1+b,\ldots,x_n+b) = S_n(x_1,\ldots,x_n). \qquad (1.3)$$

Pure scale problems are rare; in practice scale usually occurs as a nuisance parameter in location, and, more generally, in regression problems. Therefore we should tune the properties of the scale estimate to that of the location estimate to which it is subordinated. For instance, we would not want to spoil the good breakdown properties of a location estimate by an early breakdown of the scale estimate. For related reasons it appears to be more important to keep the bias of the scale estimate small than to strive for a small (asymptotic) variance.

As a result the so-called median absolute deviation (MAD) has emerged as the single most useful ancillary estimate of scale. It is defined as the median of the absolute deviations from the median:

$$\mathrm{MAD}_n = \mathrm{med}\{|x_i - M_n|\}, \qquad (1.4)$$

where

$$M_n = \mathrm{med}\{x_i\}.$$

For symmetric distributions this is asymptotically equivalent to one-half of the interquartile distance, but it has better breakdown properties under ε-contamination ($\varepsilon^* = 0.5$, as against $\varepsilon^* = 0.25$ for the interquartile distance).

Note that this clashes with the widespread opinion that, because most of the information for scale sits in the tails, we should give more consideration to the tails, and thus use a lower rejection or trimming rate in scale problems. This may be true for the pure scale problem, but is not so when scale is just a nuisance parameter.

The pure scale problem is a kind of stepping stone toward more complex estimation problems. It has the advantage that it can be converted into a location problem by taking logarithms, so the machinery of the preceding chapters is applicable. But the distributions resulting from this transformation are highly asymmetric, and there is no natural scale (corresponding to the center of symmetry). In most cases it is convenient to standardize the estimates such that they are consistent at the ideal model distribution (cf. the remarks at the end of Section 1.2). For instance, in order to make MAD consistent at the normal distribution, we must divide it by $\Phi^{-1}(\frac{3}{4}) \cong$ 0.6745.

This chapter closely follows and parallels many sections of the preceding two chapters; we again concentrate on estimates that are functionals of the empirical distribution function, $S_n = S(F_n)$, and we again exploit the heuristic approach through influence functions.

As the asymptotic variance $A(F, S)$ of $\sqrt{n}\,[S(F_n) - S(F)]$ depends on the arbitrary standardization of S, it is a poor measure of asymptotic performance. We use the relative asymptotic variance of S instead, that is, the asymptotic variance

$$A(F, \log S) = \frac{A(F, S)}{S(F)^2} \tag{1.5}$$

of

$$\sqrt{n}\,\log \frac{S(F_n)}{S(F)}. \tag{1.6}$$

Another important scale-type problem concerns the estimation of the variability of a given estimate; we have briefly touched upon this topic in Section 1.5. In the classical normal theory, the two cases are often confounded—after all, the classical estimates for the standard error of a single observation and of the sample mean differ only by a factor $\sqrt{n}$ —but we must keep them conceptually separate. We defer the discussion of this kind of problem till the end of Chapter 6.

5.2 M-ESTIMATES OF SCALE

An M-estimate S of scale is defined by an implicit relation of the form

$$\int \chi\left(\frac{x}{S(F)}\right)F(dx)=0. \tag{2.1}$$

Typically (but not necessarily), χ is an even function $\chi(-x)=\chi(x)$. From (3.2.13) we obtain the influence function

$$IC(x; F, S)=\frac{\chi(x/S(F))S(F)}{\int \chi'(x/S(F))(x/S(F))F(dx)}. \tag{2.2}$$

Example 2.1 The maximum likelihood estimate of σ for the scale family of densities $\sigma^{-1}f(x/\sigma)$ is an M-estimate with

$$\chi(x)=-x\frac{f'(x)}{f(x)}-1. \tag{2.3}$$

Example 2.2 Huber (1964) proposed the choice

$$\chi(x)=x^2-\beta, \qquad \text{for } |x|\leqslant k$$

$$=k^2-\beta, \qquad \text{for } |x|>k, \tag{2.4}$$

for some constant k, with β determined such that $S(\Phi)=1$, that is, $\int \chi(x)\Phi(dx)=0$.

Example 2.3 The choice

$$\chi(x)=\text{sign}(|x|-1) \tag{2.5}$$

yields the median absolute deviation $S=\text{med}(|X|)$, that is, that number S for which $F(S)-F(-S)=\frac{1}{2}$. (More precisely, this is the median absolute deviation from 0, to be distinguished from the median absolute deviation from the median.)

Continuity and breakdown properties can be worked out just as in the location case in Section 3.2, except that everything is slightly more complicated. Therefore we only show how the breakdown point under ε-contamination can be worked out.

Assume that χ is even and monotone increasing for positive arguments. Let $\|\chi\| = \chi(\infty) - \chi(0)$. We write (2.1) as

$$\int \left[\chi\left(\frac{x}{S(F)}\right) - \chi(0) \right] F(dx) + \chi(0) = 0. \qquad (2.6)$$

Assuming the gross error model, it is easy to see that a contamination $\varepsilon > -\chi(0)/\|\chi\|$ at $|x| = \infty$ forces the left-hand side of (2.6) to be greater than 0 for all values of $S(F)$. Similarly, a contamination $\varepsilon > 1 + \chi(0)/\|\chi\|$ at 0 forces it to be less than 0 for all values of $S(F)$. [As $0 < -\chi(0)/\|\chi\| \leqslant \frac{1}{2}$ in the more interesting cases, we can usually disregard the second contingency.] On the other hand if ε satisfies the opposite strict inequalities, then the solution $S(F)$ of (2.6) is bounded away from 0 and ∞.

We conclude that, for ε-contamination (and also for Prohorov distance), the breakdown point is given by

$$\varepsilon^* = \frac{-\chi(0)}{\|\chi\|} \leqslant \frac{1}{2}. \qquad (2.7)$$

For indeterminacy in terms of Kolmogorov or Lévy distance, this number must be halved:

$$\varepsilon^* = -\frac{1}{2} \frac{\chi(0)}{\|\chi\|} \leqslant 0.25. \qquad (2.8)$$

The reason for this different behavior is as follows. By taking away a mass ε from the central part of a distribution F and moving one-half of it to the extreme left, the other half to the extreme right, we get a distribution that is within Prohorov distance ε, but within Lévy distance $\varepsilon/2$, of the original F.

5.3 L-ESTIMATES OF SCALE

The general results of Section 3.3 apply without much change. In view of scale invariance (1.1), only the following types of functionals appear feasible:

$$S(F) = \left[\int F^{-1}(t)^q M(dt) \right]^{1/q}, \qquad \text{with integral } q \neq 0, \qquad (3.1)$$

$$S(F) = \left[\int |F^{-1}(t)|^q M(dt) \right]^{1/q}, \qquad \text{with real } q \neq 0, \qquad (3.2)$$

$$S(F) = \exp\left[\int \log|F^{-1}(t)| M(dt) \right], \qquad \text{with } M\{(0,1)\} = 1. \qquad (3.3)$$

We encounter estimates of both the first type (interquantile range, trimmed variance) and the second type (median deviation), but in what follows now we consider only (3.1).

From (3.3.11) and the chain rule, we obtain the influence function

$$IC(x; F, S) = \frac{1}{S(F)^{q-1}} \left[\int s \frac{F^{-1}(s)^{q-1}}{f(F^{-1}(s))} M(ds) - \int_{F(x)}^{1} \frac{F^{-1}(s)^{q-1}}{f(F^{-1}(s))} M(ds) \right].$$

(3.4)

Or if M has a density m, then

$$\frac{d}{dx} IC(x; F, S) = \frac{x^{q-1}}{S(F)^{q-1}} m(F(x)).$$

(3.5)

Example 3.1 The t-quantile range

$$S(F) = F^{-1}(1-t) - F^{-1}(t), \qquad 0 < t < \tfrac{1}{2}$$

(3.6)

has the influence function

$$IC(x; F, S) = \frac{1}{f(F^{-1}(t))} - c(F), \qquad \text{for } x < F^{-1}(t)$$

$$= -c(F), \qquad \text{for } F^{-1}(t) < x < F^{-1}(1-t)$$

$$= \frac{1}{f(F^{-1}(1-t))} - c(F), \qquad \text{for } x > F^{-1}(1-t),$$

(3.7)

where

$$c(F) = t \left\{ \frac{1}{f(F^{-1}(t))} + \frac{1}{f(F^{-1}(1-t))} \right\}.$$

(3.8)

If F is symmetric these formulas simplify to

$$IC(x; F, S) = \frac{1-2t}{f(F^{-1}(t))}, \qquad \text{for } x < F^{-1}(t) \quad \text{or} \quad x > F^{-1}(1-t)$$

$$= \frac{-2t}{f(F^{-1}(t))}, \qquad \text{for } F^{-1}(t) < x < F^{-1}(1-t).$$

(3.9)

Then the asymptotic variance of $\sqrt{n}\,[S(F_n)-S(F)]$ is given by

$$A(F,S)=\frac{2t(1-2t)}{f(F^{-1}(t))^2},\tag{3.10}$$

and that of $\sqrt{n}\,\log[S(F_n)/S(F)]$ is

$$A(F,\log S)=\frac{2t(1-2t)}{\left[2F^{-1}(t)f(F^{-1}(t))\right]^2}.\tag{3.11}$$

Some numerical results are given in Exhibit 5.7.3. For example, the interquartile range ($t=0.25$) has an asymptotic variance $A(\Phi,\log S)=1.361$ and an asymptotic relative efficiency (relative to the standard deviation) of $0.5/1.361=0.3674$. The same is true for the MAD.

Example 3.2 The α-*trimmed variance* is defined as the suitably scaled variance of the α-trimmed sample:

$$S(F)^2=\gamma(\alpha)\int_\alpha^{1-\alpha}F^{-1}(s)^2\,ds.\tag{3.12}$$

The normalizing factor is fixed such that $S(\Phi)=1$, that is,

$$\frac{1}{\gamma(\alpha)}=\int_{-\xi}^\xi x^2\varphi(x)\,dx=1-2\alpha-2\xi\varphi(\xi)\tag{3.13}$$

with $\xi=\Phi^{-1}(1-\alpha)$. According to (3.5) the influence function of the α-trimmed variance then satisfies

$$\frac{d}{dx}IC(x;F,S)=\gamma(\alpha)\frac{x}{S(F)},\qquad\text{for }\alpha<F(x)<1-\alpha$$

$$=0,\qquad\qquad\text{otherwise};$$

hence

$$IC(x;F,S)=\frac{\gamma(\alpha)}{2S(F)}[F^{-1}(\alpha)^2-c(F)],\qquad\text{for }x<F^{-1}(\alpha)$$

$$=\frac{\gamma(\alpha)}{2S(F)}[x^2-c(F)],\qquad\text{for }F^{-1}(\alpha)<x<F^{-1}(1-\alpha)$$

$$=\frac{\gamma(\alpha)}{2S(F)}[F^{-1}(1-\alpha)^2-c(F)],\qquad\text{for }x>F^{-1}(1-\alpha),\tag{3.14}$$

where

$$c(F) = \int_{F^{-1}(\alpha)}^{F^{-1}(1-\alpha)} x^2 \, dF + \alpha \left[F^{-1}(\alpha)^2 + F^{-1}(1-\alpha)^2 \right] \quad (3.15)$$

is the α-Winsorized variance.

Example 3.3 Define

$$S(F) = \gamma(\alpha) \int_\alpha^{1-\alpha} F^{-1}(t) \Phi^{-1}(t) \, dt, \quad (3.16)$$

with $\gamma(\alpha)$ as in (3.13). Then the influence function can be found by integrating

$$\frac{d}{dx} IC(x; F, S) = \gamma(\alpha) \Phi^{-1}(F(x)), \quad \text{for } \alpha < F(x) < 1-\alpha$$

$$= 0, \quad \text{otherwise.} \quad (3.17)$$

All of the above functionals S also have a *symmetrized version* $\tilde{S}$, which is obtained as follows. Put

$$\bar{F}(x) = 1 - F(-x+0) \quad (3.18)$$

and

$$\tilde{F}(x) = \tfrac{1}{2} \left[F(x) + \bar{F}(x) \right]. \quad (3.19)$$

We say that $\tilde{F}$ is obtained from F by symmetrizing at the origin (alternatively, we could also symmetrize at the median, etc.). Then define

$$\tilde{S}(F) = S(\tilde{F}). \quad (3.20)$$

It is immediate that

$$IC(x; F, \tilde{S}) = \tfrac{1}{2} \left[IC(x; \tilde{F}, S) + IC(-x; \tilde{F}, S) \right]. \quad (3.21)$$

Thus if S is symmetric [i.e., $S(F) = S(\bar{F})$ for all F], and if the true underlying F is symmetric ($F = \bar{F}$), then $\tilde{S}(F) = S(F)$, and S and $\tilde{S}$ have the same influence function at F. Hence for symmetric F their asymptotic properties agree.

For asymmetric F (and also for small samples from symmetric underlying distributions) the symmetrized and nonsymmetrized estimates behave quite differently. This is particularly evident in their breakdown behavior. For example, take an estimate of the form (3.1), where M is either a positive measure (q even), or positive on $[\frac{1}{2}, 1]$, negative on $[0, \frac{1}{2}]$ (q odd), and let α be the largest real number such that $[\alpha, 1-\alpha]$ contains the support of M. Then according to Theorem 3.3.1 and the remarks preceding it, for a nonsymmetrized estimate of scale, breakdown occurs at $\varepsilon^* = \alpha$ (for ε-contamination, total variation, Prohorov, Lévy, or Kolmogorov distance).

For the symmetrized version breakdown still happens at $\varepsilon^* = \alpha$ with Lévy and Kolmogorov distance, but is boosted to $\varepsilon^* = 2\alpha$ in the other three cases. It appears therefore that symmetrized scale estimates are generally preferable.

Example 3.4   Let S be one-half of the interquartile distance:

$$S(F) = \tfrac{1}{2}\left[F^{-1}\!\left(\tfrac{3}{4}\right) - F^{-1}\!\left(\tfrac{1}{4}\right) \right].$$

Then $\tilde{S}$ is the median absolute deviation (Example 2.3).

5.4   *R*-ESTIMATES OF SCALE

Rank tests for scale compare relative scale between two or more samples; there is no substitute for the left-right symmetry that makes one-sample rank-tests and estimates of location possible (though we can obtain surrogate one-sample rank tests and estimates for scale if we use a synthetic second sample, e.g., expected order statistics from a normal distribution). A brief sketch of a possible approach should suffice.

Let $(x_1, \ldots, x_m)$ and $(y_1, \ldots, y_n)$ be the two samples, and let R_i be the rank of x_i in the pooled sample of size $N = m + n$. Then form the test statistic

$$\sum_{i=1}^{m} a(R_i) \tag{4.1}$$

with $a_i = a(i)$ defined by

$$a_i = N \int_{(i-1)/N}^{i/N} J(s)\, ds \tag{4.2}$$

for some scores generating function J, just as in Section 3.4. Typically, J is a function of $|t - \frac{1}{2}|$, for example,

$$J(t) = |t - \tfrac{1}{2}| - \tfrac{1}{4} \qquad \text{(Ansari-Bradley-Siegel-Tukey)}, \qquad (4.3)$$

$$J(t) = (t - \tfrac{1}{2})^2 - \tfrac{1}{12} \qquad \text{(Mood)}, \qquad (4.4)$$

$$J(t) = \Phi^{-1}(t)^2 - 1 \qquad \text{(Klotz)}. \qquad (4.5)$$

We convert such tests into estimates of relative scale. Let $0 < \lambda < 1$ be fixed (we shall later choose $\lambda = m/N$), and define a functional $S = S(F, G)$ such that

$$\int J\left[\lambda F(x) + (1 - \lambda)G\left(\frac{x}{S}\right)\right] F(dx) = 0 \qquad (4.6)$$

or, preferably [after substituting $F(x) = t$],

$$\int J\left(\lambda t + (1 - \lambda)G\left(\frac{F^{-1}(t)}{S}\right)\right) dt = 0. \qquad (4.7)$$

If we assume that $\int J(t)\, dt = 0$, then $S(F, G)$, if well defined by (4.7), is a measure of relative scale satisfying

$$S(F_{aX}, F_X) = a, \qquad (4.8)$$

where F_{aX} denotes the distribution of the random variable aX.

We now insert $F_u = (1 - u)F + uF_1$ and $G_u = (1 - u)G + uG_1$ into (4.7) and differentiate with respect to u at $u = 0$. If $F = G$, the resulting expressions remain quite manageable; we obtain

$$\dot{S} = \left[\frac{d}{du} S(F_u, G_u)\right]_{u=0} = \frac{\int J(F(x))F_1(dx) - \int J(F(x))G_1(dx)}{\int J'(F(x))xf(x)^2\, dx}. \qquad (4.9)$$

The Gâteaux derivatives of $S(F, G)$ with respect to F and G, at $F = G$, can now be read off from (4.9).

If both samples come from the same F, and if we insert the respective empirical distributions F_m and G_n for F_1 and G_1, we obtain the Taylor expansion (with $u = 1$)

$$S(F_m, G_n) = 1 + \dot{S} + \cdots \qquad (4.10)$$

or, approximately (with $\lambda = m/N$),

$$\sqrt{N}\,(S(F_m, G_n) - 1) \cong \frac{\sqrt{\dfrac{1}{\lambda}}\,\dfrac{1}{\sqrt{m}}\sum J(F(x_i)) - \sqrt{\dfrac{1}{1-\lambda}}\,\dfrac{1}{\sqrt{n}}\sum J(F(y_j))}{\int J'(F(x))xf(x)^2\,dx}.$$

(4.11)

We thus can expect that (4.11) is asymptotically normal with mean 0 and variance

$$A(F, S) = \frac{1}{\lambda(1-\lambda)}\frac{\int J(t)^2\,dt}{\left[\int J'(F(x))xf(x)^2\,dx\right]^2}$$

(4.12)

This should hold if m and n go to ∞ comparably fast; if $m/n \to 0$, then $\sqrt{m}\,[S(F_m, G_n) - 1]$ will be asymptotically normal with the same variance (4.12), except that the factor $1/[\lambda(1-\lambda)]$ is replaced by 1.

The above derivations of course are only heuristic; for a rigorous theory we would have to refer to the extensive literature on the behavior of rank tests under alternatives, in particular Hájek (1968) and Hájek and Dupač (1969). These results on tests can be translated in a relatively straightforward way into results about the behavior of estimates; compare Section 10.6.

5.5 ASYMPTOTICALLY EFFICIENT SCALE ESTIMATES

The parametric pure scale problem corresponds to estimating σ for the family of densities

$$p(x; \sigma) = \frac{1}{\sigma}f\left(\frac{x}{\sigma}\right), \qquad \sigma > 0.$$

(5.1)

As

$$\frac{\partial}{\partial \sigma}\log p(x; \sigma) = -\frac{f'(x/\sigma)}{f(x/\sigma)}\frac{x}{\sigma^2} - \frac{1}{\sigma},$$

(5.2)

Fisher information for scale is

$$I(F; \sigma) = \frac{1}{\sigma^2} \int \left[-\frac{f'(x)}{f(x)} x - 1 \right]^2 f(x)\, dx. \tag{5.3}$$

Without loss of generality we assume now that the true scale is $\sigma = 1$. Evidently, in order to obtain full asymptotic efficiency at F, we should arrange that

$$IC(x; F, S) = \frac{1}{I(F; 1)} \left[-\frac{f'(x)}{f(x)} x - 1 \right]; \tag{5.4}$$

see Section 3.5.

Thus for M-estimates (2.1) it suffices to choose, up to a multiplicative constant,

$$\chi(x) = -\frac{f'(x)}{f(x)} x - 1. \tag{5.5}$$

For an L-estimate (3.1) the proper choice is a measure M with density m given by

$$m(F(x)) = -\frac{[f'(x)/f(x)]' x^{-q+1}}{I(F; 1)}. \tag{5.6}$$

For R-estimates of relative scale, one should choose, up to an arbitrary multiplicative constant,

$$J(F(x)) = -\frac{f'(x)}{f(x)} x - 1. \tag{5.7}$$

Example 5.1 Let $f(x) = \varphi(x)$ be the standard normal density. Then the asymptotically efficient M-estimate is, of course,

$$S(F_n)^2 = \frac{1}{n} \sum x_i^2.$$

The efficient L-estimate for $q = 2$ is exactly the same. With $q = 1$, we obtain $m(t) = \Phi^{-1}(t)$, that is,

$$S(F) = \int F^{-1}(t) \Phi^{-1}(t)\, dt;$$

thus

$$S(F_n) = \sum a_i x_{(i)}$$

with

$$a_i = \int_{(i-1)/n}^{i/n} \Phi^{-1}(t)\, dt.$$

The efficient R-estimate corresponds to the Klotz test (4.5).

5.6 DISTRIBUTIONS MINIMIZING FISHER INFORMATION FOR SCALE

Let $\mathscr{P}$ be a convex set of distribution functions that is such that, with each $F \in \mathscr{P}$, it also contains its mirror image $\bar{F}$ and thus its symmetrization $\tilde{F}$ [cf. (3.19) and (3.20)]. Assume that the observations X_i are distributed according to $F(x/\sigma)$, where σ is to be estimated.

We note that, for any parametric family of densities $f(x; \theta)$, Fisher information is convex: on any segment $f_t(x, \theta) = (1-t)f_0(x, \theta) + tf_1(x, \theta)$, $0 \leqslant t \leqslant 1$,

$$I(f_t; \theta) = \int \frac{[\partial/\partial\theta f_t(x; \theta)]^2}{f_t(x; \theta)}\, dx \tag{6.1}$$

is a convex function of t according to Lemma 4.4.4.

Clearly, F and $\bar{F}$ have the same Fisher information for scale, and it follows that

$$I(\tilde{F}; \sigma) \leqslant I(F; \sigma) = I(\bar{F}; \sigma).$$

Hence it suffices to consider symmetric distributions F when minimizing Fisher information for scale.

Then $Y_i = \log|X_i|$ is a sufficient statistic for X_i, and it has the distribution function $F^*(y - \tau)$, where

$$F^*(y) = F(e^y) - F(-e^y) \tag{6.2}$$

and

$$\tau = \log \sigma. \tag{6.3}$$

The corresponding density is

$$f^*(y) = 2e^y f(e^y).$$ (6.4)

Note that Fisher information for scale σ

$$I(F; \sigma) = \frac{1}{\sigma^2} \int \left[-x \frac{f'(x)}{f(x)} - 1 \right]^2 f(x)\, dx$$ (6.5)

agrees, apart from the factor $1/\sigma^2$, with Fisher information for location τ:

$$I(F^*; \tau) = \int \left[\frac{d}{dy} \log f^*(y) \right]^2 f^*(y)\, dy$$

$$= \int \left[-x \frac{f'(x)}{f(x)} - 1 \right]^2 f(x)\, dx.$$ (6.6)

Thus minimizing $I(F^*; \tau)$ is equivalent to minimizing $\sigma^2 I(F; \sigma)$; for reasons of scale invariance we should prefer this latter expression to $I(F; \sigma)$ anyway.

Moreover, if a set $\mathcal{P}$ of distributions is convex, then the transformed set $\mathcal{P}^* = \{F^* | F \in \mathcal{P}\}$ is also convex, and the methods and results of Sections 4.4 and 4.5 apply to $\mathcal{P}^*$. In particular, as ε-contamination in F is transformed into ε-contamination in F^*, the treatment of the gross error model goes through without change. For the other neighborhoods some more care is needed.

We consider only the ε-contaminated normal case. Let φ be the standard normal density; then

$$\varphi^*(y) = \sqrt{\frac{2}{\pi}} \, \exp\left(y - \tfrac{1}{2} e^{2y} \right);$$ (6.7)

thus

$$-\log \varphi^*(y) = \tfrac{1}{2} e^{2y} - y + \tfrac{1}{2} \log\left(\tfrac{1}{2} \pi \right)$$ (6.8)

is convex, and

$$[-\log \varphi^*(y)]' = e^{2y} - 1$$ (6.9)

is monotone.

Example 4.5.2 now shows how to find a distribution minimizing Fisher information. We have to distinguish two cases.

Case A. *Large ε.* Define two numbers $y_0 \leqslant y_1$ by

$$e^{2y_0} - 1 = -k,$$

$$e^{2y_1} - 1 = k, \tag{6.10}$$

where $k < 1$ is related to ε by

$$\int_{y_0}^{y_1} \varphi^*(y)\, dy + \frac{\varphi^*(y_0) + \varphi^*(y_1)}{k} = \frac{1}{1-\varepsilon}. \tag{6.11}$$

The least informative element of $\mathscr{P}^*$ then has the density

$$f_0^*(y) = (1-\varepsilon)\varphi^*(y_0)e^{k(y-y_0)}, \qquad \text{for } y < y_0$$

$$= (1-\varepsilon)\varphi^*(y), \qquad \text{for } y_0 \leqslant y \leqslant y_1$$

$$= (1-\varepsilon)\varphi^*(y_1)e^{-k(y-y_1)}, \qquad \text{for } y > y_1. \tag{6.12}$$

If we transform these equations back into x-space, we obtain, instead of (6.10) to (6.12),

$$x_0^2 = (1-k)^+,$$

$$x_1^2 = 1+k, \tag{6.13}$$

$$2\int_{x_0}^{x_1} \varphi(x)\, dx + \frac{2x_0\varphi(x_0) + 2x_1\varphi(x_1)}{x_1^2 - 1} = \frac{1}{1-\varepsilon}, \tag{6.14}$$

$$f_0(x) = (1-\varepsilon)\varphi(x_0)\left(\frac{x_0}{|x|}\right)^{(x_0^2)}, \qquad \text{for } |x| < x_0$$

$$= (1-\varepsilon)\varphi(x), \qquad \text{for } x_0 \leqslant |x| \leqslant x_1$$

$$= (1-\varepsilon)\varphi(x_1)\left(\frac{x_1}{|x|}\right)^{(x_1^2)}, \qquad \text{for } |x| > x_1. \tag{6.15}$$

Case B. *Small ε* In this case, the left boundary point is $y_0 = -\infty$, and correspondingly $x_0 = 0$. Nothing else is changed, and formulas (6.13) to (6.15) remain valid as they stand (with $k \geqslant 1$).

Note that Case A yields a highly pathological least informative distribution F_0; its density is ∞ at $x=0$. In Case B, F_0 corresponds to a distribution that is normal in the middle and behaves like a t-distribution with $k=x_1^2-1 \geqslant 1$ degrees of freedom in the tails. The boundary case between Cases A and B corresponds to $x_0=0$, $x_1=\sqrt{2}$, and $\varepsilon=0.205$. Exhibit 5.6.1 shows some numerical results.

We now determine the asymptotically efficient M- and L-estimates of scale for these least informative distributions (cf. Section 5.5). The efficient M-estimate (2.1) of scale is defined by

$$\chi(x)= -x\frac{f_0'(x)}{f_0(x)}-1=x_0^2-1, \quad \text{for } |x|<x_0$$

$$= x^2-1, \quad \text{for } x_0 \leqslant |x| \leqslant x_1$$

$$= x_1^2-1, \quad \text{for } |x|>x_1. \quad (6.16)$$

ε	x_0	x_1	$F_0(-x_0)$	$F_0(-x_1)$	$1/I(F_0^*)$
0	0	∞	0.5	0	0.50
0.001	0	2.88	0.5	0.002	0.52
0.002	0	2.70	0.5	0.004	0.53
0.005	0	2.46	0.5	0.009	0.56
0.01	0	2.27	0.5	0.016	0.60
0.02	0	2.07	0.5	0.029	0.66
0.05	0	1.81	0.5	0.059	0.81
0.1	0	1.62	0.5	0.098	1.02
0.15	0	1.50	0.5	0.132	1.23
0.20	0	1.42	0.5	0.162	1.45
0.205	0	1.414	0.5	0.165	1.472
0.25	0.35	1.37	0.388	0.182	1.72
0.30	0.45	1.34	0.357	0.192	1.98
0.40	0.60	1.28	0.313	0.210	2.82
0.50	0.70	1.23	0.299	0.223	4.16
0.65	0.81	1.16	0.267	0.237	8.72
0.80	0.90	1.09	0.255	0.246	28.6
1	1	1	0.25	0.25	∞

Exhibit 5.6.1 The ε-contaminated normal distributions that are least favorable for scale.

The efficient L-estimate [(3.1) with $q=2$] is a kind of trimmed variance, in Case A trimmed also on the inside; its weight density is given by

$$m(t) = \frac{2}{I(F_0^*, \tau)}, \qquad \begin{array}{l} \text{for } F_0(x_0) < t < F_0(x_1) \\ \text{and for } F_0(-x_1) < t < F_0(-x_0) \end{array}$$

$$= 0, \qquad\qquad \text{otherwise.} \qquad\qquad (6.17)$$

The limiting case $\varepsilon \to 1$ leads to an interesting nontrivial estimate, just as in the location case: the limiting M- and L-estimates of τ then coincide with the median of $\{\log|x_i|\}$, so the corresponding estimate of σ is the median of $\{|x_i|\}$. Hence the median absolute deviation [cf. (1.4), Example 2.3, and Example 3.4] is the candidate for being the "most robust estimate of scale."

Note that the above estimates (6.16) and (6.17) are biased when applied to normal data; in order to make them asymptotically unbiased at Φ, we have to divide them by a suitable constant, namely $S(\Phi)$ (see Exhibits 5.7.1 to 5.7.3 for these constants). Equivalently, we could replace the subtractive constant 1 in (6.16) by a different number β such that $E_\Phi \chi = 0$.

5.7   MINIMAX PROPERTIES

The general results of Section 4.6 show that the M-estimate determined in the preceding section is minimax with regard to asymptotic variance for the collection of ε-contaminated normal distributions satisfying $S(F) = 1$, that is,

$$\int \chi(x) F(dx) = 0. \qquad\qquad (7.1)$$

This is a rather restrictive condition, particularly in Case B, where $x_0 = 0$ and $x_1 > \sqrt{2}$; it means that those and only those distributions $F = (1 - \varepsilon)\Phi + \varepsilon H$ that put all their contaminating mass εH outside of $[-x_1, x_1]$ are admitted for competition. For any such distribution the asymptotic behavior of S is the same:

$$S(F) = S(F_0) = 1,$$

$$A(F, S) = A(F_0, S) = \frac{1}{I(F_0; 1)}.$$

Is it possible to remove this inconvenient condition (7.1)? We have the partial answer that this is indeed the case, provided ε is sufficiently small ($\varepsilon \leqslant 0.04$ and thus $x_1 \geqslant 1.88$ suffices); it would be interesting to know the precise range of ε-values for which it is true.

The point of the whole problem is, of course, that we defined our asymptotic loss as $A(F, S)/S(F)^2$. Thus if we move some contamination from the outside to the inside of $[-x_1, x_1]$, we decrease both the numerator and the denominator, and it is not evident whether the quotient decreases or increases.

From the influence function (2.2) we obtain that

$$\frac{S(F)^2}{A(F, S)} = \frac{\left[\int \chi'(x/S)(x/S)F(dx) \right]^2}{\int \chi(x/S)^2 F(dx)} \tag{7.2}$$

with the side condition (determining S)

$$\int \chi\left(\frac{x}{S}\right) f(dx) = 0, \tag{7.3}$$

where

$$\chi(x) = x^2 - 1, \qquad \text{for } |x| \leqslant x_1$$

$$= x_1^2 - 1, \qquad \text{for } |x| > x_1 \geqslant \sqrt{2}.$$

We have to show that F_0 minimizes (7.2) among *all*

$$F \in \mathcal{P}_\varepsilon = \{ F \mid F = (1-\varepsilon)\Phi + \varepsilon H, \ H \in \mathcal{M} \},$$

not only among those F satisfying (7.1).

We first note that the subsets of $\mathcal{P}_\varepsilon$ for which $S(F)$ has a given fixed value are convex, and that on each of them (7.2) is a convex function of F (Lemma 4.4.4).

Moreover, if we keep $S(F)$ fixed, then (7.2) is minimized by a contamination εH sitting on $\{0\} \cup [-x_1, x_1]^c$. The intuitive reason for this is that such a contamination evidently minimizes the numerator under the side condition $S(F) = \text{const.}$, and it also maximizes the variance of $\chi(x/S)$, that is, the denominator, as it sits on the extreme values of χ [note that $S(F) \leqslant 1$]. It is not difficult to make this intuitive reasoning precise by a variational argument (in view of convexity, local properties suffice); the details are left to the reader.

It is convenient to substitute

$$\chi(x) = \psi(x)^2 - 1; \tag{7.4}$$

then (7.2) and (7.3) can be rewritten as

$$\frac{S(F)^2}{A(F,S)} = \frac{\left[\int_{-Sx_1}^{Sx_1} (x/S)^2 F(dx) \right]^2}{\int \psi(x/S)^4 F(dx) - 1}, \tag{7.5}$$

$$\int \psi\left(\frac{x}{S}\right)^2 F(dx) = 1. \tag{7.6}$$

As it suffices to minimize (7.5) over contaminations sitting on $\{0\} \cup [-x_1, x_1]^c$, we now assume that εH puts mass $\varepsilon - \varepsilon_1$ on $\{0\}$, and mass ε_1 on $[-x_1, x_1]^c$. then (7.5) and (7.6) are further transformed into

$$\frac{S(F)^2}{A(F,S)} = \frac{\left[(1-\varepsilon) \int_{-Sx_1}^{Sx_1} (x/S)^2 \Phi(dx) \right]^2}{(1-\varepsilon) \int \psi(x/S)^4 \Phi(dx) + \varepsilon_1 x_1^4 - 1}, \tag{7.7}$$

$$(1-\varepsilon) \int \psi\left(\frac{x}{S}\right)^2 \Phi(dx) + \varepsilon_1 x_1^2 = 1. \tag{7.8}$$

We now have to find out for which values of ε the choice $\varepsilon_1 = \varepsilon$ minimizes (7.7) subject to the side condition (7.8).

From (7.8) we can determine the derivative of S with respect to ε_1. If $\varepsilon \le 0.04$, we find (with the aid of some numerical calculations) that the numerator and denominator of (7.7) have a negative and a positive derivative with respect to ε_1, respectively, and this is true over the entire range possible for S. Hence for $\varepsilon \le 0.04$ the minimum of (7.7) is reached at $\varepsilon_1 = \varepsilon$. For larger ε the situation becomes more complicated, and we do not know whether the result remains true.

Exhibits 5.7.1 to 5.7.3 compare the asymptotic performances of several estimates of scale for a normal distribution, symmetrically ε-contaminated near $\pm \infty$. To facilitate comparisons, the values of x_1 in Exhibit 5.7.1 were adjusted so that the performance at the normal distribution agrees with that of an α-trimmed standard deviation. In Exhibits 5.7.1 and 5.7.2 the value ε_{min} indicates for which least informative distribution the estimate is asymptotically efficient (cf. Exhibit 5.6.1).

x_1 (α)	ε_{min}	$S(\Phi)$	$S(F)/S(\Phi)$ and $A(F, \log S)$ for ε								
			0	0.005	0.01	0.02	0.05	0.10	0.15	0.20	0.25
2.370 (0.01)	0.0069	0.982	1.000	1.013	1.027	1.056	1.163	1.458	2.361	∞	∞
			0.530	0.566	0.605	0.697	1.138	3.677	38.14	∞	∞
2.130 (0.02)	0.016	0.964	1.000	0.011	1.022	1.046	1.128	1.323	1.690	3.045	∞
			0.557	0.581	0.607	0.665	0.909	1.854	6.059	91.43	∞
1.804 (0.05)	0.051	0.912	1.000	1.008	1.017	1.035	1.094	1.215	1.384	1.650	2.200
			0.640	0.654	0.668	0.698	0.810	1.110	1.741	3.525	12.97
1.555 (0.10)	0.123	0.824	1.000	1.007	1.014	1.028	1.075	1.165	1.276	1.419	1.615
			0.796	0.805	0.813	0.831	0.892	1.031	1.244	1.608	2.329
1.414 (0.149)	0.205	0.736	1.000	1.006	1.013	1.025	1.066	1.144	1.235	1.346	1.485
			0.989	0.995	1.001	1.014	1.056	1.146	1.269	1.449	1.728
1.311 (0.20)	—	0.642	1.000	1.006	1.012	1.024	1.061	1.131	1.212	1.308	1.422
			1.257	1.262	1.266	1.276	1.308	1.372	1.455	1.566	1.720
1.234 (0.25)	—	0.547	1.000	1.006	1.011	1.022	1.058	1.124	1.199	1.285	1.386
			1.630	1.633	1.637	1.645	1.669	1.718	1.778	1.855	1.956

Exhibit 5.7.1 Huber's scale; $\int \chi[x/S(F)]F(dx)=0$ with χ as in (6.16), $x_0=0$; asymptotic values and asymptotic variances for far-out symmetric ε-contamination.

α	ε_{min}	$S(\Phi)$	$S(F)/S(\Phi)$ and $A(F, \log S)$ for ε								
			0	0.005	0.01	0.02	0.05	0.1	0.15	0.2	0.25
0.01	0.005	0.925	1.000	1.014	1.029	∞	∞	∞	∞	∞	∞
			0.530	0.565	0.617	∞	∞	∞	∞	∞	∞
0.02	0.013	0.873	1.000	1.011	1.023	1.048	∞	∞	∞	∞	∞
			0.557	0.579	0.605	0.678	∞	∞	∞	∞	∞
0.05	0.041	0.749	1.000	1.008	1.017	1.035	1.097	∞	∞	∞	∞
			0.640	0.652	0.664	0.691	0.816	∞	∞	∞	∞
0.10	0.103	0.592	1.000	1.007	1.014	1.029	1.076	1.169	1.293	∞	∞
			0.796	0.803	0.810	0.825	0.879	1.022	1.351	∞	∞
0.15	0.180	0.466	1.000	1.006	1.013	1.025	1.067	1.145	1.238	1.356	1.513
			0.994	0.998	1.003	1.014	1.049	1.128	1.249	1.462	1.963
0.20	—	0.359	1.000	1.006	1.012	1.024	1.061	1.132	1.213	1.310	1.428
			1.257	1.261	1.264	1.272	1.298	1.352	1.425	1.530	1.693
0.25	—	0.267	1.000	1.005	1.011	1.022	1.058	1.124	1.199	1.286	1.388
			1.630	1.633	1.636	1.642	1.662	1.702	1.753	1.820	1.912

Exhibit 5.7.2 Trimmed standard deviations $S(F)=[\int_\alpha^{1-\alpha} F^{-1}(t)^2\, dt]^{1/2}$; asymptotic values and asymptotic variances for far-out symmetric ε-contamination.

		$S(F)/S(\Phi)$ and $A(F,\log S)$ for								
						ε				
α	$S(\Phi)$	0	0.005	0.01	0.02	0.05	0.1	0.15	0.2	0.25
0.01	2.327	1.000	1.045	1.106	∞	∞	∞	∞	∞	∞
		1.277	1.940	3.556	∞	∞	∞	∞	∞	∞
0.02	2.054	1.000	1.026	1.055	1.129	∞	∞	∞	∞	∞
		0.972	1.162	1.433	2.531	∞	∞	∞	∞	∞
0.05	1.645	1.000	1.014	1.028	1.059	1.178	∞	∞	∞	∞
		0.782	0.828	0.880	1.008	1.786	∞	∞	∞	∞
0.10	1.282	1.000	1.009	1.018	1.037	1.102	1.243	1.475	∞	∞
		0.791	0.808	0.827	0.867	1.026	1.548	3.465	∞	∞
0.15	1.036	1.000	1.007	1.015	1.030	1.080	1.178	1.304	1.480	1.770
		0.899	0.909	0.920	0.942	1.021	1.213	1.554	2.306	5.041
0.20	0.841	1.000	1.006	1.013	1.026	1.069	1.150	1.247	1.367	1.523
		1.081	1.088	1.095	1.110	1.160	1.268	1.425	1.672	2.109
0.25	0.674	1.000	1.006	1.012	1.024	1.062	1.134	1.217	1.316	1.435
		1.361	1.366	1.371	1.382	1.417	1.488	1.583	1.713	1.902

Exhibit 5.7.3 Interquantile distances; $S(F) = \frac{1}{2}[F^{-1}(1-\alpha) - F^{-1}(\alpha)]$; asymptotic values and asymptotic variances for far-out symmetric ε-contamination.

CHAPTER 6

Multiparameter Problems, in Particular Joint Estimation of Location and Scale

6.1 GENERAL REMARKS

We have already mentioned (Section 5.1) that M-estimates of location in practice will have to be supplemented by a simultaneous estimate of scale, since they are not scale invariant [except the median, $\psi(x) = \text{sign}(x)$]. Thus we are faced with a two-parameter problem.

The step going from one to two (or more) parameters is a troublesome one—we lose the technical advantages offered by the natural ordering of the real line, and proofs get more complicated.

L- and R-estimates of location are scale invariant, so this difficulty does not exist. On the other hand they rely so heavily on ordering that they do not generalize well beyond one-parameter location or scale problems. In fact they lose their advantages, for example the simplicity of L-estimates like the trimmed mean, or the existence of nonparametric confidence intervals for R-estimates, and their computation is quite complicated.

We therefore deal exclusively with M-estimates in this chapter. Sections 6.2 and 6.3 give some very general results (without proofs) on consistency and asymptotic normality of multiparameter M-estimates; the remaining sections discuss simultaneous estimates of location and scale (the latter being considered as a nuisance parameter).

6.2 CONSISTENCY OF M-ESTIMATES

In this section we state two theorems on consistency of M-estimates. The first one is concerned with estimates defined through a minimum property, the second one with estimates defined through a system of implicit equations. Proofs can be found in Huber (1967).

Case A. *Estimates Defined Through a Minimum Property* Assume that the parameter set Θ is a locally compact space with a countable base (e.g., an open subset of a Euclidean space), $(\mathcal{X}, \mathcal{C}, P)$ is a probability space, and $\rho(x, \theta)$ is some real-valued function on $\mathcal{X} \times \Theta$.

Assume that $x_1, x_2, \ldots$ are independent random variables with values in $\mathcal{X}$, having the common probability distribution P. Let $T_n(x_1, \ldots, x_n)$ be any sequence of functions $T_n: \mathcal{X}^n \to \Theta$, measurable or not, such that

$$\frac{1}{n} \sum_{i=1}^{n} \rho(x_i, T_n) - \inf_{\theta} \frac{1}{n} \sum_{i=1}^{n} \rho(x_i, \theta) \to 0 \qquad (2.1)$$

almost surely (or in probability). In most cases the left-hand side of (2.1)—let us denote it by Z_n—will be identically zero, but it is simpler to work with (2.1) than to add extraneous conditions only to guarantee the existence of a T_n minimizing $1/n \Sigma \rho(x_i, \theta)$. Since Z_n need not be measurable, we should more precisely speak of convergence in outer probability $[P^*(|Z_n| > \varepsilon) \to 0$ for all $\varepsilon]$ instead of convergence in probability.

We now give sufficient conditions that each sequence T_n satisfying (2.1) will converge almost surely (or in probability, respectively) to some constant θ_0, which is characterized below.

ASSUMPTIONS

(A-1)   For each fixed $\theta \in \Theta$, $\rho(x, \theta)$ is $\mathcal{C}$-measurable, and ρ is separable in the sense of Doob; that is, there is a P-null set N and a countable subset $\Theta' \subset \Theta$ such that, for every open set $U \subset \Theta$ and every closed interval A, the sets

$$\{x | \rho(x, \theta) \in A, \forall \theta \in U\}, \qquad \{x | \rho(x, \theta) \in A, \forall \theta \in U \cap \Theta'\} \quad (2.2)$$

differ by at most a subset of N.

This assumption ensures measurability of the infima and limits occurring in (A-2) and (A-5). For a fixed P, ρ might always be replaced by a separable version [see Doob (1953), p. 56 ff].

(A-2)   The function ρ is a.s. lower semicontinuous in θ, that is,

$$\inf_{\theta' \in U} (x, \theta') \to \rho(x, \theta) \qquad \text{a.s.} \qquad (2.3)$$

as the neighborhood U of θ shrinks to $\{\theta\}$.

(A-3)   There is a measurable function $a(x)$ such that

$$E\{\rho(x,\theta)-a(x)\}^- < \infty, \qquad \text{for all } \theta \in \Theta,$$

$$E\{\rho(x,\theta)-a(x)\}^+ < \infty, \qquad \text{for some } \theta \in \Theta. \tag{2.4}$$

Thus $\gamma(\theta) = E\{\rho(x,\theta)-a(x)\}$ is well defined for all θ.

(A-4)   There is a $\theta_0 \in \Theta$ such that $\gamma(\theta) > \gamma(\theta_0)$ for all $\theta \neq \theta_0$.

If Θ is not compact, let ∞ denote the point at infinity in its one-point compactification.

(A-5)   There is a continuous function $b(\theta) > 0$ such that:

(i)
$$\inf_{\theta \in \Theta} \frac{\rho(x,\theta)-a(x)}{b(\theta)} \geq h(x)$$

for some integrable function h.

(ii)
$$\liminf_{\theta \to \infty} b(\theta) > \gamma(\theta_0).$$

(iii)
$$E\left\{ \liminf_{\theta \to \infty} \frac{\rho(x,\theta)-a(x)}{b(\theta)} \right\} \geq 1.$$

If Θ is compact, then (ii) and (iii) are redundant.

Example 2.1   Let $\Theta = \mathfrak{X}$ be the real axis, and let P be any probability distribution having a unique median θ_0. Then (A-1) to (A-5) are satisfied for $\rho(x,\theta) = |x-\theta|$, $a(x) = |x|$, $b(\theta) = |\theta| + 1$, $h(x) = -1$. This will imply that the sample median is a consistent estimate of the median.

Taken together (A-2), (A-3), and (A-5) (i) imply by monotone convergence the following strengthened version of (A-2).

(A-2′)   As the neighborhood U of θ shrinks to $\{\theta\}$,

$$E \inf_{\theta' \in U} \{\rho(x,\theta')-a(x)\} \to E\{\rho(x,\theta)-a(x)\}. \tag{2.5}$$

Note that the set $\{\theta \in \Theta \mid E[|\rho(x,\theta)-a(x)|] < \infty\}$ is independent of the particular choice of $a(x)$; if there is an $a(x)$ satisfying (A-3), then we might take $a(x) = \rho(x,\theta_0)$.

For the sake of simplicity we absorb $a(x)$ into $\rho(x,\theta)$ from now on.

LEMMA 2.1 If (A-1), (A-3), and (A-5) hold, then there is a compact set $C \subset \Theta$ such that every sequence T_n satisfying (2.1) ultimately almost surely stays in C (or, with probability tending to 1, respectively).

THEOREM 2.2 If (A-1), (A-2′), (A-3), and (A-4) hold, then every sequence T_n satisfying (2.1) and the conclusion of Lemma 2.1 converges to θ_0 almost surely (or, in probability, respectively).

Quite often (A-5) is not satisfied—in particular, if location and scale are estimated simultaneously—but the conclusion of Lemma 2.1 can be verified without too much trouble by *ad hoc* methods. I do not know of any fail-safe replacement for (A-5).

In the location-scale case this problem poses itself as follows. To be specific take the maximum likelihood estimate of $\theta = (\xi, \sigma)$, $\sigma > 0$, based on a density f_0 (the true underlying distribution P may be different). Then

$$\rho(x, \theta) = \rho(x; \xi, \sigma) = \log \sigma - \log f_0\left(\frac{x - \xi}{\sigma}\right). \tag{2.6}$$

The trouble is that, if θ tends to "infinity," that is, to the boundary $\sigma = 0$ by letting $\xi = x$, $\sigma \to 0$, then $\rho \to -\infty$. If P is continuous, so that the probability of ties between the x_i's is zero, the following trick helps: take pairs $y_n = (x_{2n-1}, x_{2n})$ of the original observations as our new observations. Then the corresponding ρ_2,

$$\rho_2(y, \theta) = \rho(x_1; \xi, \sigma) + \rho(x_2; \xi, \sigma), \tag{2.7}$$

will avoid the above-mentioned difficulty. Somewhat more generally we are saved if we can show directly that the ML estimate $\hat{\theta}_n = (\hat{\xi}_n, \hat{\sigma}_n)$ ultimately satisfies $\hat{\sigma}_n \geqslant \delta > 0$ for some δ. (This again is tricky if the true underlying distribution is discontinuous and f_0 has very long tails.)

Case B. *Estimates Defined Through Implicit Equations*   Let Θ be locally compact with a countable base, let $(\mathfrak{X}, \mathfrak{A}, P)$ be a probability space, and let $\psi(x, \theta)$ be some function on $\mathfrak{X} \times \Theta$ with values in m-dimensional Euclidean space $\mathbb{R}^m$.

Assume that $x_1, x_2, \dots$ are independent random variables with values in $\mathfrak{X}$, having the common probability distribution P. We intend to give sufficient conditions that any sequence of functions $T_n : \mathfrak{X}^n \to \Theta$ such that

$$\frac{1}{n} \sum_1^n \psi(x_i; T_n) \to 0 \tag{2.8}$$

almost surely (or in probability) converges almost surely (or in probability) to some constant θ_0.

If Θ is an open subset of $\mathbb{R}^m$, and if $\psi(x,\theta)=(\partial/\partial\theta)\log f(x,\theta)$ for a differentiable parametric family of probability densities, then the ML estimate of course will satisfy (2.8). However, our ψ need not be a total differential. (This is important; for instance, it allows us to piece together joint estimates of location and scale from two essentially unrelated *M*-estimates of location and scale, respectively.)

ASSUMPTIONS

(B-1)   For each fixed $\theta\in\Theta$, $\psi(x,\theta)$ is $\mathcal{C}$-measurable in x, and ψ is separable (see A-1).

(B-2)   The function ψ is a.s. continuous in θ:

$$\lim_{\theta'\to\theta}|\psi(x,\theta')-\psi(x,\theta)|=0 \qquad \text{a.s.} \qquad (2.9)$$

(B-3)   The expected value $\lambda(\theta)=E\psi(x,\theta)$ exists for all $\theta\in\Theta$, and has a unique zero at $\theta=\theta_0$.

(B-4)   There exists a continuous function $b(\theta)$ that is bounded away from zero, $b(\theta)\geqslant b_0>0$, such that

(i) $\sup_\theta \dfrac{|\psi(x,\theta)|}{b(\theta)}$ is integrable,

(ii) $\liminf_{\theta\to\infty} \dfrac{|\lambda(\theta)|}{b(\theta)}\geqslant 1,$

(iii) $E\left[\limsup_{\theta\to\infty} \dfrac{|\psi(x,\theta)-\lambda(\theta)|}{b(\theta)}<1\right].$

In view of (B-4) (i), (B-2) can be strengthened to

(B-2')   As the neighborhood U of θ shrinks to $\{\theta\}$,

$$E\left[\sup_{\theta'\in U}|\psi(x,\theta')-\psi(x,\theta)|\right]\to 0. \qquad (2.10)$$

It follows from (B-2') that λ is continuous. Moreover, if there is a function b satisfying (B-4), we can take

$$b(\theta)=\max(|\lambda(\theta)|,b_0). \qquad (2.11)$$

LEMMA 2.3 If (B-1) and (B-4) hold, then there is a compact set $C \subset \Theta$ such that any sequence T_n satisfying (2.8) a.s. ultimately stays in C.

THEOREM 2.4 If (B-1), (B-2′), and (B-3) hold, then every sequence T_n satisfying (2.8) and the conclusion of Lemma 2.3 converges to θ_0 almost surely. An analogous statement is true for convergence in probability.

6.3 ASYMPTOTIC NORMALITY OF M-ESTIMATES

In the following Θ is an open subset of m-dimensional Euclidean space $\mathbb{R}^m$, $(\mathfrak{X}, \mathfrak{C}, P)$ is a probability space, and $\psi: \mathfrak{X} \times \Theta \to \mathbb{R}^m$ is some function.

Assume that $x_1, x_2, \ldots$ are independent random variables with values in $\mathfrak{X}$ and common distribution P. We give sufficient conditions to ensure that every sequence $T_n = T_n(x_1, \ldots, x_n)$ satisfying

$$\frac{1}{\sqrt{n}} \sum \psi(x_i, T_n) \to 0 \tag{3.1}$$

in probability is asymptotically normal; we assume that consistency of T_n has already been proved by some other means.

ASSUMPTIONS

(N-1) For each fixed $\theta \in \Theta$, $\psi(x, \theta)$ is $\mathfrak{C}$-measurable, and ψ is separable [see the preceding section, (A-1)].

Put

$$\lambda(\theta) = E\psi(x, \theta), \tag{3.2}$$

$$u(x, \theta, d) = \sup_{|\tau - \theta| \leqslant d} |\psi(x, \tau) - \psi(x, \theta)|. \tag{3.3}$$

Expectations are always taken with respect to the true underlying P.

(N-2) There is a θ_0 such that $\lambda(\theta_0) = 0$.

(N-3) There are strictly positive numbers a, b, c, d_0 such that

(i) $\qquad\qquad |\lambda(\theta)| \geqslant a|\theta - \theta_0|, \qquad$ for $|\theta - \theta_0| \leqslant d_0$.

(ii) $\qquad\qquad Eu(x, \theta, d) \leqslant bd, \qquad$ for $|\theta - \theta_0| + d \leqslant d_0$.

(iii) $\qquad\qquad E[U(x, \theta, d)^2] \leqslant cd \qquad$ for $|\theta - \theta_0| + d \leqslant d_0$.

Here, $|\theta|$ denotes any norm equivalent to Euclidean norm. Condition (iii) is somewhat stronger than needed; the proof can still be pushed through with $E[u(x, \theta, d)^2] \leq o(|\log d|^{-1})$.

(N-4) The expectation $E(|\psi(x, \theta_0)|^2)$ is nonzero and finite.

THEOREM 3.1 Assume that (N-1) to (N-4) hold and that T_n satisfies (3.1). If $P(|T_n - \theta_0| \leq d_0) \to 1$, then

$$\frac{1}{\sqrt{n}} \sum_{i=1}^{n} \psi(x_i, \theta_0) + \sqrt{n}\, \lambda(T_n) \to 0 \qquad (3.4)$$

in probability.

Proof See Huber (1967). ∎

COROLLARY 3.2 In addition to the assumptions of Theorem 3.1, assume that λ has a nonsingular derivative matrix Λ at θ_0 [i.e., $|\lambda(\theta) - \lambda(\theta_0) - \Lambda \cdot (\theta - \theta_0)| = o(|\theta - \theta_0|)$]. Then $\sqrt{n}\,(T_n - \theta_0)$ is asymptotically normal with mean 0 and covariance matrix $\Lambda^{-1}C(\Lambda^T)^{-1}$, where C is the covariance matrix of $\psi(x, \theta_0)$.

Consider now the ordinary ML estimator, that is, assume that $dP = f(x, \theta_0)\, d\mu$ and that $\psi(x, \theta) = (\partial/\partial\theta)\log f(x, \theta)$. Assume that $\psi(x, \theta)$ is jointly measurable, that (N-1), (N-3), and (N-4) hold locally uniformly in θ_0, and that the ML estimator is consistent. Assume furthermore that the Fisher information matrix

$$I(\theta) = \int \psi(x, \theta)\psi(x, \theta)^T f(x, \theta)\, d\mu \qquad (3.5)$$

is continuous at θ_0.

PROPOSITION 3.3 Under the assumptions just mentioned, we have $\lambda(\theta_0) = 0$, $\Lambda = -C = -I(\theta_0)$, and, in particular, $\Lambda^{-1}C(\Lambda^T)^{-1} = I(\theta_0)^{-1}$. That is, the ML estimator is efficient.

Proof See Huber (1967). ∎

Example 3.1 L_p-*Estimates* Define an m-dimensional estimate T_n of location by the property that it minimizes $\Sigma |x_i - T_n|^p$, where $1 \leq p \leq 2$, and $|\ |$ denotes the usual Euclidean norm. Equivalently, we could define it through $\Sigma \psi(x_i; T_n) = 0$ with

$$\psi(x, \theta) = -\frac{1}{p}\frac{\partial}{\partial\theta}(|x - \theta|^p) = |x - \theta|^{p-2}(x - \theta). \qquad (3.6)$$

Assume that $m \geq 2$.

A straightforward calculation shows that u and u^2 satisfy Lipschitz conditions of the form

$$u(x,\theta,d) \leqslant c_1 \cdot d \cdot |x-\theta|^{p-2} \tag{3.7}$$

$$u^2(x,\theta,d) \leqslant c_2 \cdot d \cdot |x-\theta|^{p-2} \tag{3.8}$$

for $0 \leqslant d \leqslant d_0 < \infty$. Thus assumptions (N-3) (ii) and (iii) are satisfied, provided

$$E(|x-\theta|^{p-2}) \leqslant K < \infty \tag{3.9}$$

in some neighborhood of θ_0. This certainly holds if the true underlying distribution has a density with respect to Lebesgue measure. Furthermore under the same condition (3.9), we have

$$\frac{\partial}{\partial\theta}\lambda(\theta) = E\frac{\partial\psi(x,\theta)}{\partial\theta}. \tag{3.10}$$

Thus

$$\operatorname{tr}\frac{\partial\lambda}{\partial\theta} = E\operatorname{tr}\frac{\partial\psi}{\partial\theta} = -(m+p-2)E(|x-\theta|^{p-2}) < 0;$$

hence (N-3) (i) is also satisfied.

Assumption (N-1) is immediate, (N-2) and (N-4) hold if $E(|x|^{2p-2}) < \infty$, and consistency follows either from verifying (B-1) to (B-4) [with $b(\theta) = \max(1,|\theta|^{p-1})$] or from an easy *ad hoc* proof using convexity of $\rho(x,\theta) = |x-\theta|^p$.

Occasionally, the theorems of this and of the preceding section are also useful in the one-dimensional case.

Example 3.2   Let $\mathfrak{X} = \Theta = \mathbb{R}$, and let

$$\rho(x,\theta) = \tfrac{1}{2}(x-\theta)^2, \qquad \text{for} \quad |x-\theta| \leqslant k$$

$$= \tfrac{1}{2}k^2 \qquad \text{for} \quad |x-\theta| > k.$$

Assumption (A-4) of Section 6.2, namely unicity of θ_0, imposes a restriction on the true underlying distribution; the other assumptions (A-1), (A-2), (A-3), and (A-5) are trivially satisfied [with $a(x) \equiv 0$, $b(\theta) \equiv \tfrac{1}{2}k^2$, $h(x) \equiv 0$]. Then the T_n minimizing $\Sigma\rho(x_i, T_n)$ is a consistent estimate of θ_0.

Under slightly more stringent conditions, it is also asymptotically normal. Assume for simplicity that $\theta_0 = 0$, and assume that the true underlying distribution function F has a density F' in some neighborhoods of the points $\pm k$, and that F' is continuous at these points. Assumptions (N-1), (N-2), (N-3) (ii), (iii), and (N-4) are obviously satisfied with $\psi(x, \theta) = (\partial/\partial\theta)\rho(x, \theta)$. If

$$\int_{-k}^{k} F(dx) - kF'(-k) - kF'(k) > 0,$$

then (N-3) (i) is also satisfied. We can easily check that Corollary 3.2 is applicable; hence T_n is asymptotically normal.

6.4 SIMULTANEOUS *M*-ESTIMATES OF LOCATION AND SCALE

In order to make an *M*-estimate of location scale invariant, we must couple it with an estimate of scale.

If the underlying distribution F is symmetric, location estimates T and scale estimates S typically are asymptotically independent, and the asymptotic behavior of T depends on S only through the asymptotic value $S(F)$. We can therefore afford to choose S on criteria other than low statistical variability.

Consider the simultaneous maximum likelihood estimates of θ and σ for a family of densities

$$\frac{1}{\sigma} f\left(\frac{x-\theta}{\sigma}\right), \tag{4.1}$$

that is, the values $\hat{\theta}$ and $\hat{\sigma}$ maximizing

$$\prod_{i \leqslant n} \frac{1}{\sigma} f\left(\frac{x_i - \theta}{\sigma}\right). \tag{4.2}$$

Evidently, these satisfy the following system of equations [with $\psi(x) = -(d/dx)\log f(x)$]

$$\sum \psi\left(\frac{x_i - \theta}{\sigma}\right) = 0, \tag{4.3}$$

$$\sum \left[\psi\left(\frac{x_i - \theta}{\sigma}\right)\frac{x_i - \theta}{\sigma} - 1\right] = 0. \tag{4.4}$$

We generalize this and call *simultaneous M-estimate of location and scale* any pair of statistics (T_n, S_n) determined by two equations of the form

$$\sum \psi\left(\frac{x_i - T_n}{S_n}\right) = 0, \tag{4.5}$$

$$\sum \chi\left(\frac{x_i - T_n}{S_n}\right) = 0. \tag{4.6}$$

Evidently, $T_n = T(F_n)$ and $S_n = S(F_n)$ can be expressed in terms of functionals T and S, defined by

$$\int \psi\left(\frac{x - T(F)}{S(F)}\right) F(dx) = 0, \tag{4.7}$$

$$\int \chi\left(\frac{x - T(F)}{S(F)}\right) F(dx) = 0. \tag{4.8}$$

Neither ψ nor χ need be determined by a probability density as in (4.3) and (4.4). In most cases, however, ψ will be an odd and χ an even function.

As before the influence functions can be found straightforwardly by inserting $F_t = (1-t)F + t\delta_x$ for F into (4.7) and (4.8), and then taking the derivative with respect to t at $t = 0$. We obtain that the two influence curves $IC(x; F, T)$ and $IC(x; F, S)$ satisfy the system of equations

$$IC(x; F, T)\int \psi'(y)F(dx) + IC(x; F, S)\int \psi'(y)yF(dx) = \psi(y)S(F),$$

$$\tag{4.9}$$

$$IC(x; F, T)\int \chi'(y)F(dx) + IC(x; F, S)\int \chi'(y)yF(dx) = \chi(y)S(F),$$

$$\tag{4.10}$$

where y is short for $y = [x - T(F)]/S(F)$.

If F is symmetric, ψ is odd, and χ is even, some integrals vanish for reasons of symmetry and there are considerable simplifications:

$$IC(x; F, T) = \frac{\psi\left(\dfrac{x}{S(F)}\right)S(F)}{\int \psi'\left(\dfrac{x}{S(F)}\right)F(dx)}, \tag{4.11}$$

$$IC(x; F, S) = \frac{\chi\left(\dfrac{x}{S(F)}\right) S(F)}{\int \chi'\left(\dfrac{x}{S(F)}\right) \dfrac{x}{S(F)} F(dx)}. \qquad (4.12)$$

Example 4.1   Let

$$\psi(x) = \max[-k, \min(k, x)] \qquad (4.13)$$

and

$$\chi(x) = \min(c^2, x^2) - \beta, \qquad (4.14)$$

where $0 < \beta < c^2$. With $\beta = \beta(c)$,

$$\beta(c) = \int \min(c^2, x^2) \Phi(dx), \qquad (4.15)$$

we obtain consistency of the scale estimate at the normal model.

This example is a combination of the asymptotic minimax estimates of location (Section 4.6) and of scale (Section 5.7); k and $c = x_1$ might be determined from (4.5.21) and (5.6.14), respectively. A simplified version of this estimate uses $c = k$ [Huber (1964), p. 96 "Proposal 2"], that is,

$$\chi(x) = \psi(x)^2 - \beta(k). \qquad (4.16)$$

Example 4.2   *Median and Median Absolute Deviation*   Let

$$\psi(x) = \text{sign}(x), \qquad (4.17)$$

$$\chi(x) = \text{sign}(|x| - 1). \qquad (4.18)$$

A (formal) evaluation of (4.9) and (4.10) gives

$$IC(x; F, T) = \frac{\text{sign}(x - T(F))}{2 f(T(F))}, \qquad (4.19)$$

and

$$IC(x; F, S) = \frac{\text{sign}(|x - T| - S) - \dfrac{f(T+S) - f(T-S)}{f(T)} \text{sign}(x - T)}{2[f(T+S) + f(T-S)]}. \qquad (4.20)$$

If F is symmetric, (4.20) simplifies to

$$IC(x; F, S) = \frac{\text{sign}(|x| - S(F))}{4f(S(F))}. \tag{4.21}$$

Existence and Uniqueness of the solutions of (4.7) and (4.8)

We follow Scholz (1971). Assume that ψ and χ are differentiable, that $\psi' > 0$, that ψ has a zero at $x = 0$ and χ has a minimum at $x = 0$, and that χ'/ψ' is strictly monotone. [In the particular case $\chi(x) = \psi(x)^2 - \beta$, this last assumption follows from $\psi' > 0$]. F is indifferently either the true or the empirical distribution.

The Jacobian of the map

$$(t, s) \rightarrow \left(\int \psi\left(\frac{x-t}{s}\right) F(dx), \int \chi\left(\frac{x-t}{s}\right) F(dx) \right) \tag{4.22}$$

is

$$-\frac{1}{s} \begin{vmatrix} \int \psi'(y) \, dF & \int y \psi'(y) \, dF \\ \int \chi'(y) \, dF & \int y \chi'(y) \, dF \end{vmatrix} \tag{4.23}$$

with $y = (x - t)/s$. We define a new probability measure F^* by

$$F^*(dy) = \frac{\psi'(y)}{E_F[\psi'(y)]} F(dx); \tag{4.24}$$

then the Jacobian can be written as

$$-\frac{1}{s} E_F[\psi'(y)] \begin{vmatrix} 1 & E_{F^*}(y) \\ E_{F^*}\left(\frac{\chi'}{\psi'}\right) & E_{F^*} y\left(\frac{\chi'}{\psi'}\right) \end{vmatrix}. \tag{4.25}$$

Its determinant

$$\left[\frac{E_F \psi'(y)}{s} \right]^2 \text{cov}_{F^*}\left(y, \frac{\chi'}{\psi'} \right)$$

is strictly positive unless F is concentrated at a single point. To prove this, let f and g be any two strictly monotone functions, and let Y_1 and Y_2 be two independent, identically distributed random variables. As $[f(Y_1)-f(Y_2)][g(Y_1)-g(Y_2)]>0$ unless $Y_1=Y_2$, we have

$$\text{cov}\left[f(Y_1), g(Y_1)\right] = \tfrac{1}{2}E\{\left[f(Y_1)-f(Y_2)\right]\left[g(Y_1)-g(Y_2)\right]\}>0$$

unless $P(Y_1=Y_2)=1$.

Thus as the diagonal elements of the Jacobian are strictly negative, and its determinant is strictly positive, we conclude [cf. Gale and Nikaidô (1965), Theorem 4] that (4.22) is a one-to-one map.

The existence of a solution now follows from the observations (1) that, for each fixed s, the first component of (4.22) has a unique zero at some $t=t(s)$ that depends continuously on s, and (2) that the second component $\int \chi\{[x-t(s)]/s\}F(dx)$ ranges from $\chi(0)$ to (at least) $(1-\eta)\chi(\pm\infty)+\eta\chi(0)$, where η is the largest pointmass of F, when s varies from ∞ to 0. We now conclude from the intermediate value theorem for continuous functions that $[T(F),T(S)]$ exists uniquely, provided $\chi(0)<0<\chi(\pm\infty)$ and F does not have pointmasses that are too large; the largest one should satisfy $\eta<\chi(\pm\infty)/[\chi(\pm\infty)-\chi(0)]$.

The special case of Example 4.1 is not covered by this proof, as ψ is not strictly monotone, but the result remains valid [approximate ψ by strictly monotone functions; for a direct proof see Huber (1964), p. 98; cf. also Section 7.7].

It is intuitively obvious (and easy to check rigorously) that the map $F\rightarrow(T(F),S(F))$ is not only well defined but also weakly continuous, provided ψ and χ are bounded; hence T and S are qualitatively robust in Hampel's sense. The Glivenko-Cantelli theorem then implies consistency of (T_n,S_n). The monotonicity and differentiability properties of ψ and χ make it relatively easy to check assumptions (N-1) to (N-4) of Section 6.3, and, since the map (4.22) is differentiable by assumption, (T_n,S_n) is asymptotically normal in virtue of Corollary 3.2. The special case of Example 4.1 is again not quite covered; if F puts pointmasses on the discontinuities of ψ', asymptotic normality is destroyed just as in the case of location alone (Section 3.2), but for finite n the case is now milder, because the random fluctuations in the scale estimate smooth away these discontinuities.

If F is symmetric, and ψ and χ skew symmetric and symmetric, respectively, the location and scale estimates are uncorrelated for symmetry reasons, and hence asymptotically independent.

6.5 M-ESTIMATES WITH PRELIMINARY ESTIMATES OF SCALE

The simultaneous solution of two equations (4.5) and (4.6) is perhaps unnecessarily complicated. A somewhat simplified variant is an *M-estimate of location with a preliminary estimate of scale*: take *any* estimate $S_n = S(F_n)$ of scale, and determine location from (4.5) or (4.7), respectively. If the influence function of the scale estimate is known, then the influence function of the location estimate can be determined from (4.9), or, in the symmetric case, simply from (4.11). Note that, in the symmetric case, only the limiting value $S(F)$, but neither the influence function nor the asymptotic variance of S, enters into the expression for the influence function of T.

Another, even simpler, variant is the so-called *one-step M-estimate*. Here, we start with some preliminary estimates $T_0(F)$ and $S_0(F)$ of location and scale, and then we solve (4.7) approximately for T by applying Newton's rule just *once*. Since the Taylor expansion of (4.7) with respect to T at $T_0 = T_0(F)$ begins with

$$\int \psi\left(\frac{x-T}{S_0}\right)F(dx) = \int \psi\left(\frac{x-T_0}{S_0}\right)F(dx) - \frac{T-T_0}{S}\int \psi'\left(\frac{x-T_0}{S_0}\right)F(dx) + \cdots,$$

this estimate can be formally defined by the functional

$$T(F) = T_0(F) + \frac{S_0 \int \psi\left(\frac{x-T_0}{S_0}\right)F(dx)}{\int \psi'\left(\frac{x-T_0}{S_0}\right)F(dx)}. \tag{5.1}$$

The influence function corresponding to (5.1) can be calculated straightforwardly, if those of T_0 and S_0 are known. In the general asymmetric case this leads to unpleasantly complicated expressions:

$$IC(x; F, T) = \frac{S_0}{\int \psi'}\psi - \frac{S_0 \int \psi}{\left(\int \psi'\right)^2}\psi' + \frac{\int \psi \int \psi''}{\left(\int \psi'\right)^2}IC(x; F, T_0)$$

$$+ \left[\frac{\int \psi}{\int \psi'} - \frac{\int y\psi'}{\int \psi'} + \frac{\int \psi \int y\psi''}{\left(\int \psi'\right)^2}\right]IC(x; F, S_0), \tag{5.2}$$

where the argument of ψ, ψ', and ψ'' in all instances is $y = [x - T_0(F)]/S_0(F)$, and all integrals are with respect to dF.

If we assume that T_0 is translation invariant and odd,

$$T(F_{X+c}) = T(F_X) + c,$$

$$T(F_{-X}) = -T(F_X),$$

that ψ is odd, and that F is symmetric, then all terms except the first vanish, and the formula simplifies again to (4.11):

$$IC(x; F, T) = \frac{\psi\left(\dfrac{x}{S_0(F)}\right) S_0(F)}{\displaystyle\int \psi'\left(\dfrac{x}{S_0(F)}\right) F(dx)}. \tag{5.3}$$

It is intuitively clear from the influence functions that the estimate with preliminary scale and the corresponding one-step estimate will both be asymptotically normal and asymptotically equivalent to each other if T_0 is consistent. Asymptotic normality proofs utilizing one-step estimates as auxiliary devices are usually relatively straightforward to construct.

6.6 QUANTITATIVE ROBUSTNESS PROPERTIES OF SIMULTANEOUS ESTIMATES FOR LOCATION AND SCALE

The breakdown properties of the estimates considered in the preceding two sections are mainly determined by the breakdown of the scale part. Thus they can differ considerably from those of fixed-scale M-estimates of location.

To be specific consider first joint M-estimates, assume that both ψ and χ are continuous, that ψ is odd and χ is even, and that both are monotone increasing for positive arguments. We only consider ε-contamination (the results for Prohorov-ε-neighborhoods are the same). We regard scale as a nuisance parameter and concentrate on the location aspects.

Let ε_S^* and ε_T^* be the infima of the set of ε-values for which $S(F)$, or $T(F)$, respectively, can become infinitely large. We first note that $\varepsilon_S^* \leqslant \varepsilon_T^*$. Otherwise $T(F)$ would break down while $S(F)$ stays bounded; therefore we would have $\varepsilon_T^* = 0.5$ as in the fixed-scale case, but $\varepsilon_S^* > 0.5$ is impossible (5.2.7). Scale breakdown by "implosion," $S \to 0$, is uninteresting in the present context because then the location estimate is converted into the highly robust sample median.

Now let $\{F\}$ be a sequence of ε-contaminated distributions, $F = (1-\varepsilon)F_0 + \varepsilon H$, such that $T(F) \to \infty$, $S(F) \to \infty$, and $\varepsilon \to \varepsilon_T^* = \varepsilon^*$. Without loss of generality we assume that the limit

$$0 \leqslant \lim \frac{T(F)}{S(F)} = y \leqslant \infty \tag{6.1}$$

exists (if necessary we pass to a subsequence).

We write the defining equations (4.7) and (4.8) as

$$(1-\varepsilon)\int \psi\left(\frac{x-T}{S}\right) F_0(dx) + \varepsilon \int \psi\left(\frac{x-T}{S}\right) H(dx) = 0, \tag{6.2}$$

$$(1-\varepsilon)\int \chi\left(\frac{x-T}{S}\right) F_0(dx) + \varepsilon \int \chi\left(\frac{x-T}{S}\right) H(dx) = 0. \tag{6.3}$$

If we replace the coefficients of ε by their upper bounds $\psi(\infty)$ and $\chi(\infty)$, respectively, we obtain from (6.2) and (6.3), respectively,

$$(1-\varepsilon)\int \psi\left(\frac{x-T}{S}\right) F_0(dx) + \varepsilon\psi(\infty) \geqslant 0,$$

$$(1-\varepsilon)\int \chi\left(\frac{x-T}{S}\right) F_0(dx) + \varepsilon\chi(\infty) \geqslant 0.$$

In the limit we have

$$(1-\varepsilon^*)\psi(-y) + \varepsilon^*\psi(\infty) \geqslant 0, \tag{6.4}$$

$$(1-\varepsilon^*)\chi(-y) + \varepsilon^*\chi(\infty) \geqslant 0; \tag{6.5}$$

hence using the symmetry and monotonicity properties of ψ and χ,

$$\chi^{-1}\left(-\frac{\varepsilon^*}{1-\varepsilon^*}\chi(\infty)\right) \leqslant y \leqslant \psi^{-1}\left(\frac{\varepsilon^*}{1-\varepsilon^*}\psi(\infty)\right). \tag{6.6}$$

It follows that the solution ε_0 of

$$\chi^{-1}\left(-\frac{\varepsilon}{1-\varepsilon}\chi(\infty)\right) = \psi^{-1}\left(\frac{\varepsilon}{1-\varepsilon}\psi(\infty)\right) \tag{6.7}$$

is a lower bound for ε^* (assume for simplicity that ε_0 is unique).

It is not difficult to check that this is also an upper bound for ε^*. Assume that ε is small enough so that the solution $[T(F), S(F)]$ of (6.2) and (6.3) stays bounded for all H. In particular, if we let H tend to a pointmass at $+\infty$, (6.2) and (6.3) then converge to

$$(1-\varepsilon)\int\psi\left(\frac{x-T}{S}\right)F_0(dx)+\varepsilon\psi(\infty)=0, \tag{6.8}$$

$$(1-\varepsilon)\int\chi\left(\frac{x-T}{S}\right)F_0(dx)+\varepsilon\chi(\infty)=0. \tag{6.9}$$

Now let ε increase until the solutions $T(F)$ and $S(F)$ of (6.8) and (6.9) begin to diverge. We can again assume that (6.1) holds for some y. The limiting ε must be at least as large as the breakdown point, and it will satisfy (6.4) and (6.5), with equality signs. It follows that the solution ε_0 of (6.7) is an upper bound for ε^*, and that it is the common breakdown point of T and S.

Example 6.1 *Continuation of Example 4.1*   In this case we have $\psi(\infty)=k$,

$$\psi^{-1}\left[\frac{\varepsilon}{1-\varepsilon}\psi(\infty)\right]=\frac{\varepsilon}{1-\varepsilon}k;$$

hence (6.7) can be written

$$\left[\left(\frac{\varepsilon}{1-\varepsilon}\right)^2k^2-\beta(c)\right]+\frac{\varepsilon}{1-\varepsilon}\left[c^2-\beta(c)\right]=0. \tag{6.10}$$

If $c=k$, the solution of (6.10) is simply

$$\varepsilon^*=\frac{\beta(k)}{\beta(k)+k^2}. \tag{6.11}$$

For symmetric contamination the variance of the location estimate breaks down $[(v(\varepsilon)\to\infty]$ for

$$\varepsilon^{**}=\frac{\beta(k)}{k^2}. \tag{6.12}$$

These values should be compared quantitatively to the corresponding breakdown points $\varepsilon^*=\alpha$ and $\varepsilon^{**}=2\alpha$ of the α-trimmed mean. To facilitate this comparison, the following table of breakdown points (Exhibit 6.6.1) also gives the "equivalent trimming rate" $\alpha_\Phi=\Phi(-k)$, for which the corresponding α-trimmed mean has the same influence function and the same asymptotic performance at the normal model.

| | Example 6.1 "Proposal 2" | | Example 6.2 | | | | Trimmed Mean Equivalent for Φ, $\alpha = \Phi(-k)$ | |
| | | | Scale: Interquartile Range | | Scale: Median Deviation | | | |
k	ε^*	ε^{**}	ε^*	ε^{**}	ε^*	ε^{**}	$\varepsilon^* = \alpha$	$\varepsilon^{**} = 2\alpha$
3.0	0.100	0.111					0.001	0.003
2.5	0.135	0.156					0.006	0.012
2.0	0.187	0.230					0.023	0.046
1.7	0.227	0.294					0.045	0.090
1.5	0.257	0.346					0.067	0.134
1.4	0.273	0.375	0.25	0.5	0.5	0.5	0.081	0.162
1.3	0.290	0.407					0.097	0.194
1.2	0.307	0.441					0.115	0.230
1.1	0.324	0.478					0.136	0.272
1.0	0.340	0.516					0.159	0.318
0.7	0.392	0.645					0.242	0.484

Exhibit 6.6.1 Breakdown points for the estimates of Examples 6.1 and 6.2, and for the trimmed mean with equivalent performance at the normal distribution.

Also the breakdown of M-estimates with preliminary estimates of scale is governed by the breakdown of the scale part, but the situation is much simpler. The following example will suffice to illustrate this.

Example 6.2 With the same ψ as in Example 6.1, but with the interquartile range as scale [normalized such that $S(\Phi) = 1$], we have $\varepsilon^* = 0.25$ and $\varepsilon^{**} = 0.5$. For the symmetrized version $\tilde{S}$ (the median absolute deviation, cf. Sections 5.1 and 5.3) breakdown is pushed up to $\varepsilon^* = \varepsilon^{**} = 0.5$. See Exhibit 6.6.1.

As a further illustration Exhibit 6.6.2 compares the suprema $v_s(\varepsilon)$ of the asymptotic variances for *symmetric* ε-contamination, for various estimates whose finite sample properties had been investigated by Andrews et al. (1972). Among these estimates:

* H14, H10, and H07 are Huber's "Proposal 2" with $k = 1.4$, 1.0, and 0.7 respectively; see Examples 4.1 and 6.1.
* A14, A10, and A07 have the same ψ as the corresponding H-estimates, but use MAD/0.6745 as a preliminary estimate of scale (cf. Section 6.5).

	0	0.001	0.002	0.005	0.01	0.02	0.05	0.1	0.15	0.2	0.25	0.3	0.4	0.5
Normal scores estimate	1.000	1.014	1.026	1.058	1.106	1.197	1.474	2.013	2.714	3.659	4.962	6.800	13.415	29.161
Hodges-Lehmann	1.047	1.051	1.056	1.068	1.090	1.135	1.286	1.596	2.006	2.557	3.310	4.361	8.080	16.755
H 14	1.047	1.050	1.054	1.065	1.084	1.123	1.258	1.554	1.992	2.698	4.003	7.114	∞	∞
A 14	1.047	1.050	1.054	1.065	1.084	1.124	1.257	1.539	1.928	2.482	3.31	4.61	10.51	∞
10% trimmed mean	1.061	1.064	1.067	1.077	1.095	1.131	1.256	1.541	2.030	∞	∞	∞	∞	∞
H 10	1.107	1.110	1.113	1.123	1.138	1.170	1.276	1.490	1.770	2.150	2.690	3.503	7.453	64.4
A 14	1.107	1.110	1.113	1.123	1.138	1.170	1.276	1.490	1.768	2.140	2.660	3.434	6.752	∞
15%	1.100	1.103	1.106	1.115	1.131	1.163	1.270	1.492	1.797	2.253	3.071	∞	∞	∞
H 07	1.187	1.189	1.192	1.201	1.215	1.244	1.339	1.525	1.755	2.046	2.423	2.926	4.645	9.09
A 07	1.187	1.189	1.192	1.201	1.215	1.244	1.339	1.524	1.754	2.047	2.431	2.954	4.915	∞
25%	1.195	1.198	1.201	1.209	1.223	1.252	1.346	1.530	1.758	2.046	2.422	2.930	4.808	∞
25A (2.5, 4.5, 9.5)	1.025	1.031	1.036	1.053	1.082	1.143	1.356	1.843	2.590	3.790	5.825	9.530	33.70	∞
21A (2.1, 4.0, 8.2)	1.050	1.055	1.060	1.075	1.101	1.155	1.342	1.759	2.376	3.333	4.904	7.676	25.10	∞
17A (1.7, 3.4, 8.5)	1.092	1.096	1.100	1.113	1.135	1.180	1.331	1.653	2.099	2.743	3.715	5.280	13.46	∞
12A (1.2, 3.5, 8,0)	1.166	1.170	1.174	1.185	1.205	1.247	1.383	1.661	2.025	2.516	3.201	4.201	8.47	∞
Minimax Bound	1.000	1.010	1.017	1.037	1.065	1.116	1.256	1.490	1.748	2.046	2.397	2.822	3.996	5.928

Exhibit 6.6.2 Suprema $v_s(\epsilon)$ of the asymptotic variance for symmetrically ϵ-contaminated normal distributions, for various estimates of location.

145

• 25A, 21A, 17A, and 12A are descending Hampel estimates, with the constants (a, b, c) given in parentheses [cf. Section 4.8, especially (4.8.8)]; they use MAD as a preliminary estimate of scale.

6.7 THE COMPUTATION OF M-ESTIMATES

We describe several variants, beginning with some where the median absolute deviation is used as an auxiliary estimate of scale.

Variant 1 *Modified Residuals* Let

$$T^{(0)} = \mathrm{med}\{x_i\}, \tag{7.1}$$

$$S^{(0)} = \mathrm{med}\{|x_i - T^{(0)}|\}. \tag{7.2}$$

Perform at least one Newton step, that is, one iteration of

$$T^{(m+1)} = T^{(m)} + \frac{\dfrac{1}{n}\sum \psi\left(\dfrac{x_i - T^{(m)}}{S^{(0)}}\right)S^{(0)}}{\dfrac{1}{n}\sum \psi'\left(\dfrac{x_i - T^{(m)}}{S^{(0)}}\right)}. \tag{7.3}$$

Compare Section 6.5 and note that the one-step estimate $T^{(1)}$ is asymptotically ($n \to \infty$) equivalent to the iteration limit $T^{(\infty)}$, provided that the underlying distribution is symmetric and ψ is skew symmetric. The denominator in (7.3) is not very critical, and it might be replaced by a constant. If $0 \leqslant \psi' \leqslant 1$, then any constant denominator $> \frac{1}{2}$ will give convergence (for a proof, see Section 7.8). However, if ψ is piecewise linear, then (7.3) will lead to the *exact* solution of

$$\sum \psi\left(\frac{x_i - T}{S^{(0)}}\right) = 0 \tag{7.4}$$

in a *finite* number of steps (if it converges at all).

Variant 2 *Modified Weights* Let $T^{(0)}$ and $S^{(0)}$ be defined as above. Perform a few iterations of

$$T^{(m+1)} = \frac{\sum w_i^{(m)} x_i}{\sum w_i^{(m)}}, \tag{7.5}$$

with

$$w_i^{(m)} = \frac{\psi[(x_i - T^{(m)})/S^{(0)}]}{(x_i - T^{(m)})/S^{(0)}}. \tag{7.6}$$

A convergence proof is also given in Section 7.8; the iteration limit $T^{(\infty)}$, of course, is a solution of (7.4).

Variant 3 *Joint M-Estimates of Location and Scale* Assume that we want to solve the system

$$\sum \psi\left(\frac{x_i - T}{S}\right) = 0, \tag{7.7}$$

$$\sum \psi^2\left(\frac{x_i - T}{S}\right) = (n-1)\beta, \tag{7.8}$$

with

$$\beta = E_\Phi(\psi^2),$$

and where ψ is assumed to be skew symmetric and monotone, $0 \leqslant \psi' \leqslant 1$.

Start with $T^{(0)}$ and $S^{(0)}$ as above. Let

$$[S^{(m+1)}]^2 = \frac{1}{(n-1)\beta} \sum \psi^2\left(\frac{x_i - T^{(m)}}{S^{(m)}}\right)[S^{(m)}]^2, \tag{7.9}$$

$$T^{(m+1)} = T^{(m)} + \frac{\frac{1}{n}\sum \psi\left(\frac{x_i - T^{(m)}}{S^{(m)}}\right)S^{(m)}}{\frac{1}{n}\sum \psi'\left(\frac{x_i - T^{(m)}}{S^{(m)}}\right)}. \tag{7.10}$$

For a convergence proof [with a constant denominator in (7.10)] see Section 7.8.

Variant 4 *Joint M-Estimates of Location and Scale, Continued* Assume that $\psi(x) = \max[-c, \min(c, x)]$. Let m_1, m_2, and m_3 be the number of observations satisfying $x_i \leqslant T - cS$, $T - cS < x_i < T + cS$, and $T + cS \leqslant x_i$, respectively. Then (7.7) and (7.8) can be written

$$\sum' x_i - m_2 T + (m_3 - m_1) cS = 0, \tag{7.11}$$

$$\sum' (x_i - T)^2 + (m_1 + m_3) c^2 S^2 - (n-1)\beta S^2 = 0. \tag{7.12}$$

Here, the primed summation sign indicates that the sum is extended only over the observations for which $|x_i - T| < cS$. If we determine T from (7.11) and insert it into (7.12), we obtain the equivalent system

$$\bar{x}' = \frac{\sum' x_i}{m_2}, \tag{7.13}$$

$$S^2 = \frac{\sum' (x_i - \bar{x}')^2}{(n-1)\beta - \left[m_1 + m_2 + \frac{(m_3 - m_1)^2}{m_2} \right] c^2}, \tag{7.14}$$

$$T = \bar{x}' + cS \frac{(m_3 - m_1)}{m_2}. \tag{7.15}$$

These three last equations now are used to calculate T and S. Assume that we have already determined $T^{(m)}$ and $S^{(m)}$. Find the corresponding partition of the sample according to $T^{(m)} \pm cS^{(m)}$, then evaluate (7.13) and (7.14) to find $S^{(m+1)}$, and finally find $T^{(m+1)}$ through (7.15), using $S^{(m+1)}$.

The convergence of this procedure has not yet been proved, and in fact, there are counterexamples for small values of c. But in practice it converges extremely fast and reaches the exact solution in a finite number of steps.

6.8 STUDENTIZING

As a matter of principle each estimate $T_n = T_n(x_1, \ldots, x_n)$ of any parameter θ should be accompanied by an estimate $D_n = D_n(x_1, \ldots, x_n)$ of its own variability. Since T_n will often be asymptotically normal, D_n should be standardized such that it estimates the (asymptotic) standard deviation of T_n, that is

$$\mathcal{L}\{\sqrt{n}\,[T_n - T(F)]\} \to \mathcal{N}[0, A(F, T)], \tag{8.1}$$

and

$$nD_n^2 \rightarrow A(F,T). \tag{8.2}$$

Most likely D_n will be put to either of two uses:

(1) For finding confidence intervals $(T_n - cD_n, T_n + cD_n)$ for the un-known true parameter estimated by T_n.

(2) For finding (asymptotic) standard deviations for functions of T_n by the so-called Δ-method:

$$\sigma(h(T_n)) \approx |h'(T_n)| \sigma(T_n). \tag{8.3}$$

In Section 1.2 we have proposed standardizing an estimate T of θ such that it would be Fisher consistent at the model, that is, $T(F_\theta) = \theta$, and otherwise defining the estimand in terms of the limiting value of the estimate.

For D we do not have this freedom; the estimand is asymptotically fixed by (8.2). If $A(F_n, T) \rightarrow A(F, T)$, our estimate should therefore satisfy

$$\sqrt{n}\, D_n \approx A(F_n, T)^{1/2}, \tag{8.4}$$

and we might in fact define D_n by this relation, that is,

$$D_n^2 = \frac{1}{n(n-1)} \sum IC(x_i, F_n, T)^2. \tag{8.5}$$

The factor $n-1$ (instead of n) was substituted to preserve equivalence with the classical formula for the estimated standard deviation of the sample mean.

Almost equivalently, we can use the jackknife method (Section 1.5).

In some cases both (8.5) and the jackknife fail, for instance for the sample median. In this particular case we can take recourse to the well-known nonparametric confidence intervals for the median, given by the interval between two selected order statistics $(x_{(i)}, x_{(n+1-i)})$. If we then divide $x_{(n+1-i)} - x_{(i)}$ by a suitable constant $2c$, we may also get an estimate D_n satisfying (8.2). In view of the central limit theorem, the proper choice is, asymptotically,

$$c = \Phi^{-1}\left(\tfrac{1}{2} + \tfrac{1}{2}\alpha\right), \tag{8.6}$$

where α is the level of the confidence interval.

If T_n and D_n are jointly asymptotically normal, they will be asymptotically independent in the symmetric case (for reasons of symmetry, their covariance is 0). We can expect that the quotient

$$\frac{T_n - T(F)}{D_n} \tag{8.7}$$

will behave very much like a t-statistic, but with how many degrees of freedom?

This question is tricky and probably does not have a satisfactory answer. The difficulties are connected with the following points: (1) we intend to use (8.5) not only for normal data; and (2) the answer is interesting only for relatively small sample sizes, where the asymptotic approximations are poor and depend very much on the actual underlying F.

The common opinion is that the appropriate number of degrees of freedom is somewhat smaller than the classical $n-1$, but by how much is anybody's guess. Since we are typically interested in a 95 or 99% confidence interval, it is really the tail behavior of (8.7) that matters. For small n this is overwhelmingly determined by the density of D_n near 0. Huber's approach (1970), which determined an equivalent number of degrees of freedom by matching the asymptotic moments of D_n^2 with those of a χ^2-distribution, might therefore be rather misleading.

All this notwithstanding, (8.5) and (8.7) work remarkably well for M-estimates; compare the extensive Monte Carlo study by Shorack (1976). [Shorack's definition and the use of his number of degrees of freedom df^* in formula (5) are unsound—df^* is not only unstable under small perturbations of ψ, but even gives wrong asymptotic results when used in (5). But for his favorite Hampel estimate, the difference between df^* and $n-1$ is negligible.]

Example 8.1 For an M-estimate T of location we obtain, from (8.5) and the influence function (4.11),

$$nD_n^2 = \frac{\frac{1}{n-1} \sum \psi \left(\frac{x_i - T_n}{S_n} \right)^2 S_n^2}{\left[\frac{1}{n} \sum \psi' \left(\frac{x_i - T_n}{S_n} \right) \right]^2}. \tag{8.8}$$

Example 8.2 In the case of the α-trimmed mean $\bar{x}_\alpha$, an instructive, explicit comparison between the scatter estimates derived from the jackknife and from the influence function is possible. Assume that the sample is ordered, $x_1 \leqslant x_2 \leqslant \cdots \leqslant x_n$. We distinguish two cases:

Case A $g-1 \leqslant (n-1)\alpha < n\alpha \leqslant g$, g *integral*   Then, with $p=g-n\alpha$, $q=g-(n-1)\alpha$, we have

$$(1-2\alpha)n\bar{x}_{\alpha,n} = px_g + x_{g+1} + \cdots + x_{n-g} + px_{n-g+1}.$$

The jackknifed pseudo-observations can be represented as

$$T_{ni}^* = \frac{1}{1-2\alpha}(x_i^W - \Delta) \qquad\qquad (8.9)$$

where $\{x_i^W\}$ is the α'-Winsorized sample [with $\alpha'=\alpha(n-1)/n$]

$$x_i^W = qx_g + (1-q)x_{g+1}, \qquad\qquad \text{for } i \leqslant g$$

$$= x_i, \qquad\qquad\qquad\qquad \text{for } g < i < n-g+1$$

$$= (1-q)x_{n-g} + qx_{n-g+1} \qquad \text{for } i \geqslant n-g+1, \qquad\qquad (8.10)$$

and

$$\Delta = \alpha(x_g + x_{n-g+1}). \qquad\qquad (8.11)$$

Thus

$$T_n^* = \frac{1}{n} \sum T_{ni}^* = \bar{x}_{\alpha,n} + \frac{g(1-q)}{(1-2\alpha)n}\left[-x_g + x_{g+1} + x_{n-g} - x_{n-g+1}\right],$$

$$(8.12)$$

and we obtain the jackknifed variance

$$nD_n^2 = \frac{1}{n-1}\sum(T_{ni}^* - T_n^*)^2 = \frac{1}{n-1}\frac{1}{(1-2\alpha)^2}\sum(x_i^W - \bar{x}^W)^2. \quad (8.13)$$

Case B $(n-1)\alpha = g - q \leqslant g \leqslant g + p = n\alpha$, g *integral*   Then

$$(1-2\alpha)n\bar{x}_{\alpha,n} = (1-p)x_{g+1} + x_{g+2} + \cdots + x_{n-g-1} + (1-p)x_{n-g}.$$

$$(8.14)$$

Formulas (8.9) to (8.13) remain valid with the following changes:

$$\Delta = qx_g + px_{g+1} + px_{n-g} + qx_{n-g+1}, \tag{8.15}$$

and in (8.12) for T_n^* the factor in front of the square bracket is changed into $(n-g)q/(1-2\alpha)n$.

The influence function approach (8.5) works as follows. The influence function of the α-trimmed mean is given by (3.3.18). There is a question of taste whether we should define $F_n^{-1}(\alpha) = x_{\lceil n\alpha \rceil} = x_g$ or, by linear interpolation, $F_n^{-1}(\alpha) = px_g + (1-p)x_{g+1}$, with g and p as in Case A. For either choice we obtain the representation

$$nD_n^2 = \frac{n}{n-1} \int IC_n^2 \, dF_n = \frac{1}{n-1} \frac{1}{(1-2\alpha)^2} \sum (x_i^W - \bar{x}^W)^2, \tag{8.16}$$

where the Winsorizing parameter used in the definition of x_i^W is g/n or α, respectively. The difference to the jackknifed variance clearly is negligible and has mostly to do with the fine print in the definition of the estimates and sample distribution functions. But obviously, the influence function approach, when available, is cheaper to calculate.

CHAPTER 7

Regression

7.1 GENERAL REMARKS

Regression poses some peculiar and difficult robustness problems. Consider the following example. Assume that a straight line is to be fitted through six points, whose coordinates are given in Exhibit 7.1.1. A least squares fit (fit 1) yields the line shown in Exhibit 7.1.2a. A casual scan of the values in Exhibit 7.1.1 leaves the impression that everything is fine; in particular, none of the residuals $r_i = y_i - \hat{y}_i$ is exceptionally large when compared to the estimated standard deviation $\hat{\sigma}$ of the observations. A closer scrutiny, in particular, a closer look at Exhibit 7.1.2a, may, however, lead to the suspicion that there could be something wrong either with point 1 (which has the largest residual), or, perhaps, with point 6. If we drop point 6 from the fit, we obtain fit 2 (shown in Exhibit 7.1.2b). But, possibly, a linear model was inappropriate to start with, and we should have fitted a parabola (fit 3, Exhibit 7.1.2c). It is fairly clear that the available data do not suffice to distinguish between these three possibilities. Because of the low residual error $\hat{\sigma}$, we might perhaps lean towards

			Fit 1		Fit 2		Fit 3	
Point	x	y	$\hat{y}$	$y - \hat{y}$	$\hat{y}$	$y - \hat{y}$	$\hat{y}$	$y - \hat{y}$
1	-4	2.48	0.39	2.09	2.04	0.44	2.23	0.25
2	-3	0.73	0.31	0.42	1.06	-0.33	0.99	-0.26
3	-2	-0.04	0.23	-0.27	0.08	-0.12	-0.09	-0.13
4	-1	-1.44	0.15	-1.59	-0.90	-0.54	-1.00	-0.44
5	0	-1.32	0.07	-1.39	-1.87	0.55	-1.74	0.42
6	10	0.	-0.75	0.75	-11.64	(11.64)	0.01	-0.01
	e.s.d.			$\hat{\sigma} = 1.55$		$\hat{\sigma} = 0.55$		$\hat{\sigma} = 0.41$
				$r_{max}/\hat{\sigma} = 1.35$		$r_{max}/\hat{\sigma} = 1.00$		$r_{max}/\hat{\sigma} = 1.08$

Exhibit 7.1.1

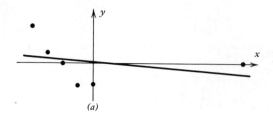

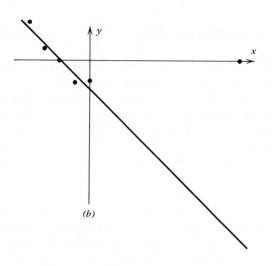

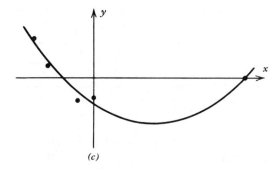

Exhibit 7.1.2 (*a*) Fit 1. (*b*) Fit 2. (*c*) Fit 3.

the third variant. In actual fact the example is synthetic and the points have been generated by taking the line $y = -2 - x$, adding random normal errors (with mean 0 and standard error 0.6) to points 1 to 5, and a gross error of 12 to point 6. Thus fit 2 is the appropriate one, and it happens to hit the true line almost perfectly.

In this case, only two parameters were involved, and a graph like Exhibit 7.1.2 helped to spot the potentially troublesome point number 6, even if the corresponding residual was quite unobtrusive. But what can be done in more complicated multiparameter problems? The difficulty is of course that a gross error does not necessarily show up through a large residual; by causing an overall increase in the size of other residuals, it can even hide behind a veritable smokescreen. In order to disentangle the issues, we must:

(1) Find analytical methods for identifying so-called leverage points, that is, points in the design space where an observation, by virtue of its position, has an overriding influence on the fit and in particular on its own fitted value. Such an observation may be the most important one in the sample (e.g., an isolated astronomical observation from remote antiquity), but on the other hand its value is difficult or impossible to crosscheck.

(2) Find routine methods for robust estimation of regression coefficients when there are no leverage points.

(3) Find estimation methods that work reasonably well and robustly also in the presence of moderately bad leverage points.

Not surprisingly, the treatment of the last part is the least satisfactory of the three.

We leave aside all issues connected with ridge regression, Stein estimation, and the like. These questions and robustness seem to be sufficiently orthogonal to each other that it should be possible to superimpose them without very serious interactions.

7.2   THE CLASSICAL LINEAR LEAST SQUARES CASE

The main purpose of this section is to discuss and clarify some of the issues connected with the leverage points.

Assume that p unknown parameters $\theta_1, \ldots, \theta_p$ are to be estimated from n observations $y_1, \ldots, y_n$ to which they are related by

$$y_i = \sum_{j=1}^{p} x_{ij}\theta_j + u_i, \tag{2.1}$$

where the x_{ij} are known coefficients and the u_i are independent random variables with (approximately) identical distributions. We also use matrix notation,

$$\mathbf{y} = X\boldsymbol{\theta} + \mathbf{u}. \tag{2.2}$$

Classically, the problem is solved by minimizing the sum of squares

$$\sum_i \left(y_i - \sum x_{ij}\theta_j \right)^2 = \min! \tag{2.3}$$

or, equivalently, by solving the system of p equations obtained by differentiating (2.3),

$$\sum_i \left(y_i - \sum x_{ik}\theta_k \right) x_{ij} = 0, \tag{2.4}$$

or, in matrix notation,

$$X^T X \boldsymbol{\theta} = X^T \mathbf{y}. \tag{2.5}$$

We assume that X has full rank p, so the solution can be written

$$\hat{\boldsymbol{\theta}} = (X^T X)^{-1} X^T \mathbf{y}, \tag{2.6}$$

and, in particular, the fitted values [the least squares estimates $\hat{y}_i$ of the expected values $Ey_i = (X\boldsymbol{\theta})_i$ of the observations] are given by

$$\hat{\mathbf{y}} = X(X^T X)^{-1} X^T \mathbf{y} = H\mathbf{y} \tag{2.7}$$

with

$$H = X(X^T X)^{-1} X^T. \tag{2.8}$$

The matrix H is often called "hat matrix" (since it puts the hat on y). We note that H is a symmetric $n \times n$ projection matrix, that is, $HH = H$, and that it has p eigenvalues equal to 1 and $n - p$ eigenvalues equal to 0. Its diagonal elements, denoted by $h_i = h_{ii}$, satisfy

$$0 \leqslant h_i \leqslant 1, \tag{2.9}$$

and the trace of H is

$$\text{tr}(H) = p. \tag{2.10}$$

We now assume that the errors u_i are independent and have a common distribution F with mean $Eu_i = 0$ and variance $Eu_i^2 = \sigma^2 < \infty$. Assume that our regression problem is imbedded in an infinite sequence of similar problems, such that the number n of observations, and possibly also the number p of parameters, tend to infinity; we suppress the index that gives the position of our problems in this sequence.

Questions When is a fitted value $\hat{y}_i$ consistent, in the sense that

$$\hat{y}_i - E(y_i) \to 0 \qquad (2.11)$$

in probability? When are all fitted values consistent?

Since $E\mathbf{u} = 0$, $\hat{\mathbf{y}}$ is unbiased:

$$E\hat{\mathbf{y}} = E\mathbf{y} = X\boldsymbol{\theta}. \qquad (2.12)$$

We have

$$\hat{y}_i = \sum h_{ik} y_k \qquad (2.13)$$

and thus

$$\operatorname{var}(\hat{y}_i) = \sum_k h_{ik}^2 \sigma^2 = h_i \sigma^2 \qquad (2.14)$$

(note that $\sum_k h_{ik}^2 = h_i$, since H is symmetric and idempotent). Hence by Chebyshev's inequality,

$$P\big[|\hat{y}_i - E(y_i)| \geq \varepsilon\big] \leq \frac{h_i \sigma^2}{\varepsilon^2}, \qquad (2.15)$$

and we have proved the sufficiency part of the following proposition.

PROPOSITION 2.1 Assume that the errors u_i are independent with mean 0 and common variance $\sigma^2 < \infty$. Then $\hat{y}_i$ is consistent iff $h_i \to 0$, and the fitted values $\hat{y}_i$ are all consistent iff

$$h = \max_{1 \leq i \leq n} h_i \to 0.$$

Proof We have to show necessity of the condition. This follows easily from

$$\hat{y}_i - E\hat{y}_i = h_i u_i + \sum_{k \neq i} h_{ik} u_k$$

and the remark that, for independent random variables X and Y, we have

$$P(|X+Y| \geqslant \varepsilon) \geqslant P(X \geqslant \varepsilon)P(Y \geqslant 0) + P(X \leqslant -\varepsilon)P(Y < 0),$$

$$\geqslant \min[P(X \geqslant \varepsilon), P(X \leqslant -\varepsilon)].$$

Thus

$$P(|\hat{y}_i - E\hat{y}_i| \geqslant \varepsilon) \geqslant \min\left[P\left(u_i \geqslant \frac{\varepsilon}{h_i}\right), P\left(u_i \leqslant -\frac{\varepsilon}{h_i}\right)\right]. \quad \blacksquare$$

Note that $h = \max h_i \geqslant \operatorname{ave} h_i = \operatorname{tr}(H)/n = p/n$; hence h cannot converge to 0 unless $p/n \to 0$.

The following formulas are straightforward to establish (under the assumptions of the preceding proposition):

$$\operatorname{var} \hat{y}_i = h_i \sigma^2, \tag{2.16}$$

$$\operatorname{var}(y_i - \hat{y}_i) = (1 - h_i)\sigma^2, \tag{2.17}$$

$$\operatorname{cov}(\hat{y}_i, \hat{y}_k) = h_{ik}\sigma^2, \tag{2.18}$$

$$\operatorname{cov}(y_i - \hat{y}_i, y_k - \hat{y}_k) = (\delta_{ik} - h_{ik})\sigma^2, \tag{2.19}$$

$$\operatorname{cov}(\hat{y}_i, y_k - \hat{y}_k) = 0, \qquad \text{for all } i, k. \tag{2.20}$$

Now let

$$\hat{\alpha} = \sum a_j \hat{\theta}_j = \mathbf{a}^T \hat{\boldsymbol{\theta}} \tag{2.21}$$

be the least squares estimate of an arbitrary linear combination $\alpha = \mathbf{a}^T \boldsymbol{\theta}$. If F is normal, then $\hat{\alpha}$ is automatically normal.

Question Assume that F is not normal. Under which conditions is $\hat{\alpha}$ asymptotically normal (as $p, n \to \infty$)?

Without restricting generality we can choose the coordinate system in the parameter space such that $X^T X = I$ is the $p \times p$ identity matrix. Furthermore assume $\mathbf{a}^T\mathbf{a} = 1$. Then $\hat{\boldsymbol{\theta}} = X^T\mathbf{y}$, and

$$\hat{\alpha} = \mathbf{a}^T\hat{\boldsymbol{\theta}} = \mathbf{a}^T X^T \mathbf{y} = \mathbf{s}^T\mathbf{y}, \tag{2.22}$$

with

$$\mathbf{s} = X\mathbf{a} \tag{2.23}$$

and

$$\mathbf{s}^T\mathbf{s} = \mathbf{a}^T X^T X \mathbf{a} = \mathbf{a}^T\mathbf{a} = 1. \tag{2.24}$$

Thus

$$\text{var}(\hat{\alpha}) = \sigma^2. \tag{2.25}$$

PROPOSITION 2.2 $\hat{\alpha}$ is asymptotically normal iff $\max_i |s_i| \to 0$.

Proof   If $\max_i |s_i| \not\to 0$, then $\hat{\alpha}$ either does not have a limiting distribution at all, or, if it has, the limiting distribution can be written as a convolution of two parts one of which is F (apart from a scale factor); hence it cannot be normal [see, e.g., Feller (1966), p. 498]. If $\gamma = \max_i |s_i| \to 0$, then we can easily check Lindeberg's condition:

$$\frac{1}{\sigma^2} \sum_i E\left\{ s_i^2 u_i^2 1_{[|s_i u_i| > \epsilon\sigma]} \right\} \leq \frac{1}{\sigma^2} \sum_i s_i^2 E\left\{ u_i^2 1_{[|u_i| > \epsilon\sigma/\gamma]} \right\}$$

$$= \frac{1}{\sigma^2} E\left\{ u^2 1_{[|u| > \epsilon\sigma/\gamma]} \right\} \to 0.$$

This finishes the proof of the proposition.   ∎

Note that the Schwarz inequality gives

$$s_i^2 = \left(\sum_k x_{ik} a_k \right)^2 \leq \sum_k x_{ik}^2 \sum_k a_k^2 = h_i.$$

Hence we obtain, as a corollary, the following theorem.

THEOREM 2.3   If $h = \max_i h_i \to 0$, then all least squares estimates $\hat{\alpha} = \sum a_j \hat{\theta}_j = \mathbf{a}^T\hat{\boldsymbol{\theta}}$ are asymptotically normal. If not, then, in particular, some of the fitted values are not asymptotically normal.

Proof The direct part is an immediate consequence of the preceding proposition. For the converse recall that $\hat{y}_i = \Sigma h_{ik} y_k$. For each n choose i such that $h_i = h$. Then the standardized sequence $[(\hat{y}_i - E(\hat{y}_i))] / \sqrt{h_i}$ has expectation 0 and variance σ^2, but cannot be asymptotically normal. ∎

Residuals and Outliers

The ith residual can be written

$$r_i = y_i - \hat{y}_i = (1 - h_i) y_i - \sum_{k \neq i} h_{ik} y_k. \tag{2.26}$$

Hence if h_i is close to 1, a gross error in y_i will not necessarily show up in r_i. But it might show up elsewhere, say in r_k, if h_{ki} happens to be large. For instance, in the introductory example of Section 7.1 (fit 1), we have $h_6 = 0.936$. Points with large h_i are, by definition, *leverage points*.

We may say that $1/h_i$ is the equivalent number of observations entering into the determination of $\hat{y}_i$. We show later that, if $h_i = 1/k$ and if we duplicate the ith row of X (make an additional observation there), then h_i is changed to $1/(k+1)$. In other words if h_i is large, it can easily be decreased by duplication or triplication of the observation y_i. (In practice approximate duplication, i.e., observing under slightly varied conditions, is to be preferred over exact duplication, since this helps to avoid repetition of systematic errors.)

We now work this out in detail. We have

$$H = X(X^T X)^{-1} X^T. \tag{2.27}$$

What happens if we add another row vector $\mathbf{x}^T$ to X:

$$\tilde{X} = \begin{pmatrix} X \\ \mathbf{x}^T \end{pmatrix}? \tag{2.28}$$

Without loss of generality we assume $X^T X = I$; then

$$\tilde{X}^T \tilde{X} = I + \mathbf{x} \mathbf{x}^T. \tag{2.29}$$

We can easily check that

$$(\tilde{X}^T \tilde{X})^{-1} = I - \frac{\mathbf{x} \mathbf{x}^T}{1 + \mathbf{x}^T \mathbf{x}}. \tag{2.30}$$

The modified hat matrix $\tilde{H}$ is also easy to work out:

$$\tilde{H} = \tilde{X}(\tilde{X}^T\tilde{X})^{-1}\tilde{X}^T$$

$$= \left(\frac{X}{\mathbf{x}^T}\right)\left(I - \frac{\mathbf{x}\mathbf{x}^T}{1+\mathbf{x}^T\mathbf{x}}\right)(X^T|\mathbf{x})$$

$$= \left[\begin{array}{c|c} XX^T - \dfrac{(X\mathbf{x})(X\mathbf{x})^T}{1+\mathbf{x}^T\mathbf{x}} & \dfrac{X\mathbf{x}}{1+\mathbf{x}^T\mathbf{x}} \\ \hline \dfrac{(X\mathbf{x})^T}{1+\mathbf{x}^T\mathbf{x}} & \dfrac{\mathbf{x}^T\mathbf{x}}{1+\mathbf{x}^T\mathbf{x}} \end{array}\right]. \tag{2.31}$$

Example 2.1 Duplication of a row, say row n. Then (still assuming $X^TX = I$) we have $\mathbf{x}^T\mathbf{x} = h_n$, and thus

$$\tilde{h}_{n+1} = \frac{h_n}{1+h_n}. \tag{2.32}$$

Since there is no possibility of confusion, we omit the tilde on h_{n+1} from now on. In particular, if $h_n = 1/k$, we obtain $h_{n+1} = 1/(k+1)$.

Example 2.2 Leaving out a row (say row $n+1$, after we have added it):

(1) With row $n+1$ in we obtain, from (2.31),

$$\operatorname{var}(\hat{y}_{n+1}) = h_{n+1}\sigma^2 = \frac{\mathbf{x}^T\mathbf{x}}{1+\mathbf{x}^T\mathbf{x}}\sigma^2. \tag{2.33}$$

(2) With row $n+1$ out, let $\hat{\alpha}_{n+1}$ be the estimate of $E(y_{n+1})$ based on the remaining observations $y_1, \ldots, y_n$:

$$\hat{\alpha}_{n+1} = \mathbf{x}^T\hat{\theta}. \tag{2.34}$$

$$\operatorname{var}(\hat{\alpha}_{n+1}) = \mathbf{x}^T\mathbf{x}\sigma^2 = \frac{h_{n+1}}{1-h_{n+1}}\sigma^2. \tag{2.35}$$

Note that $\operatorname{var}(\hat{\alpha}_{n+1})$ is larger than $\operatorname{var}(y_{n+1})$ if $h_{n+1} > \frac{1}{2}$.

We have

$$\hat{y}_{n+1} = (1-h_{n+1})\hat{\alpha}_{n+1} + h_{n+1}y_{n+1}, \tag{2.36}$$

that is the $(n+1)$th fitted value is a convex linear combination of the "predicted" value $\hat{\alpha}_{n+1}$ (which disregards y_{n+1}) and the observation y_{n+1}, with weights $1-h_{n+1}$ and h_{h+1}, respectively. This is shown by a look at the last row of the matrix $\tilde{H}$.

In terms of residuals the above formula reads

$$r_{n+1}=y_{n+1}-\hat{y}_{n+1}=(1-h_{n+1})(y_{n+1}-\hat{\alpha}_{n+1}). \qquad (2.37)$$

This relation is important; it connects the ordinary residual $y_{n+1}-\hat{y}_{n+1}$ with the residual $y_{n+1}-\hat{\alpha}_{n+1}$ relative to the "interpolated" value $\hat{\alpha}_{n+1}$ ignoring y_{n+1}.

Of course, all these relations hold for arbitrary indices, not only for $i=n+1$.

We may conclude from this discussion that the diagonal of the hat matrix contains extremely useful information. In particular, large values of h_i should serve as warning signals that the ith observation may have a decisive, yet hardly checkable, influence. Values $h_i \leq 0.2$ appear to be safe, values between 0.2 and 0.5 are risky, and if we can control the design at all, we had better avoid values above 0.5.

7.3   ROBUSTIZING THE LEAST SQUARES APPROACH

The classical equations (2.3) and (2.4) can be robustized in a straightforward way; instead of minimizing a sum of squares, we minimize a sum of less rapidly increasing functions of the residuals:

$$\sum_{i=1}^{n} \rho\left(y_i - \sum x_{ij}\theta_j\right)=\min!, \qquad (3.1)$$

or, after taking derivatives, we solve

$$\sum_{i=1}^{n} \psi\left(y_i - \sum x_{ij}\theta_j\right)x_{ik}=0, \qquad k=1,\ldots,p, \qquad (3.2)$$

with $\psi=\rho'$. If ρ is convex, the two approaches are essentially equivalent; otherwise the selection of the "best" solution of (3.2) may create problems.

We denote the ith residual by

$$r_i=r_i(\theta)=y_i - \sum_j x_{ij}\theta_j. \qquad (3.3)$$

Note that (3.2) can be viewed as a robustized version of the cross product of the residual vector **r** with the kth column vector of X; the residuals r_i have been replaced by metrically Winsorized versions $\psi(r_i)$.

Ordinarily, scale will not be known, so it will be necessary to make (3.2) scale invariant by introducing some estimate s of scale:

$$\sum \psi\left(\frac{r_i}{s}\right)x_{ik}=0. \qquad (3.4)$$

Possibly we might use individual scales s_i, and even more generally, we might use different functions ρ_i and ψ_i for different observations.

We are acting under the assumption that the x_{ij} are known and error-free. If this is not so, it might be advisable to modify not only the residual vector **r**, but also the column vectors $\mathbf{x}_j$ in order to gain robustness also with regard to errors in the coefficients x_{ij}. There are several obvious proposals for doing so; they may look plausible, but they hitherto lack a sound theoretical underpinning. Conceivably they might do more harm (by introducing bias) than good.

We obtain R-estimates of regression if we minimize, instead of (3.1),

$$\sum_i a_n(R_i)r_i=\min!. \qquad (3.5)$$

Here, R_i is the rank of r_i in $(r_1,\dots,r_n)$, and $a_n(\cdot)$ is some monotone scores function satisfying $\sum_i a_n(i)=0$ [see Jaeckel (1972)]. Note, however, that these estimates are unable to estimate an additive main effect and thus do not contain estimates of location as particular cases. On the contrary the additive main effect has to be estimated by applying an estimate of location to the residuals.

If we differentiate (3.5), which is a piecewise linear convex function of $\boldsymbol{\theta}$, we obtain the following approximate equalities at the minimum:

$$\sum_i a_n(R_i)x_{ik}\approx 0, \qquad k=1,\dots,p. \qquad (3.6)$$

These approximate equations in turn can be reconverted into a minimum problem, for example,

$$\sum_k \left|\sum_i a_n(r_i)x_{ik}\right|=\min!. \qquad (3.7)$$

This last variant was investigated by Jurečková (1971), and asymptotic equivalence between (3.6) and (3.7) was shown by Jaeckel (1972). The task of solving (3.6) or (3.7) by linear programming techniques appears to be very formidable, however, unless p and n are quite small.

All of the regression estimates allow one-step versions: start with some reasonably good preliminary estimate θ^*, and then apply one step of Newton's method to (3.2), and so on, just as in the location case. A one-step L-estimate of regression has been investigated by Bickel (1973). However, in the regression case it is very difficult to find a good starting value. We know from the location case that the least squares estimate, that is, the sample mean, will not do for one-step estimates, and the analogue of the sample median, which gives an excellent starting point for location, would be the so-called L_1-estimate [corresponding to $\rho(X) = |X|$], which itself may be harder to compute than most of the robust regression estimates we want to use.

It appears that M-estimates offer enough flexibility and are by far the easiest to cope with, simultaneously, with regard to computation, asymptotic theory, and intuitive interpretation; moreover, the step from (2.3) to (3.1) is easily explainable to nonstatisticians also. We therefore restrict ourselves to M-estimates of regression.

7.4 ASYMPTOTICS OF ROBUST REGRESSION ESTIMATES

The obvious approach to asymptotics is: keep the number p of parameters fixed, let the number n of observations go to infinity. However, in practice p and n tend to become large simultaneously; in crystallography, where some of the largest least squares problems occur (with hundreds or thousands of parameters) we find the explicit recommendation that there should be at least five observations per parameter (Hamilton 1970). This suggests that a meaningful asymptotic theory should be in terms of $p/n \rightarrow 0$, or, perhaps better, in terms of $h = \max h_i \rightarrow 0$. The point we make here, which is given some technical substantiation later, is that, if the asymptotic theory requires, say, $p^3/n \rightarrow 0$, and if it is able to give a useful approximation for $n = 20$ if $p = 1$, then, for $p = 10$, we would need $n = 20,000$ to get an equally good approximation! In an asymptotic theory that keeps p fixed such distinctions do not become visible at all.

We begin with a short discussion of the overall regularity conditions. They separate into three parts, conditions on: the design matrix X, the estimate, and the error laws.

Conditions on the Design Matrix X

X has full rank p, and the diagonal elements of the hat matrix

$$H = X(X^TX)^{-1}X^T \qquad (4.1)$$

are assumed to be uniformly small:

$$\max_{i \leq i \leq n} h_i = h \ll 1. \qquad (4.2)$$

The precise order of smallness will be specified from case to case. Without loss of generality we choose the coordinate system in the parameter space such that the true parameter point is $\theta^0 = 0$, and such that X^TX is the $p \times p$ identity matrix.

Conditions on the Estimate

The function ρ is assumed to be convex and nonmonotone and to possess bounded derivatives of sufficiently high order (approximately four). In particular, $\psi(x) = (d/dx)\rho(x)$ should be continuous and bounded. Convexity of ρ serves to guarantee equivalence between (3.1) and (3.2) and asymptotic uniqueness of the solution. If we are willing to forego this and are satisfied with local uniqueness, the convexity assumption can be omitted. Higher order derivatives are technically convenient, since they make Taylor expansions possible, but their existence does not seem to be essential for the results to hold.

Conditions on the Error Laws

We assume that the errors u_i are independent, identically distributed, such that

$$E[\psi(u_i)] = 0. \qquad (4.3)$$

We require this in order that the expectation of (3.1) reaches its minimum and the expectation of (3.2) vanishes, at the true value θ^0.

The assumption of independence is a serious restriction. The assumption that the errors are identically distributed simplifies notations and calculations, but could easily be relaxed: "random" deviations (i.e., not related to the structure of X) can be modeled by identical distributions (take the "averaged" cumulative distribution). Nonrandom deviations (e.g., changes in scale that depend on X in a systematic fashion) can be handled by a minimax approach if the deviations are small; if they are large, they transgress our notion of robustness.

The Case $hp^2 \to 0$ and $hp \to 0$

A simple but rigorous treatment is possible if $hp^2 \to 0$, or, with slightly weaker results, if $hp \to 0$. Note that this implies $p^3/n \to 0$ and $p^2/n \to 0$, respectively. Thus quite moderate values of p already lead to very large and impractical values for n.

The idea is to compare the zeros of two vector-valued random functions Φ and Ψ of θ:

$$\Phi_j(\theta) = \frac{-1}{E(\psi')} \sum_i \psi \left(y_i - \sum x_{ik}\theta_k \right) x_{ij}, \tag{4.4}$$

$$\Psi_j(\theta) = \theta_j - \frac{1}{E(\psi')} \sum_i \psi(y_i) x_{ij}. \tag{4.5}$$

The zero $\hat{\theta}$ of Φ is our estimate. The zero $\tilde{\theta}$ of Ψ,

$$\tilde{\theta}_j = \frac{1}{E(\psi')} \sum_i \psi(y_i) x_{ij} \tag{4.6}$$

of course is not a genuine estimate, but according to Theorem 2.3 all linear combinations $\tilde{\alpha} = \sum a_j \tilde{\theta}_j$ are asymptotically normal if $h \to 0$. So we can prove asymptotic normality of $\hat{\theta}$ (or, better, of $\hat{\alpha} = \sum a_j \hat{\theta}_j$) by showing that the difference between $\hat{\theta}$ and $\tilde{\theta}$ is small.

Let a_j be indeterminate coefficients satisfying $\sum a_j^2 = 1$ and write for short

$$s_i = \sum x_{ij} a_j, \tag{4.7}$$

$$t_i = \sum x_{ij} \theta_j. \tag{4.8}$$

Since $X^T X = I$, we have

$$\|t\|^2 = (X\theta)^T X\theta = \|\theta\|^2, \tag{4.9}$$

$$\|s\|^2 = 1. \tag{4.10}$$

We expand $\sum a_j \Phi_j(\boldsymbol{\theta})$ into a Taylor series with remainder term

$$\sum a_j \Phi_j(\boldsymbol{\theta}) = \frac{-1}{E(\psi')} \left[\sum \psi(y_i)s_i - \sum \psi'(y_i)t_i s_i + \tfrac{1}{2} \sum \psi''(y_i - \eta t_i)t_i^2 s_i \right]$$

$$(4.11)$$

with $0 < \eta < 1$. This can be rearranged to give

$$\sum a_j [\Phi_j(\boldsymbol{\theta}) - \Psi_j(\boldsymbol{\theta})] = \sum_{jk} \Delta_{jk} a_j \theta_k - \frac{1}{2E(\psi')} \sum_i \psi''(y_i - \eta t_i)t_i^2 s_i,$$

$$(4.12)$$

where

$$\Delta_{jk} = \frac{1}{E(\psi')} \sum_i [\psi'(y_i) - E\psi'(y_i)] x_{ij} x_{ik}.$$

$$(4.13)$$

We now intend to show that $\boldsymbol{\Phi} - \boldsymbol{\Psi}$ is uniformly small in a neighborhood of $\boldsymbol{\theta} = 0$, or more precisely, that (4.12) is uniformly small on sets of the form

$$\{ (\boldsymbol{\theta}, \mathbf{a}) | \, \|\boldsymbol{\theta}\|^2 \leqslant Kp, \|\mathbf{a}\| = 1 \}.$$

$$(4.14)$$

By the Schwarz inequality the first term on the right-hand side of (4.12) can be bounded as follows:

$$\left(\sum \Delta_{jk} a_j \theta_k \right)^2 \leqslant \sum \Delta_{jk}^2 \sum a_j^2 \sum \theta_k^2 = \sum \Delta_{jk}^2 \|\boldsymbol{\theta}\|^2.$$

$$(4.15)$$

We have

$$E\left(\sum \Delta_{jk}^2 \right) = \sum E(\Delta_{jk}^2) = \sum_{jki} x_{ij}^2 x_{ik}^2 \cdot \frac{\text{var}(\psi')}{(E\psi')^2}$$

$$(4.16)$$

and

$$\sum_{jki} x_{ij}^2 x_{ik}^2 = \sum h_i^2 \leqslant \max(h_i) \sum h_i = hp.$$

$$(4.17)$$

Now let $\delta > 0$ be given. Markov's inequality then yields that there is a constant K_1, namely

$$K_1 = \frac{\text{var}(\psi')}{(E\psi')^2} \cdot \frac{1}{\delta}, \tag{4.18}$$

such that

$$P\left\{ \sum \Delta_{jk}^2 \geqslant K_1 hp \right\} \leqslant \delta. \tag{4.19}$$

We conclude that, with probability greater than $1 - \delta$,

$$\left(\sum \Delta_{jk} a_j \theta_k \right)^2 \leqslant KK_1 hp^2 \tag{4.20}$$

holds simultaneously for all $(\mathbf{a}, \boldsymbol{\theta})$ in (4.14).

Assume that ψ'' is bounded, say $|\psi''(x)| \leqslant 2|E(\psi')|M$ for some M; then

$$\left| \frac{1}{2E(\psi')} \sum \psi''(y_i - \eta t_i) t_i^2 s_i \right| \leqslant M \max |s_i| \sum t_i^2 \leqslant M h^{1/2} \|\boldsymbol{\theta}\|^2 \tag{4.21}$$

[see (4.9) and recall that $s_i^2 \leqslant \sum x_{ij}^2 \sum a_j^2 = h_i$].

If we put things together, we obtain that, with probability $> 1 - \delta$, (4.12) is bounded in absolute value by

$$r = \left[(KK_1)^{1/2} + MK \right] (hp^2)^{1/2}, \tag{4.22}$$

and this uniformly on the set (4.14). Since the results hold simultaneously for all $\mathbf{a}$ with $\|\mathbf{a}\| = 1$, we have in fact shown that, with probability greater than $1 - \delta$,

$$\|\boldsymbol{\Phi}(\boldsymbol{\theta}) - \boldsymbol{\Psi}(\boldsymbol{\theta})\| \leqslant r, \qquad \text{for } \|\boldsymbol{\theta}\|^2 \leqslant Kp. \tag{4.23}$$

If K is chosen large enough, and since

$$E\left(\|\tilde{\boldsymbol{\theta}}\|^2 \right) = \frac{E(\psi^2)}{\left[E(\psi') \right]^2} p, \tag{4.24}$$

it follows from Markov's inequality that

$$P\left\{ \|\tilde{\boldsymbol{\theta}}\|^2 \leqslant Kp/4 \right\} \tag{4.25}$$

can be made arbitrarily large. Moreover, then

$$\|\Phi(\theta) - \theta\| \leq \|\Phi(\theta) - \Psi(\theta)\| + \|\tilde{\theta}\| \leq r + \tfrac{1}{2}(Kp)^{1/2}, \qquad (4.26)$$

on the set $\|\theta\|^2 \leq Kp$.

If $hp \to 0$, then r can be made smaller than $\tfrac{1}{2}(Kp)^{1/2}$, so that (4.26) implies

$$\|\theta - \Phi(\theta)\| < (Kp)^{1/2} \qquad (4.27)$$

on the set $\|\theta\| \leq (Kp)^{1/2}$.

But this is precisely the premiss of Brouwer's fixed point theorem: we conclude that the map $\theta \to \theta - \Phi(\theta)$ has a fixed point $\hat{\theta}$, which necessarily then is a zero of $\Phi(\theta)$, with $\|\hat{\theta}\| < (Kp)^{1/2}$.

If we substitute $\hat{\theta}$ for θ into (4.23), we obtain

$$\|\hat{\theta} - \tilde{\theta}\| \leq r. \qquad (4.28)$$

We thus obtain the following proposition.

PROPOSITION 4.1

(1) If $hp^2 \to 0$, then

$$\|\hat{\theta} - \tilde{\theta}\| \to 0 \qquad (4.29)$$

in probability.

(2) If $hp \to 0$, then

$$\frac{\|\hat{\theta} - \tilde{\theta}\|}{p^{1/2}} \to 0 \qquad (4.30)$$

in probability. [Note that $\|\tilde{\theta} - \theta^0\| \sim p^{1/2}$ in view of (4.24).]

Now let $\hat{\alpha} = \Sigma a_j \hat{\theta}_j$ and $\tilde{\alpha} = \Sigma a_j \tilde{\theta}_j$, with $\|\mathbf{a}\| = 1$. Recall that $\hat{\alpha}$ is the estimate to be investigated, while $\tilde{\alpha}$ is a sum of independent random variables and is asymptotically normal if $h \to 0$.

PROPOSITION 4.2

(1) If $hp^2 \to 0$, then

$$\sup_{\|\mathbf{a}\| = 1} |\hat{\alpha} - \tilde{\alpha}| \to 0 \qquad (4.31)$$

in probability.

(2) If **a** is chosen at random with respect to the invariant measure on the sphere $\|\mathbf{a}\| = 1$, and if $hp \to 0$, then

$$\hat{\alpha} - \tilde{\alpha} \to 0 \qquad\qquad (4.32)$$

in probability.

Both (1) and (2) imply that $\hat{\alpha}$ is asymptotically normal.

Proof (1) is an immediate consequence of part (1) of the preceding proposition; similarly, (2) follows from part (2) and the fact that the average of $|\hat{\alpha} - \tilde{\alpha}|^2$ over the unit sphere $\|\mathbf{a}\| = 1$ is $\|\hat{\boldsymbol{\theta}} - \tilde{\boldsymbol{\theta}}\|^2/p$.

Remark 1 In essence we have shown that $\boldsymbol{\Phi}(\boldsymbol{\theta})$ is asymptotically linear in a neighborhood of the true parameter point $\boldsymbol{\theta}^0$. Actually, the assumption that $\boldsymbol{\theta}^0 = 0$ was used only once, namely, in (4.27). If $\boldsymbol{\theta}^*$ is any estimate satisfying $\|\boldsymbol{\theta}^* - \boldsymbol{\theta}^0\| = O_p(p^{1/2})$, then we can show in the same way that just one step of Newton's method for solving $\boldsymbol{\Phi}(\boldsymbol{\theta}) = 0$, with trial value $\boldsymbol{\theta}^*$, leads to an estimate $\hat{\boldsymbol{\theta}}^*$ satisfying

$$\|\hat{\boldsymbol{\theta}}^* - \tilde{\boldsymbol{\theta}}\| \to 0, \qquad \|\hat{\boldsymbol{\theta}}^* - \hat{\boldsymbol{\theta}}\| \to 0$$

in probability, provided $hp^2 \to 0$.

Remark 2 Recently Yohai and Maronna (1979) have improved this result and shown that $\hat{\alpha}$ is asymptotically normal for arbitrary choices of **a**, assuming only $hp^{3/2} \to 0$, instead of $hp^2 \to 0$. My conjecture is that $hp \to 0$ is sufficient for (4.32) to hold for arbitrary **a**, and that $hp^{1/2} \to 0$ is necessary, if either the distribution of the u_i or ρ are allowed to be asymmetric. If both the distribution of the u_i and ρ are symmetric, then perhaps already $h \to 0$ is sufficient, as in the classical least squares case.

7.5 CONJECTURES AND EMPIRICAL RESULTS

An asymptotic theory that requires $hp^2 \to 0$ (and hence _a fortiori_ $p^3/n \to 0$) is for all practical purposes worthless and the situation is only a little more favorable for $hp \to 0$; already for a moderately large number of parameters we would need an impossibly large number of observations. Of course, my inability to prove theorems assuming only $h \to 0$ does not imply that then the robustized estimates fail to be consistent and asymptotically normal, but what if they should fail? In order to get some insight into what is going on, we may resort to asymptotic expansions. These are without remainder terms, so the results are nonrigorous, but they can be verified by Monte

Carlo simulations. The expansions are reported in some detail in Huber (1973a); here we only summarize the salient points.

The Question of Bias

Assume that either the distribution of the errors u_i or the function ρ, or both, are asymmetric. Then the parameters to be estimated are not intrinsically defined by symmetry considerations; we chose to fix them through the convention that $E\psi(u_i)=0$. Then, for instance, a location estimate T_n, defined by

$$\sum_{i=1}^{n} \psi(u_i - T_n) = 0,$$

is asymptotically normal with mean 0, but its distribution for finite n is asymmetric and not exactly centered at 0.

Take now the following simple regression design (which actually represents the worst possible case). Assume that we have p unknown parameters $\theta_1, \ldots, \theta_p$; we take r independent observations on each of them, and as an overall check we take one observation y_n of

$$\sqrt{\frac{r}{n-p}} \ (\theta_1 + \cdots + \theta_p).$$

Here, $n = rp + 1$ is the total number of observations, and the corresponding hat matrix happens to be balanced, that is, all its diagonal elements h_i are equal to p/n.

It is intuitively obvious that any robust regression estimate of $(\theta_1, \ldots, \theta_p)$, for all practical purposes, is equivalent to estimating the p parameters separately from $\sum_1^r \psi(y_i - \hat\theta_1) = 0$, and so on, since the single observation y_n of the scaled sum should have only a negligible influence. So the predicted value of this last observation is

$$\hat\alpha_n = \sqrt{\frac{r}{n-p}} \ (\hat\theta_1 + \cdots + \hat\theta_p),$$

where the $\hat\theta_i$ have been estimated from the r observations of each separately. [The definition of g in Huber (1973a), p. 810, should read $g^2 = r/(n-p)$.] But the distributions of the $\hat\theta_i$ are slightly asymmetric and not quite centered at their "true" values, and if we work things out in detail, we find that $\hat\alpha_n$ is affected by a bias of the order $p^{3/2}/n$. Note that the asymptotic variance of $\hat{\alpha}_n$ is of the order p/n, so that the bias measured in units of the standard deviation is $p/n^{1/2}$. The asymptotic behavior of the fitted value $\hat{y}_n$ is the same as that of $\hat{\alpha}_n$, of course.

In other words, if $h=p/n\to0$, but $p^{3/2}/n\to\infty$, it can happen that the residual $r_n=y_n-\hat{y}_n$ tends to infinity, not because of a gross error in y_n, but because the small biases in the $\hat{\theta}_i$ have added up to a large bias in $\hat{y}_n$! However, we should hasten to add that this bias of the order $\sqrt{p}\ p/n$ is asymptotically negligible against the bias

$$\sqrt{\frac{r}{n-p}}\ p\delta\approx\sqrt{p}\ \delta$$

caused by a systematic error $+\delta$ in all the observations.

Moreover, the quantitative aspects are such that it is far from easy to verify the effect by Monte Carlo simulation; with $p/n=\frac{1}{8}$ we need $p\cong100$ to make the bias of $\hat{y}_n$ approximately equal to $(\operatorname{var}\hat{y}_n)^{1/2}$, and this with highly asymmetric error distributions (χ^2 with two to four degrees of freedom).

From these remarks we derive the following practical conclusions:

(1) The biases caused by asymmetric error distributions exist and can cause havoc within the asymptotic theory but, for most practical purposes, they will be so small that they can be neglected.

(2) The biases are largest in situations that should also be avoided for another reason (robustness of design), namely, situations where the estimand is interpolated between observations that are widely separated in the design space—as in our example where $\alpha=\Sigma\theta_i$ is estimated from observed values for the individual θ_i. In such cases relatively minor deviations from the linear model may cause large deviations in the fitted values.

7.6 ASYMPTOTIC COVARIANCES AND THEIR ESTIMATION

The covariance matrix of the classical least squares estimate $\hat{\boldsymbol{\theta}}_{\mathrm{LS}}$ is traditionally estimated by

$$\operatorname{cov}(\hat{\boldsymbol{\theta}}_{\mathrm{LS}})\approx\frac{1}{n-p}\left(\sum r_i^2\right)(X^TX)^{-1}. \tag{6.1}$$

By what should this be replaced in the robustized case?

The limiting expression for the covariance of the robust estimate, derived from Proposition 4.1, is

$$\text{cov}(\hat{\boldsymbol{\theta}}) = \frac{E(\psi)^2}{(E\psi')^2}(X^T X)^{-1}, \tag{6.2}$$

which can be translated straightforwardly into the estimate

$$\text{cov}(\hat{\boldsymbol{\theta}}) \approx \frac{(1/n)\sum \psi(x_i)^2}{\left[(1/n)\sum \psi'(x_i)\right]^2}(X^T X)^{-1}. \tag{6.3}$$

If we want to recapture the classical formula (6.1) in the classical case $[\psi(x) = x]$, we should multiply the right-hand side of (6.3) by $n/(n-p)$, and perhaps some other corrections of the order $h = p/n$ are needed. Also the matrix $X^T X$ should perhaps be replaced by something like the matrix

$$W_{jk} = \sum \psi'(r_i)x_{ij}x_{ik}. \tag{6.4}$$

The second, and perhaps even more important, goal of the asymptotic expansions mentioned in the preceding sections is to find proposals for correction terms of the order h.

The general expressions are extremely unwieldy, but in the balanced case (i.e., $h_i = h = p/n$), with symmetric error distributions and skew symmetric ψ, assuming that $1 \ll p \ll n$, and if we neglect terms of the orders $h^2 = (p/n)^2$ or $1/n$, the following three expressions all are unbiased estimates of $\text{cov}(\hat{\boldsymbol{\theta}})$:

$$K^2 \frac{[1/(n-p)]\sum \psi(r_i)^2}{\left[(1/n)\sum \psi'(r_i)\right]^2}(X^T X)^{-1}, \tag{6.5}$$

$$K\frac{[1/(n-p)]\sum \psi(r_i)^2}{(1/n)\sum \psi'(r_i)}W^{-1}, \tag{6.6}$$

$$K^{-1}\frac{1}{n-p}\sum \psi(r_i)^2 W^{-1}(X^T X)W^{-1}. \tag{6.7}$$

The correction factors are expressed in terms of

$$K = 1 + \frac{p}{n}\frac{\text{var}(\psi')}{(E\psi')^2}. \tag{6.8}$$

In practice $E(\psi')$ and var(ψ') are unknown and will be estimated by

$$E(\psi') \approx m = \frac{1}{n} \sum \psi'(r_i), \tag{6.9}$$

$$\text{var}(\psi') \approx \frac{1}{n} \sum [\psi'(r_i) - m]^2. \tag{6.10}$$

In the special case

$$\psi(x) = \min[c, \max(-c, x)],$$

(6.8) simplifies to

$$K = 1 + \frac{p}{n} \frac{1-m}{m}, \tag{6.11}$$

where m is the relative frequency of the residuals satisfying $-c < r_i < c$.

Note that, in the classical case, all three expressions (6.5), (6.6), and (6.7) reduce to (6.1).

In the simple location case ($p = 1$, $x_{ij} = 1$), the three expressions agree exactly if we put $K = 1$ (the derivation of K neglected terms of the order $1/n$ anyhow).

For details and a comparison with Monte Carlo results, see Huber (1973a). (For normal errors, the agreement between the expansions and the Monte Carlo results was excellent up to $p/n = \frac{1}{4}$; for Cauchy errors excellent up to $p/n = \frac{1}{16}$, and still tolerable for $p/n = \frac{1}{8}$.)

NOTE   Since $\hat{\theta}$ can also be characterized formally as the solution of the weighted least squares problem

$$\sum_i w_i r_i x_{ij} = 0, \qquad j = 1, \ldots, p, \tag{6.12}$$

with weights $w_i = \psi(r_i)/r_i$ depending on the sample, a further variant to $X^T X$ and (6.4), namely,

$$\sum w_i x_{ij} x_{ik}, \tag{6.13}$$

looks superficially attractive, together with

$$\frac{1}{n-p} \sum w_i r_i^2 \tag{6.14}$$

in place of

$$\frac{1}{n-p}\sum \psi(r_i)^2.$$

However, (6.14) is *not robust* in general [$w_i r_i^2 = \psi(r_i)r_i$ is unbounded unless ψ is redescending] and is not a consistent estimate of $E(\psi^2)$. So we should be strongly advised against the use of (6.14).

It would be feasible to use the suitably scaled matrix (6.13), namely

$$\frac{\sum w_i x_{ij} x_{ik}}{(1/n)\sum w_i} \qquad (6.15)$$

in place of $X^T X$, just as we did with W in (6.6) and (6.7), but the bias correction factors then seem to become discouragingly complicated.

7.7 CONCOMITANT SCALE ESTIMATES

For the sake of simplicity, we have so far assumed that scale was known and fixed. In practice we would have to estimate also a scale parameter σ, and we would have to solve

$$\sum \psi \left(\frac{y_i - \sum x_{ij}\theta_j}{\sigma} \right) x_{ij} = 0, \qquad j = 1, \ldots, p \qquad (7.1)$$

in place of (3.2).

This introduces some technical complications similar to those in Chapter 6, but does not change the asymptotic results. The reason is that, if we have *some* estimate for which the fitted values $\hat{y}_i$ are consistent, we can estimate scale $\hat{\sigma}$ consistently from the corresponding residuals $r_i = y_i - \hat{y}_i$, and then use this $\hat{\sigma}$ in (7.1) for calculating the final estimate $\hat{\theta}$.

In practice we calculate the estimates $\hat{\theta}$ and $\hat{\sigma}$ by simultaneous iterations (which may cause difficulties with the convergence proofs).

Which scale estimate should we use? In the simple location case the unequivocal answer is given by the results of the Princeton study (Andrews et al., 1972); the estimates using the median absolute deviation, that is

$$\hat{\sigma} = \text{med}\{|r_i|\}, \qquad (7.2)$$

when expressed in terms of residuals relative to the sample median, fared best. This result is theoretically underpinned by the facts that the median absolute deviation (1) is minimax with respect to bias (Section 5.7), and (2) has the highest possible breakdown point ($\varepsilon^* = \frac{1}{2}$).

In regression the case for the median absolute residual (7.2) is less well founded. First, it is not feasible to calculate it beforehand (the analogue to the sample median, the L_1-estimate, may take more time to calculate than our intended estimate $\hat{\theta}$). Second, we still lack a convergence proof for procedures simultaneously iterating (7.1) and (7.2) (the empirical evidence is, however, good).

For the following we assume, somewhat more generally, that $\hat{\theta}$ and $\hat{\sigma}$ are estimated by solving the simultaneous equations

$$\sum \psi\left(\frac{y_i - f_i(\boldsymbol{\theta})}{\sigma}\right) \frac{\partial f_i(\boldsymbol{\theta})}{\partial \theta_j} = 0, \qquad j = 1, \ldots, p, \tag{7.3}$$

$$\sum \chi\left(\frac{y_i - f_i(\boldsymbol{\theta})}{\sigma}\right) = 0 \tag{7.4}$$

for $\boldsymbol{\theta}$ and σ; the functions f_i are not necessarily linear.

Note that this contains, in particular:

(1) Maximum likelihood estimation. Assume that the observations have a probability density of the form

$$\frac{1}{\sigma} g\left(\frac{y_i - f_i(\boldsymbol{\theta})}{\sigma}\right); \tag{7.5}$$

then (7.3) and (7.4) give the maximum likelihood estimates if

$$\psi(x) = -\frac{g'(x)}{g(x)} \tag{7.6}$$

$$\chi(x) = x\psi(x) - 1. \tag{7.7}$$

(2) Median absolute residuals as the scale estimate, if

$$\chi(x) = \operatorname{sign}(|x| - 1). \tag{7.8}$$

Some problems with existence and convergence proofs arise when ψ and χ are totally unrelated. For purely technical reasons we therefore introduce the following minimum problem:

$$Q(\boldsymbol{\theta}, \sigma) = \frac{1}{n} \sum \rho\left(\frac{y_i - f_i(\boldsymbol{\theta})}{\sigma}\right) \sigma = \min!, \tag{7.9}$$

where ρ is a convex function that has a strictly positive minimum at 0. If we take partial derivatives of (7.9) with respect to θ_j and σ, we obtain the following characterization of the minimum:

$$\sum \psi\left(\frac{y_i - f_i(\boldsymbol{\theta})}{\sigma}\right)\frac{\partial f_i}{\partial \theta_i} = 0, \tag{7.10}$$

$$\sum \chi\left(\frac{y_i - f_i(\boldsymbol{\theta})}{\sigma}\right) = 0, \tag{7.11}$$

with

$$\psi(x) = \rho'(x), \tag{7.12}$$

$$\chi(x) = x\psi(x) - \rho(x). \tag{7.13}$$

Note that $\chi'(x) = x\psi'(x)$ then is negative for $x \leqslant 0$ and positive for $x \geqslant 0$; hence χ has an absolute minimum at $x = 0$, namely, $\chi(0) = -\rho(0) < 0$.

In particular, with

$$\rho(x) = \tfrac{1}{2}x^2 + \tfrac{1}{2}\beta, \qquad \text{for } |x| < c$$

$$= c|x| - \tfrac{1}{2}c^2 + \tfrac{1}{2}\beta, \quad \text{for } |x| \geqslant c, \tag{7.14}$$

we obtain

$$\psi(x) = -c, \quad \text{for } x \leqslant -c$$

$$= x, \qquad \text{for } -c < x < c$$

$$= c, \qquad \text{for } x \geqslant c, \tag{7.15}$$

and

$$\chi(x) = \tfrac{1}{2}\left[\psi(x)^2 - \beta\right]. \tag{7.16}$$

Note that this is a ψ, χ pair suggested by minimax considerations both for location and for scale (cf. Example 6.4.1), and that both ψ and χ are bounded [whereas, with the maximum likelihood approach, the χ corresponding to a monotone ψ would always be unbounded; cf. (7.7)].

If the f_i are linear, then $Q(\boldsymbol{\theta}, \sigma)$ in fact is a convex function not only of $\boldsymbol{\theta}$, but of $(\boldsymbol{\theta}, \sigma)$. In order to demonstrate this, we assume that $(\boldsymbol{\theta}, \sigma)$

depends linearly on some real parameter t and calculate the second derivative with respect to t of the summands of (7.9):

$$q_i(t) = \rho\left(\frac{y_i - f_i(\boldsymbol{\theta})}{\sigma}\right)\sigma. \tag{7.17}$$

Denote differentiation with respect to t by a superscript dot, then (omitting the index i)

$$\dot{q} = \rho\left(\frac{y-f}{\sigma}\right)\dot{\sigma} + \rho'\left(\frac{y-f}{\sigma}\right)\left(-\frac{y-f}{\sigma}\dot{\sigma} - \dot{f}\right), \tag{7.18}$$

and

$$\ddot{q} = \rho''\left(\frac{y-f}{\sigma}\right)\left(\frac{y-f}{\sigma}\dot{\sigma} + \dot{f}\right)^2 \frac{1}{\sigma} \geq 0. \tag{7.19}$$

Thus Q is convex. If ρ is not twice differentiable, the result still holds (prove this by approximating ρ differentiably).

Assume now that

$$0 < \lim_{|x|\to\infty} \frac{\rho(x)}{|x|} = c \leq \infty. \tag{7.20}$$

If $c < \infty$, Q can be extended by continuity:

$$Q(\boldsymbol{\theta}, 0) = c\frac{1}{n}\sum |y_i - f_i(\boldsymbol{\theta})|. \tag{7.21}$$

Hence the limiting case $\sigma = 0$ corresponds to L_1-estimation. Of course, on the boundary $\sigma = 0$, the characterization of the minimum by (7.10) and (7.11) breaks down, but in any case, the set of solutions $(\boldsymbol{\theta}, \sigma)$ of (7.9) is a convex subset of $(p+1)$-space. Often it reduces to a single point. For this it suffices, for instance, that ρ is strictly convex, that the f_i are linear, and that the columns of the design matrix $x_{ij} = \partial f_i/\partial \theta_j$ and the residual vector $x_i - f_i$ are linearly independent (that is, the design matrix has full rank, and there is no exact solution with vanishing residuals). Then also Q is strictly convex [cf. (7.19)], and the solution $(\boldsymbol{\theta}, \sigma)$ is necessarily unique.

Even if ρ is not strictly convex everywhere, but contains a strictly convex piece, the solution is usually unique when n/p is large (because then enough residuals will fall into the strictly convex region of ρ for the above argument to carry through).

7.8   COMPUTATION OF REGRESSION M-ESTIMATES

We now describe some simple algorithms. They alternate between improving trial values for $\hat{\boldsymbol{\theta}}$ and $\hat{\sigma}$, and they decrease (7.9). We prefer to write the latter expression in the form

$$Q(\boldsymbol{\theta}, \sigma) = \frac{1}{n} \sum \left[\rho_0 \left(\frac{y_i - f_i(\boldsymbol{\theta})}{\sigma} \right) + a \right] \sigma, \qquad (8.1)$$

where $\rho_0(0) = 0$ and $a > 0$. The equations (7.10) and (7.11) can then be written

$$\frac{1}{n} \sum \psi_0 \left(\frac{r_i}{\sigma} \right) \frac{\partial f_i}{\partial \theta_j} = 0, \qquad (8.2)$$

$$\frac{1}{n} \sum \chi_0 \left(\frac{r_i}{\sigma} \right) = a, \qquad (8.3)$$

with

$$\psi_0(x) = \rho_0'(x), \qquad (8.4)$$

$$\chi_0(x) = x \psi_0(x) - \rho_0(x). \qquad (8.5)$$

Note that χ_0 has an absolute minimum at $x = 0$, namely, $\chi_0(0) = 0$. We assume throughout that ψ_0 and χ_0 are continuous.

In order to obtain consistency of the scale estimate at the normal model and to recapture the classical estimates for the classical choice $\rho_0(x) = \frac{1}{2} x^2$, we propose to take

$$a = \frac{n - p}{n} E_\Phi(\chi_0). \qquad (8.6)$$

The Scale Step

Let $\boldsymbol{\theta}^{(m)}$ and $\sigma^{(m)}$ be trial values for $\boldsymbol{\theta}$ and σ, and put $r_i = y_i - f_i(\boldsymbol{\theta}^{(m)})$. Define

$$(\sigma^{(m+1)})^2 = \frac{1}{na} \sum \chi_0 \left(\frac{r_i}{\sigma^{(m)}} \right) (\sigma^{(m)})^2. \qquad (8.7)$$

Remarks For the classical choice $\rho_0(x) = \frac{1}{2} x^2$, with a as in (8.6), we obtain

$$(\sigma^{(m+1)})^2 = \frac{1}{n-p} \sum r_i^2. \tag{8.8}$$

For the choice (7.14) we obtain

$$(\sigma^{(m+1)})^2 = \frac{1}{(n-p)\beta} \sum \psi\left(\frac{r_i}{\sigma^{(m)}}\right)^2 (\sigma^{(m)})^2, \tag{8.9}$$

with

$$\beta = E_\Phi(\psi^2). \tag{8.10}$$

In the latter case we may say that $\sigma^{(m+1)}$ is an ordinary variance estimate (8.8), but calculated from *metrically Winsorized residuals*

$$r_i^w = \psi\left(\frac{r_i}{\sigma^{(m)}}\right) \sigma^{(m)} = -c\sigma^{(m)}, \qquad \text{for } r_i < -c\sigma^{(m)},$$

$$= r_i, \qquad\qquad \text{for } |r_i| \leqslant c\sigma^{(m)},$$

$$= c\sigma^{(m)}, \qquad \text{for } r_i > c\sigma^{(m)}, \tag{8.11}$$

and corrected for bias by the factor β.

LEMMA 8.1 Assume that $\rho_0 \geqslant 0$ is convex, that $\rho_0(0) = 0$, and that $\rho_0(x)/x$ is convex for $x < 0$, concave for $x > 0$. Then

$$Q(\boldsymbol{\theta}^{(m)}, \sigma^{(m)}) - Q(\boldsymbol{\theta}^{(m)}, \sigma^{(m+1)}) \geqslant \frac{a(\sigma^{(m+1)} - \sigma^{(m)})^2}{\sigma^{(m)}}. \tag{8.12}$$

In particular, unless (8.3) is already satisfied, Q is strictly decreased.

Proof The idea is to construct a simple "comparison function" $U(\sigma)$ that agrees with $Q(\boldsymbol{\theta}^{(m)}, \sigma)$ at $\sigma = \sigma^{(m)}$, that lies wholly above $Q(\boldsymbol{\theta}^{(m)}, \cdot)$, and that reaches its minimum at $\sigma^{(m+1)}$, namely,

$$U(\sigma) = Q(\boldsymbol{\theta}^{(m)}, \sigma^{(m)}) + a(\sigma - \sigma^{(m)}) + \frac{1}{n} \sum \chi_0\left(\frac{r_i}{\sigma^{(m)}}\right)\left[\frac{(\sigma^{(m)})^2}{\sigma} - \sigma^{(m)}\right]. \tag{8.13}$$

Obviously, $U(\sigma^{(m)}) = Q(\theta^{(m)}, \sigma^{(m)})$. The derivatives with respect to σ are

$$U'(\sigma) = -\frac{1}{n} \sum \chi_0\left(\frac{r_i}{\sigma^{(m)}}\right)\left(\frac{\sigma^{(m)}}{\sigma}\right)^2 + a, \qquad (8.14)$$

$$Q'(\theta^{(m)}, \sigma) = -\frac{1}{n} \sum \chi_0\left(\frac{r_i}{\sigma}\right) + a; \qquad (8.15)$$

hence they agree at $\sigma = \sigma^{(m)}$. Define

$$f(z) = U\left(\frac{1}{z}\right) - Q\left(\theta^{(m)}, \frac{1}{z}\right), \qquad z > 0. \qquad (8.16)$$

This function is convex, since it can be written

$$f(z) = -\frac{1}{n} \sum \frac{\rho_0(zr_i)}{z} + b_0 + b_1 z \qquad (8.17)$$

with some constants b_0 and b_1; it has a horizontal tangent at $z = 1/\sigma^{(m)}$, and it vanishes there. It follows that $f(z) \geqslant 0$ for all $z > 0$; hence

$$U(\sigma) \geqslant Q(\theta^{(m)}, \sigma) \qquad (8.18)$$

for all $\sigma > 0$. Note that U reaches its minimum at $\sigma^{(m+1)}$. A simple calculation, using (8.7) to eliminate $\sum \chi_0$, gives

$$U(\sigma^{(m+1)}) = Q(\theta^{(m)}, \sigma^{(m)}) + a(\sigma^{(m+1)} - \sigma^{(m)}) + a\left(\frac{\sigma^{(m+1)}}{\sigma^{(m)}}\right)^2\left[\frac{(\sigma^{(m)})^2}{\sigma^{(m+1)}} - \sigma^{(m)}\right]$$

$$= Q(\theta^{(m)}, \sigma^{(m)}) - a\frac{(\sigma^{(m+1)} - \sigma^{(m)})^2}{\sigma^{(m)}}.$$

The assertion of the lemma now follows from (8.18). ∎

For the location step, we have two variants: one modifies the residuals, the other the weights.

The Location Step with Modified Residuals

Let $\theta^{(m)}$ and $\sigma^{(m)}$ be trial values for θ and σ. Put

$$r_i = y_i - f_i(\theta^{(m)}), \qquad (8.19)$$

$$r_i^* = \psi\left(\frac{r_i}{\sigma^{(m)}}\right)\sigma^{(m)}, \qquad (8.20)$$

$$x_{ik} = \frac{\partial}{\partial \theta_k} f_i(\theta^{(m)}). \qquad (8.21)$$

Solve

$$\sum \left(r_i^* - \sum_k x_{ik} \tau_k \right)^2 = \min! \tag{8.22}$$

for τ, that is, determine the solution $\tau = \hat{\tau}$ of

$$X^T X \tau = X^T \mathbf{r}^*. \tag{8.23}$$

Put

$$\theta^{(m+1)} = \theta^{(m)} + q\hat{\tau}, \tag{8.24}$$

where $0 < q < 2$ is an arbitrary relaxation factor.

Remark Except that the residuals r_i have been replaced by their metrically Winsorized versions r_i^*, this is just the ordinary iterative Gauss-Newton step one uses to solve nonlinear least squares problems (if the f_i are linear, it gives the least squares solution in one step).

LEMMA 8.2 Assume that $\rho_0 \geqslant 0$, $\rho_0(0) = 0$, $0 \leqslant \rho_0'' \leqslant 1$, and that the f_i are linear. Without loss of generality choose the coordinate system such that $X^T X = I$. Then

$$Q(\theta^{(m)}, \sigma^{(m)}) - Q(\theta^{(m+1)}, \sigma^{(m)}) \geqslant \frac{q(2-q)}{2\sigma^{(m)}n} \sum_j \left(\sum_i r_i^* x_{ij} \right)^2$$

$$= \frac{q(2-q)}{2\sigma^{(m)}n} \| \hat{\tau} \|^2$$

$$= \frac{2-q}{2\sigma^{(m)}nq} \| \theta^{(m+1)} - \theta^{(m)} \|^2. \tag{8.25}$$

In particular, unless (8.2) is already satisfied, Q is strictly decreased.

Proof As in the scale step, we use a comparison function that agrees with Q at $\theta^{(m)}$, that lies wholly above Q, and that reaches its minimum at $\theta^{(m+1)}$. Put

$$W(\tau) = Q(\theta^{(m)}, \sigma^{(m)}) + \frac{1}{2\sigma^{(m)}n} \sum \left[\left(r_i^* - \sum_k x_{ik} \tau_k \right)^2 - (r_i^*)^2 \right]. \tag{8.26}$$

The functions $W(\tau)$ and $Q(\theta^{(m)}+\tau, \sigma^{(m)})$ then have the same value and the same first derivative at $\tau=0$, as we can easily check. The matrix of second order derivatives of the difference,

$$\frac{\partial^2}{\partial \tau_j \partial \tau_k} \left[W(\tau) - Q(\theta^{(m)}+\tau, \sigma^{(m)}) \right] = \frac{1}{\sigma^{(m)} n} \sum_i \left[1 - \psi'\left(\frac{r_i - \Sigma x_{il}\tau_l}{\sigma^{(m)}} \right) \right] x_{ij} x_{ik},$$

(8.27)

is positive semidefinite; hence

$$W(\tau) \geqslant Q(\theta^{(m)}+\tau, \sigma^{(m)})$$

(8.28)

for all τ. The minimum of $W(\tau)$ occurs at $\hat{\tau}=X^T r^*$, and we easily check that it has the value

$$W(\hat{\tau}) = Q(\theta^{(m)}, \sigma^{(m)}) - \frac{1}{2\sigma^{(m)} n} \|\hat{\tau}\|^2.$$

(8.29)

As a function of q,

$$W(q\hat{\tau}) - Q(\theta^{(m)}, \sigma^{(m)})$$

is quadratic, vanishes at $q=0$, has a minimum at $q=1$, and for reasons of symmetry must vanish again at $q=2$. Hence we obtain, by quadratic interpolation,

$$W(q\hat{\tau}) - Q(\theta^{(m)}, \sigma^{(m)}) = -\frac{q(2-q)}{2\sigma^{(m)} n} \|\hat{\tau}\|,$$

(8.30)

and the assertion of the lemma now follows from (8.28). ■

Remark The relaxation factor q had originally been introduced because theoretical considerations had indicated that $q \approx 1/E\psi' \geqslant 1$ should give faster convergence than $q=1$. The empirical experience shows hardly any difference.

The Location Step with Modified Weights

Instead of (8.2) we can equivalently write

$$\sum w_i r_i \frac{\partial f_i}{\partial \theta_j} = 0, \qquad j = 1, \ldots, p,$$

(8.31)

with weights depending on the current residuals r_i, determined by

$$w_i = \frac{\psi(r_i/\sigma^{(m)})}{r_i/\sigma^{(m)}}. \tag{8.32}$$

Let $\theta^{(m)}$ and $\sigma^{(m)}$ be trial values, then find $\theta^{(m+1)}$ by solving the weighted least squares problem (8.31), that is, find the solution $\tau = \hat{\tau}$ of

$$X^T W X \tau = X^T W \mathbf{r}, \tag{8.33}$$

where W is the diagonal matrix with diagonal elements w_i, and put

$$\theta^{(m+1)} = \theta^{(m)} + \hat{\tau}. \tag{8.34}$$

LEMMA 8.3 (Dutter 1975) Assume that ρ_0 is convex and symmetric, that $\psi(x)/x$ is bounded and monotone decreasing for $x > 0$, and that the f_i are linear. Then, for $\sigma > 0$, we have $Q(\theta^{(m+1)}, \sigma^{(m)}) < Q(\theta^{(m)}, \sigma^{(m)})$, unless $\theta^{(m)}$ already minimizes $Q(\cdot, \sigma^{(m)})$. The decrease in Q exceeds that of the corresponding modified residuals step.

Proof To simplify notation assume $\sigma^{(m)} = 1$. We also use a comparison function U here, and we define it as follows:

$$U(\theta) = \sum_i U_i(y_i - f_i(\theta)), \tag{8.35}$$

where each U_i is a quadratic function

$$U_i(x) = a_i + \tfrac{1}{2} b_i x^2 \tag{8.36}$$

with a_i and b_i determined such that

$$U_i(x) \geqslant \rho_0(x), \qquad \text{for all } x, \tag{8.37}$$

and

$$U_i(r_i) = \rho_0(r_i), \tag{8.38}$$

with $r_i = y_i - f_i(\theta^{(m)})$; see Exhibit 7.8.1.

These conditions imply that U_i and ρ have a common tangent at r_i:

$$U_i'(r_i) = b_i r_i = \psi_0(r_i); \tag{8.39}$$

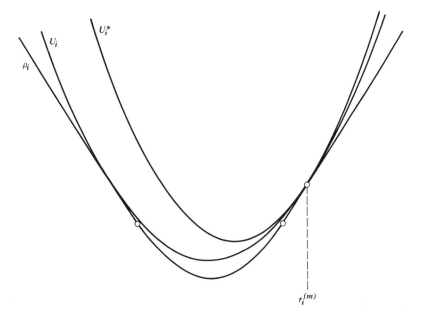

ρ_i U_i U_i^* $r_i^{(m)}$

Exhibit 7.8.1

hence

$$b_i = \frac{\psi(r_i)}{r_i} = w_i \qquad (8.40)$$

and

$$a_i = \rho_0(r_i) - \tfrac{1}{2} r_i \psi(r_i). \qquad (8.41)$$

We have to check that (8.37) holds. Write r instead of r_i. The difference

$$z(x) = U_i(x) - \rho_0(x)$$

$$= \rho_0(r) - \tfrac{1}{2} r \psi_0(r) + \frac{1}{2} \frac{\psi_0(r)}{r} x^2 - \rho_0(x) \qquad (8.42)$$

satisfies

$$z(r) = z(-r) = 0, \qquad (8.43)$$

$$z'(r) = z'(-r) = 0, \qquad (8.44)$$

and

$$z'(x) = \frac{\psi(r)}{r} x - \psi(x).$$ (8.45)

Since $\psi(x)/x$ is decreasing for $x > 0$, this implies that

$$z'(x) \leqslant 0, \qquad \text{for } 0 < x \leqslant r$$

$$\geqslant 0, \qquad \text{for } x \geqslant r;$$ (8.46)

hence $z(x) \geqslant z(r) = 0$ for $x \geqslant 0$, and the same holds for $x \leqslant 0$ because of symmetry.

In view of (8.40) we have

$$U(\boldsymbol{\theta}) = \tfrac{1}{2} \sum w_i [y_i - f_i(\boldsymbol{\theta})]^2 + \text{const.},$$ (8.47)

and this is, of course, minimized by $\boldsymbol{\theta}^{(m+1)}$. This proves the first part of the lemma. The second part follows from the remark that, if we had used comparison functions of the form

$$U_i^*(x) = a_i + c_i x + \tfrac{1}{2} x^2$$ (8.48)

instead of (8.36), we would have recaptured the proof of Lemma 8.2, and that

$$U_i^*(x) \geqslant U_i(x), \qquad \text{for all } x$$ (8.49)

provided $0 \leqslant \rho'' \leqslant 1$ (if necessary rescale ψ to achieve this). Hence

$$W(\tau) \geqslant U(\boldsymbol{\theta}^{(m)} + \tau) \geqslant \rho(\boldsymbol{\theta}^{(m)} + \tau).$$

In fact the same argument shows that U is the best possible quadratic comparison function. ∎

Remark 1 If we omit the convexity assumption and only assume that $\rho(x)$ increases for $x > 0$, the above proof still goes through and shows that the modified weights algorithm converges to a (local) minimum if the scale is kept fixed.

Remark 2 The second part of Lemma 8.3 implies that the modified weights approach should give a faster convergence than the modified residuals approach. However the empirically observed convergence rates show only small differences. Since the modified residuals approach (for linear f_i) can use the same matrices over all iterations, it even seems to have a slight advantage in total computing costs [cf. Dutter (1977a, b)].

If we alternate between location and scale steps (using either of the two versions for the location step), we obtain a sequence $(\boldsymbol{\theta}^{(m)}, \sigma^{(m)})$, which is guaranteed to decrease Q at each step. We now want to prove that the sequence converges toward a solution of (8.2) and (8.3).

THEOREM 8.4

(1) The sequence $(\boldsymbol{\theta}^{(m)}, \sigma^{(m)})$ has at least one accumulation point $(\hat{\boldsymbol{\theta}}, \hat{\sigma})$.

(2) Every accumulation point $(\hat{\boldsymbol{\theta}}, \hat{\sigma})$ with $\hat{\sigma} > 0$ is a solution of (8.2) and (8.3) and minimizes (8.1).

Proof The sets of the form

$$A_b = \{(\boldsymbol{\theta}, \sigma) | \sigma \geqslant 0, Q(\boldsymbol{\theta}, \sigma) \leqslant b\} \tag{8.50}$$

are compact. First, they are obviously closed, since Q is continuous. We have $\sigma \leqslant b/a$ on A_b. Since the f_i are linear and the matrix of the $x_{ij} = \partial f_i / \partial \theta_j$ is assumed to have full rank, $\|\boldsymbol{\theta}\|$ must also be bounded (otherwise at least one of the $f_i(\boldsymbol{\theta})$ would be unbounded on A_b; hence $\sigma \rho\{[y_i - f_i(\boldsymbol{\theta})]/\sigma\}$ would be unbounded). Compactness of the sets A_b obviously implies (1). To prove (2), assume $\hat{\sigma} > 0$ and let $(\boldsymbol{\theta}^{(m_l)}, \sigma^{(m_l)})$ be a subsequence converging toward $(\hat{\boldsymbol{\theta}}, \hat{\sigma})$. Then

$$Q(\boldsymbol{\theta}^{(m_l)}, \sigma^{(m_l)}) \geqslant Q(\boldsymbol{\theta}^{(m_l)}, \sigma^{(m_l+1)}) \geqslant Q(\boldsymbol{\theta}^{(m_{l+1})}, \sigma^{(m_{l+1})})$$

(see Lemma 8.1); the two outer members of this inequality tend to $Q(\hat{\boldsymbol{\theta}}, \hat{\sigma})$; hence (see Lemmas 8.2 and 8.3)

$$Q(\boldsymbol{\theta}^{(m_l)}, \sigma^{(m_l)}) - Q(\boldsymbol{\theta}^{(m_l)}, \sigma^{(m_l+1)}) \geqslant a \frac{(\sigma^{(m_l+1)} - \sigma^{(m_l)})^2}{\sigma^{(m_l)}}$$

converges to 0. In particular, it follows that

$$\left(\frac{\sigma^{(m_l+1)}}{\sigma^{(m_l)}}\right)^2 = \frac{1}{na} \sum \chi_0\left(\frac{y_i - f_i(\boldsymbol{\theta}^{(m_l)})}{\sigma^{(m_l)}}\right)$$

converges to 1; hence in the limit

$$\frac{1}{n} \sum \chi_0 \left(\frac{y_i - f_i(\hat{\boldsymbol{\theta}})}{\hat{\sigma}} \right) = a.$$

Thus (8.3) is satisfied.

In the same way, we obtain from Lemma 8.2 that

$$Q(\boldsymbol{\theta}^{(m_l)}, \sigma^{(m_l)}) - Q(\boldsymbol{\theta}^{(m_l+1)}, \sigma^{(m_l)}) \geq \frac{q(2-q)}{2\sigma^{(m_l)}n} \sum_j \left(\sum_i r_i^* x_{ij} \right)^2$$

tends to 0; in particular

$$\sum_i r_i^* x_{ij} = \sigma^{(m_l)} \sum_i \psi \left(\frac{y_i - f_i(\boldsymbol{\theta}^{(m_l)})}{\sigma^{(m_l)}} \right) x_{ij} \to 0.$$

Hence in the limit

$$\sum \psi \left(\frac{y_i - f_i(\hat{\boldsymbol{\theta}})}{\hat{\sigma}} \right) x_{ij} = 0,$$

and thus (8.2) also holds. In view of the convexity of Q, every solution of (8.2) and (8.3) minimizes (8.1). ∎

We now intend to give conditions sufficient for there to be no accumulation points with $\hat{\sigma} = 0$. The main condition is one ensuring that the maximum number p' of residuals that can be made simultaneously 0 is not too big; assume that χ_0 is symmetric and bounded, and that

$$n - p' > (n - p) \frac{E_\Phi(\chi_0)}{\chi_0(\infty)}. \tag{8.51}$$

Note that $p' = p$ with probability 1 if the error distribution is absolutely continuous with respect to Lebesgue measure, so (8.51) is then automatically satisfied, since $E_\Phi(\chi_0) < \max(\chi_0) = \chi_0(\infty)$.

We assume that the iteration is started with $(\boldsymbol{\theta}^{(0)}, \sigma^{(0)})$ where $\sigma^{(0)} > 0$. Then $\sigma^{(m)} > 0$ for all finite m. Moreover, we note that, for all m, $(\boldsymbol{\theta}^{(m)}, \sigma^{(m)})$ then is contained in the compact set A_b, with $b = Q(\boldsymbol{\theta}^{(0)}, \sigma^{(0)})$. Hence it suffices to restrict $(\boldsymbol{\theta}, \sigma)$ to A_b for all the following arguments.

Clearly, (8.51) is equivalent to the following: for sufficiently small σ we have

$$\frac{1}{n}\sum \chi_0\left(\frac{r_i}{\sigma}\right) > a = \frac{n-p}{n} E_\Phi(\chi_0). \tag{8.52}$$

This is strengthened in the following lemma.

LEMMA 8.5   Assume that (8.51) holds. Then there is a $\sigma_0 > 0$ and a $d > 1$ such that for all $(\boldsymbol{\theta}, \sigma) \in A_b$ with $\sigma \leqslant \sigma_0$

$$\frac{1}{n}\sum \chi_0\left(\frac{r_i}{\sigma}\right) \geqslant d^2 a. \tag{8.53}$$

Proof   For each $\boldsymbol{\theta}$ order the corresponding residuals according to increasing absolute magnitude, and let $h(\boldsymbol{\theta}) = |r_{(p'+1)}|$ be the $(p'+1)$st smallest. Then $h(\boldsymbol{\theta})$ is a continuous (in fact piecewise linear) strictly positive function. Since A_b is compact, the minimum h_0 of $h(\boldsymbol{\theta})$ is attained and hence must be strictly positive. It follows that

$$\frac{1}{n}\sum \chi_0\left(\frac{r_i}{\sigma}\right) \geqslant \frac{n-p'}{n}\chi_0\left(\frac{h_0}{\sigma}\right). \tag{8.54}$$

In the limit $\sigma \to 0$ the right-hand side becomes

$$\frac{n-p'}{n}\chi_0(\infty) > \frac{n-p}{n} E_\Phi(\chi_0) = a \tag{8.55}$$

in view of (8.51). Clearly strict inequality must already hold for some nonzero σ_0, and the assertion of the lemma follows. ∎

PROPOSITION 8.6   Assume (8.51), that χ_0 is symmetric and bounded, and that $\sigma^{(0)} > 0$. Then the sequence $(\boldsymbol{\theta}^{(m)}, \sigma^{(m)})$ cannot have an accumulation point on the boundary $\sigma = 0$.

Proof   Lemma 8.5 implies that $\sigma^{(m+1)} \geqslant d\sigma^{(m)}$. It follows that the sequence $\sigma^{(m)}$ cannot indefinitely stay below σ_0 and that there must be infinitely many m for which $\sigma^{(m)} > \sigma_0$. Hence $(\boldsymbol{\theta}^{(m)}, \sigma^{(m)})$ has an accumulation point $(\hat{\boldsymbol{\theta}}, \hat{\sigma})$ with $\hat{\sigma} > 0$, and, by Theorem 8.4, $(\hat{\boldsymbol{\theta}}, \hat{\sigma})$ minimizes $Q(\boldsymbol{\theta}, \sigma)$. It follows from (8.51) that, on the boundary, $Q(\boldsymbol{\theta}, 0) > Q(\hat{\boldsymbol{\theta}}, \hat{\sigma}) = b_0$. Furthermore $(\boldsymbol{\theta}^{(m)}, \sigma^{(m)})$ ultimately stays in $A_{b_0 + \varepsilon}$ for every $\varepsilon > 0$, and, for sufficiently small ε, $A_{b_0 + \varepsilon}$ does not intersect the boundary.

THEOREM 8.7 Assume (8.51). Then, with the location step using modified residuals, the sequence $(\boldsymbol{\theta}^{(m)}, \sigma^{(m)})$ always converges to some solution of (8.2) and (8.3).

Proof If the solution $(\hat{\boldsymbol{\theta}}, \hat{\sigma})$ of the minimum problem (8.1), or of the simultaneous equations (8.2) and (8.3), is unique, then Theorem 8.4 and Proposition 8.6 together imply that $(\hat{\boldsymbol{\theta}}, \hat{\sigma})$ must be the unique accumulation point of the sequence, and there is nothing to prove. Assume now that the (necessarily convex) solution set S contains more than one point.

A look at Exhibit 7.8.2 helps us to understand the following arguments; the diagram shows S and some of the surfaces $Q(\boldsymbol{\theta}, \sigma) = \text{const.}$

Clearly, for $m \to \infty$, $Q(\boldsymbol{\theta}^{(m)}, \sigma^{(m)}) \to \inf Q(\boldsymbol{\theta}, \sigma)$, that is, $(\boldsymbol{\theta}^{(m)}, \sigma^{(m)})$ converge to the *set* S. The idea is to demonstrate that the iteration steps succeeding $(\boldsymbol{\theta}^{(m)}, \sigma^{(m)})$ will have to stay inside an approximately conical region (the shaded region in the picture). With increasing m the base of the respective cone will get smaller and smaller, and since each cone is contained in the preceding one, this will imply convergence. The details of the proof are messy; we only sketch the main idea.

We standardize the coordinate system such that $X^T X = 2anI$. Then

$$\Delta\theta_j = \frac{1}{2an} \sum \psi_0\left(\frac{r_i}{\sigma^{(m)}}\right) x_{ij} \sigma^{(m)}.$$

$$\Delta\sigma \cong \frac{1}{2}\sigma^{(m)}\left[\left(\frac{\sigma^{(m+1)}}{\sigma^{(m)}}\right)^2 - 1\right] = \frac{\sigma^{(m)}}{2a}\left[\frac{1}{n}\sum \chi_0\left(\frac{r_i}{\sigma^{(m)}}\right) - a\right],$$

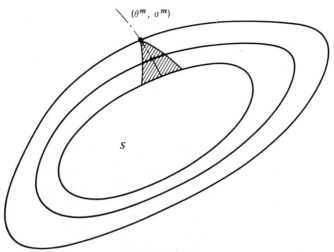

(θ^m, σ^m)

S

Exhibit 7.8.2

whereas the gradient of Q is given by

$$g_j = \frac{\partial}{\partial \theta_j} Q(\boldsymbol{\theta}^{(m)}, \sigma^{(m)}) = -\tfrac{1}{2} \sum \psi_0\left(\frac{r_i}{\sigma^{(m)}}\right) x_{ij}, \qquad j=1,\ldots,p,$$

$$g_{p+1} = \frac{\partial}{\partial \sigma} Q(\boldsymbol{\theta}^{(m)}, \sigma^{(m)}) = -\frac{1}{n} \sum \chi_0\left(\frac{r_i}{\sigma^{(m)}}\right) - a.$$

In other words in this particular coordinate system, the step

$$\Delta \theta_j = -\frac{\sigma^{(m)}}{2a} g_j,$$

$$\Delta \sigma = -\frac{\sigma^{(m)}}{2a} g_{p+1},$$

is in the direction of the negative gradient at the point $(\boldsymbol{\theta}^{(m)}, \sigma^{(m)})$. ∎

It is not known to me whether this theorem remains true with the location step with modified weights.

Of course, the algorithms described so far have to be supplemented with a stopping rule, for example, stop iterations when the shift of every linear combination $\alpha = \mathbf{a}^T \boldsymbol{\theta}$ is smaller than ε times its own estimated standard deviation [using (6.5)], with $\varepsilon = 0.001$ or the like. Our experience is that on the average this will need about 10 iterations [for ρ as in (7.14), with $c = 1.5$], with relatively little dependence on p and n.

If ψ is piecewise linear, it is possible to devise algorithms that reach the *exact* solution in a finite (and usually small, mostly under 10) number of iterations, if they converge at all: partition the residuals according to the linear piece of ψ on which they sit and determine the algebraically exact solution under the assumption that the partitioning of the residuals stays the same for the new parameter values. If this assumption turns out to be true, we have found the exact solution $(\hat{\boldsymbol{\theta}}, \hat{\sigma})$; otherwise iterate. In the one-dimensional location case, this procedure seems to converge without fail; in the general regression case, some elaborate safeguards against singular matrices and other mishaps are needed. See Huber (1973a) and Dutter (1975, 1977a, b).

As starting values $(\boldsymbol{\theta}^{(0)}, \sigma^{(0)})$ we usually take the ordinary least squares estimate, despite its known poor properties [cf. Andrews et al. (1972), for the simple location case].

Descending ψ-functions are tricky, especially when the starting values for the iterations are nonrobust. Residuals that are accidentally large because of the poor starting parameters then may stay large forever because they exert zero pull. It is therefore preferable to start with a monotone ψ, iterate to death, and then append a few (1 or 2) iterations with the nonmonotone ψ.

7.9 MODERATE LEVERAGE POINTS

We now return to the example of Section 7.1 and to the formulas derived in Section 7.2. In particular, we recall that in the classical least squares case, with $\mathrm{var}(y_i) = \sigma^2$,

$$\mathrm{var}(\hat{y}_i) = h_i \sigma^2, \tag{9.1}$$

$$\mathrm{var}(y_i - \hat{y}_i) = (1 - h_i)\sigma^2, \tag{9.2}$$

$$\mathrm{var}(\hat{\alpha}_i) = \frac{h_i}{1 - h_i} \sigma^2, \tag{9.3}$$

$$y_i - \hat{y}_i = (1 - h_i)(y_i - \hat{\alpha}_i), \tag{9.4}$$

$$\mathrm{var}(y_i - \hat{\alpha}_i) = \frac{1}{1 - h_i} \sigma^2. \tag{9.5}$$

Here $\hat{y}_i$ denotes the fitted value, and $\hat{\alpha}_i$ the "interpolated" value, estimated without using y_i.

If we robustize least squares by using (7.1), that is,

$$\sum \psi\left(\frac{y_i - \hat{y}_i}{\sigma}\right) x_{ij} = 0, \qquad j = 1, \dots, p, \tag{9.6}$$

with ψ as in (7.15), for instance, and if point i happens to be a leverage point with a high h_i, then y_i can be grossly aberrant, but $(y_i - \hat{y}_i)/\sigma$ will still remain on the linear part of ψ. This is, of course, undesirable.

We can try to correct this by cutting down the overall influence of observation i by introducing a weight factor γ; we can shorten the linear part of ψ by cutting down scale by a factor δ; and we can do both (γ and δ, of course, depend on h_i and possibly other variables). This means that we replace the term $\psi[(y_i - \hat{y}_i)/\sigma]$ in (9.6) by

$$\gamma\psi\left(\frac{y_i - \hat{y}_i}{\delta\sigma}\right). \tag{9.7}$$

We claim there are strong heuristic arguments in favor of choosing $\gamma = \delta$, and somewhat weaker ones suggest choosing

$$\gamma = \delta = \sqrt{1 - h_i} \ . \tag{9.8}$$

These arguments run as follows. Assume first that y_i and the interpolated value $\hat{\alpha}_i$ differ only slightly, so that the residual sits on the linear part of ψ. Then (9.7) means, essentially, that the ith observation is treated with weight γ/δ. Clearly, if h_i is small, we should have $\gamma \approx \delta \approx 1$. On the other hand if h_i is large, say $h_i > \frac{1}{2}$, then according to (9.3) a "good" observation y_i is more accurate than $\hat{\alpha}_i$. If the underlying distribution is a moderately contaminated normal one ($\varepsilon = 1$ to 10%), then the likelihood is quite high that y_i is a "good" observation if it does not deviate too much from $\hat{\alpha}_i$. But then we would not want the extrapolated value $\hat{\alpha}_i$ to take precedence over y_i, that is, we would not want to downweight y_i. [Note that $\hat{\alpha}_i$ is more influential than y_i anyway, cf. (2.36).] Thus we are induced to put $\gamma = \delta$.

Now imagine that, for most observations, the residuals sit on the linear part of ψ, so that, essentially, the parameters are determined by ordinary least squares. Move one observation y_i from $-\infty$ to $+\infty$, and, for each value of y_i, let y_i^* be the corresponding observational value for which the least squares solution would agree with the robustized version based on (9.7). Using (9.4) we find that

$$y_i^* = y_i, \qquad \text{for } |y_i - \hat{\alpha}_i| \leqslant \frac{\delta}{1 - h_i} c\sigma,$$

$$= \hat{\alpha}_i \pm \frac{\delta}{1 - h_i} c\sigma, \qquad \text{for } |y_i - \hat{\alpha}_i| \geqslant \frac{\delta}{1 - h_i} c\sigma. \tag{9.9}$$

In view of (9.5) it would seem natural to choose $\delta = \sqrt{1 - h_i}$, so that the changeover in (9.9) would be related to a natural measure of scale of $|y_i - \hat{\alpha}_i|$.

In other words we propose to modify (9.6) to

$$\sum_i \sigma \sqrt{1 - h_i} \ \psi\left[\frac{y_i - \hat{y}_i}{\sigma \sqrt{1 - h_i}}\right] x_{ij} = 0 \tag{9.10}$$

with scale σ determined simultaneously from

$$\sum_i \left[\sqrt{1-h_i} \, \psi \left(\frac{y_i - \hat{y}_i}{\sigma \sqrt{1-h_i}} \right) \right]^2 = (n-p) E_\Phi \psi^2. \qquad (9.11)$$

Equivalently, this corresponds to minimizing, simultaneously with respect to $\boldsymbol{\theta}$ and σ, the expression

$$\sum \rho \left[\frac{y_i - \hat{y}_i}{\sigma \sqrt{1-h_i}} \right] (1-h_i)\sigma, \qquad (9.12)$$

with ρ as in (7.14) and $\beta = E_\Phi \psi^2$.

Computationally, this does not introduce any new problems; instead of modifying the residual $r_i = y_i - \hat{y}_i$ to $r_i^* = \pm c\sigma$ whenever $|r_i| > c\sigma$, we now modify it to $r_i^* = \pm \sqrt{1-h_i}\, c\sigma$ whenever $|r_i| > \sqrt{1-h_i}\, c\sigma$ (cf. the location step with modified residuals).

If we look at these matters quantitatively, then it is clear that, for $h_i \leqslant 0.2$, the change from (9.6) to (9.10) is hardly noticeable (and the effort hardly worthwhile); on the other hand, for $h_i \geqslant 0.8$, it may not give a good enough protection from the ill effects of outliers.

Therefore, several researchers have proposed more drastic downweighting of leverage points in order to bound, simultaneously for all possible positions in the design space, the influence of any observational value toward any estimable quantity. This is an outgrowth of Hampel's approach (cf. Section 11.1). Much of the material is unpublished; the most systematic treatment so far, together with a comprehensive account of earlier work done by Hampel, Mallows, Schweppe, and others, can be found in Krasker and Welsch (1980). The last-named authors find estimates that are asymptotically efficient at the model, subject to an overall bound on the gross-error sensitivity, both with regard to value and position.

However, some basic difficulties remain unresolved. The most fundamental one probably is that we are confronted with a small sample problem—the fitted value at a high leverage point is essentially determined by a single observation. Therefore, we cannot really rely on tools derived from asymptotic considerations, like the influence function, and we should check everything against insights gained from finite sample theory (cf. Chapter 10). Some preliminary explorations along these lines seem to suggest that the Krasker-Welsch approach may be overly pessimistic with regard to outliers at leverage points, but not pessimistic enough with regard to small systematic errors at nonleverage points, and, somewhat obnox-iously, the first effect becomes particularly severe when there are no real leverage points (i.e., when max h_i is small, but large against p/n). Clearly, much work still remains to be done in this area.

To avoid possible misunderstandings, we should add that the preceding discussion has little to do with robustness relative to outliers among the independent variables (the rows of X). Although several studies have been devoted to this important problem in the past few years, it does not seem that an adequate conceptual penetration has been achieved. In particular, unless we have an (approximate) model underlying the generation of the rows of X, the concept of robustness would seem to be ill defined. In some cases a treatment via robust covariance/correlation matrices would seem to make more sense than the regression approach.

7.10   ANALYSIS OF VARIANCE

Geometrically speaking, analysis of variance is concerned with nested models, say a larger p-parameter model and a smaller q-parameter model, $q < p$, and with orthogonal projections of the observational vector y into the linear subspaces $V_q \subset V_p$ spanned by the columns of the respective design matrices; see Exhibit 7.10.1. Let $\hat{\mathbf{y}}_{(p)}$ and $\hat{\mathbf{y}}_{(q)}$ be the fitted values.

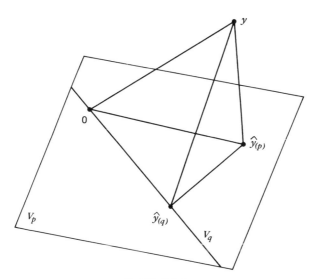

Exhibit 7.10.1

If the experimental errors are independent normal with (say) unit variance, then the differences squared,

$$\|\mathbf{y} - \hat{\mathbf{y}}_{(q)}\|^2, \qquad \|\mathbf{y} - \hat{\mathbf{y}}_{(p)}\|^2, \qquad \|\hat{\mathbf{y}}_{(p)} - \hat{\mathbf{y}}_{(q)}\|^2$$

are χ^2-distributed with $n-q$, $n-p$, $p-q$ degrees of freedom, and the latter two are independent, so that

$$\frac{1/(p-q)\|\hat{\mathbf{y}}_{(p)} - \hat{\mathbf{y}}_{(q)}\|^2}{1/(n-p)\|\mathbf{y} - \hat{\mathbf{y}}_{(p)}\|^2} \tag{10.1}$$

has an F-distribution, on which we can then base a test of the adequacy of the smaller model.

What of this can be salvaged if the errors are no longer normal? Of course, the distributional assumptions behind (10.1) are then violated, and worse, the power of the tests may be severely impaired.

If we try to improve by estimating $\hat{\mathbf{y}}_{(p)}$ and $\hat{\mathbf{y}}_{(q)}$ robustly, then these two quantities at least will be asymptotically normal under fairly general assumptions (cf. Sections 7.4 and 7.5). Since the projections are no longer orthogonal, but defined in a somewhat complicated nonlinear fashion, we do not obtain the same result by first projecting to V_p, then to V_q as by directly projecting to V_q (even though the two results are asymptotically equivalent). For the sake of internal consistency, when more than two nested models are concerned, the former variant (project via V_p) is preferable.

It follows from Proposition 4.1 that, under suitable regularity conditions, $\|\hat{\mathbf{y}}_{(p)} - \hat{\mathbf{y}}_{(q)}\|^2$ for the robust estimates still is asymptotically χ^2, when suitably scaled, with $p-q$ degrees of freedom. The denominator of (10.1), however, is nonrobust and useless as it stands. We must replace it by something that is a robust and consistent estimate of the expected value of the numerator. The obvious choice for the denominator, suggested by (6.5), is of course

$$\frac{1}{n-p} \frac{K^2 \sum \psi(r_i/\sigma)^2 \sigma^2}{\left[(1/n)\sum \psi'(r_i/\sigma)\right]^2}, \tag{10.2}$$

where

$$r_i = y_i - f_i\left(\hat{\boldsymbol{\theta}}_{(p)}\right).$$

Since the asymptotic approximations will not work very well unless p/n is reasonably small (say $p/n \leqslant 0.2$), and since $p \geqslant 2$, $n-p$ will be much larger than $p-q$, and the numerator

$$\frac{1}{p-q} \|\hat{\mathbf{y}}_{(p)} - \hat{\mathbf{y}}_{(q)}\|^2 \tag{10.3}$$

will always be much more variable than (10.2). Thus the quotient of (10.3) divided by (10.2),

$$\frac{\dfrac{1}{p-q} \|\hat{\mathbf{y}}_{(p)} - \hat{\mathbf{y}}_{(q)}\|^2}{\dfrac{1}{n-p} \dfrac{K^2 \sum \psi(r_i/\sigma)^2 \sigma^2}{\left[(1/n) \sum \psi'(r_i/\sigma)\right]^2}} \tag{10.4}$$

will be approximated quite well by a χ^2-variable with $p-q$ degrees of freedom, divided by $p-q$, and presumably even better by an F-distribution with $p-q$ degrees of freedom in the numerator and $n-p$ degrees in the denominator. We might argue that the latter value—but *not* the factor $n-p$ occurring in (10.4)—should be lowered somewhat; however, since the exact amount depends on the underlying distribution and is not known anyway, we may just as well stick to the classical value $n-p$.

Thus we end up with the following proposal for doing analysis of variance. Unfortunately, it is only applicable when there is a considerable excess of observations over parameters, say $p/n \leqslant 0.2$. First, fit the largest model under consideration, giving $\hat{\mathbf{y}}_{(p)}$. Make sure that there are no leverage points (an erroneous observation at a leverage point of the larger model may cause an erroneous rejection of the smaller model), or at least, be aware of the danger. Then estimate the dispersion of the "unit weight" fitted value by (10.2). Estimate the parameters of smaller models using $\hat{\mathbf{y}}_{(p)}$ (*not* y) by ordinary least squares. Then proceed in the classical fashion [but replace $[1/(n-p)]\|\mathbf{y} - \mathbf{y}_{(p)}\|^2$ by (10.2)].

Incidentally, the above procedure can also be described as follows. Let

$$r_k^* = \frac{K\psi(r_k/\sigma)\sigma}{\dfrac{1}{n} \sum \psi'(r_i/\sigma)}. \tag{10.5}$$

Put

$$\mathbf{y}^* = \hat{\mathbf{y}}_{(p)} + \mathbf{r}^*. \tag{10.6}$$

Then proceed classically, using the pseudo-observations $\mathbf{y}^*$ instead of y.

At first sight the following approach also might look attractive. First, fit the largest model, yielding $\hat{y}_{(p)}$. This amounts to an ordinary weighted least squares fit with modified weights (8.32). Now freeze the weights w_i and proceed in the classical fashion, using y_i and the same weights w_i for all models. However, this gives improper (inconsistent) values for the denominator of (10.1), and for monotone ψ-functions it is not even outlier resistant.

CHAPTER 8

Robust Covariance and Correlation Matrices

8.1 GENERAL REMARKS

The classical covariance and correlation matrices are used for a variety of different purposes. We mention just a few:

— They (or better: the associated ellipsoids) give a simple description of the overall shape of a pointcloud in p-space. This is an important aspect with regard to discriminant analysis as well as principal component and factor analysis.

— They allow us to calculate variances in arbitrary directions: $\mathrm{var}(\mathbf{a}^T\mathbf{x}) = \mathbf{a}^T\mathrm{cov}(\mathbf{x})\mathbf{a}$.

— For a multivariate Gaussian distribution, the sample covariance matrix, together with the sample mean, is a sufficient statistic.

— They can be used for tests of independence.

Unfortunately, sample covariance matrices are excessively sensitive to outliers. All too often a principal component or factor analysis "explains" a structure that, on closer scrutiny, has been created by a mere one or two outliers (cf. Exhibit 8.1.1).

The robust alternative approaches can be roughly classified into the following three categories:

(1) Robust estimation of the individual matrix elements of the covariance/correlation matrix.

(2) Robust estimation of variances in sufficiently many selected directions (to which a quadratic form is then fitted).

(3) Direct (maximum likelihood) estimation of the shape matrix of some elliptical distribution.

The third of these approaches is affinely invariant; the first one certainly is not. The second one is somewhere in between, depending on how the

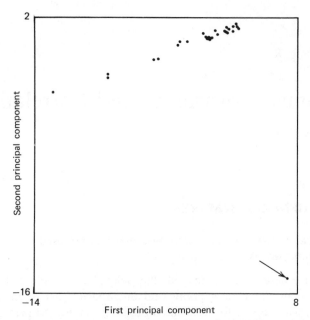

Exhibit 8.1.1 From Devlin, Gnanadesikan, and Kettenring (1979); see also Chen, H., Gnanadesikan, R., and Kettenring, J. R. (1974), with permission of the authors.

directions are selected. For example, we can choose them in relation to the coordinate axes and determine the matrix elements as in (1), or we can mimic an eigenvector/eigenvalue determination and find the direction with the smallest or largest robust variance, leading to an orthogonally invariant approach.

The coordinate dependent approaches are more germane to the estimation of correlation matrices (which are coordinate dependent anyway); the affinely invariant ones are better matched to covariance matrices.

Exhibits 8.1.1 and 8.1.2 illustrate the severity of the effects.

Exhibit 8.1.1 shows a principal component analysis of 14 economic characteristics of 29 chemical companies, namely the projection of the data on the plane of the first 2 components. The sample correlation between the two principal components is zero, as it should be, but there is a maverick company in the bottom right-hand corner, invalidating the analysis.

Exhibit 8.1.2 compares the influence of outliers on classical and on robust covariance estimates. The solid lines show ellipses derived from the classical sample covariance, theoretically containing 80% of the total normal mass; the dotted lines derive from the maximum likelihood estimate based on (10.26) with $\kappa = 2$ and correspond to the ellipse $|\mathbf{y}| = b = 2$,

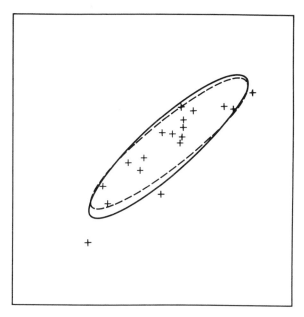

(a)

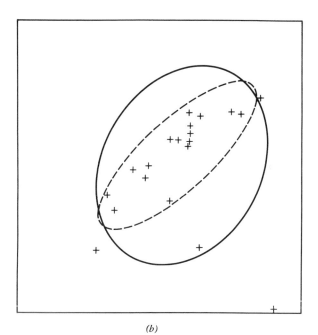

(b)

Exhibit 8.1.2

which, asymptotically, also contains about 80% of the total mass, if the underlying distribution is normal. The observations in Exhibit 8.1.2a are a random sample of size 18 from a bivariate normal distribution with covariance matrix

$$\begin{pmatrix} 1 & 0.9 \\ 0.9 & 1 \end{pmatrix}.$$

In Exhibit 8.1.2b, two contaminating observations with covariance matrix

$$\begin{pmatrix} 4 & -3.6 \\ -3.6 & 4 \end{pmatrix}$$

were added to the sample.

8.2   ESTIMATION OF MATRIX ELEMENTS THROUGH ROBUST VARIANCES

This approach is based on the following identity, valid for square integrable random variables X and Y:

$$\text{cov}(X, Y) = \frac{1}{4ab} \left[\text{var}(aX + bY) - \text{var}(aX - bY) \right]. \qquad (2.1)$$

It has been utilized by Gnanadesikan and Kettenring (1972).

Assume that S is a robust scale functional; we write for short $S(X) = S(F_X)$ and assume that

$$S(aX + b) = |a| S(X). \qquad (2.2)$$

If we replace $\text{var}(\cdot)$ by $S(\cdot)^2$, then (2.1) is turned into the definition of a robust alternative $C(X, Y)$ to the covariance between X and Y:

$$C(X, Y) = \frac{1}{4ab} \left[S(aX + bY)^2 - S(aX - bY)^2 \right]. \qquad (2.3)$$

The constants a and b can be chosen arbitrarily, but (2.3) will have awkward and unstable properties if aX and bY are on an entirely different scale. Gnanadesikan and Kettenring therefore recommend to take, for a and b, the inverses of some robust scale estimates for X and Y, respectively; for example, take

$$a = \frac{1}{S(X)},$$

$$b = \frac{1}{S(Y)}. \qquad (2.4)$$

Then

$$\tfrac{1}{4}\left[S(aX+bY)^2 - S(aX-bY)^2 \right] \tag{2.5}$$

becomes a kind of robust correlation. However, it is not necessarily confined to the interval $[-1, +1]$, and the expression

$$R^*(X,Y) = \frac{S(aX+bY)^2 - S(aX-bY)^2}{S(aX+bY)^2 + S(aX-bY)^2} \tag{2.6}$$

will therefore be preferable to (2.5) as a definition of a robust correlation coefficient. "Covariances" then can be reconstructed as

$$C^*(X,Y) = R^*(X,Y)S(X)S(Y). \tag{2.7}$$

It is convenient to standardize S such that $S(X)=1$ if X is normal $\mathcal{N}(0,1)$. Then if the joint distribution of (X,Y) is bivariate normal, we have

$$C(X,Y) = C^*(X,Y) = \operatorname{cov}(X,Y). \tag{2.8}$$

Proof (2.8) Note that $aX \pm bY$ is normal with variance

$$a^2 \operatorname{var}(X) \pm 2ab\operatorname{cov}(X,Y) + b^2 \operatorname{var}(Y). \tag{2.9}$$

Now (2.8) follows from (2.2). ∎

If X and Y are independent, but not necessarily normal, and if the distribution of either X or Y is symmetric, then, clearly, $C(X,Y) = C^*(X,Y) = 0$.

Now let $S_n(X)$ and $C_n(X,Y)$ be the finite sample versions based on $(x_1, y_1), \ldots, (x_n, y_n)$. We can expect that the asymptotic distribution of $\sqrt{n}\,[C_n(X,Y) - C(X,Y)]$ will be normal, but already for a normal parent distribution we obtain some quite complicated expressions. For a nonnormal parent the situation seems to become almost intractably messy.

This approach has another and even more serious drawback: when it is applied to the components of a p-vector $\mathbf{X} = (X_1, \ldots, X_p)$, it does not automatically produce a positive definite robust correlation or covariance matrix $[C(X_i, X_j)]$ and thus these matrices may cause both computational and conceptual trouble (the shape ellipsoid may be a hyperboloid!). The schemes proposed by Devlin et al. (1975) to enforce positive definiteness would seem to be very difficult to analyze theoretically.

There is an intriguing and, as far as I know, not yet explored variant of this approach, that avoids this drawback. It directly determines the eigenvalues λ_i and eigenvectors $\mathbf{u}_i$ of a robust covariance matrix; namely find that unit vector $\mathbf{u}_1$ for which $\lambda_1 = S(\mathbf{u}_1^T \mathbf{X})^2$ is maximal (or minimal), then do the same for unit vectors $\mathbf{u}_2$ orthogonal to $\mathbf{u}_1$, and so on. This will automatically give a positive definite matrix.

8.3   ESTIMATION OF MATRIX ELEMENTS THROUGH ROBUST CORRELATION

This approach is based on a remarkable distribution-free property of the ordinary sample correlation coefficient

$$r_n(\mathbf{x}, \mathbf{y}) = \frac{\sum (x_i - \bar{x})(y_i - \bar{y})}{\left[\sum (x_i - \bar{x})^2 \cdot \sum (y_i - \bar{y})^2 \right]^{1/2}}. \tag{3.1}$$

THEOREM 3.1 If the two vectors $\mathbf{x}^T = (x_1, \ldots, x_n)$ and $\mathbf{y}^T = (y_1, \ldots, y_n)$ are independent, and either the distribution of $\mathbf{y}$ or the distribution of $\mathbf{x}$ is invariant under permutation of the components, then

$$E(r_n) = 0,$$

$$E(r_n^2) = \frac{1}{n-1}.$$

Proof It suffices to calculate the above expectations conditionally, $\mathbf{x}$ given, and $\mathbf{y}$ given up to a random permutation.   ∎

Despite this distribution-free result, r_n obviously is not robust—one single, sufficiently bad outlying pair (x_i, y_i) can shift r_n to any value in $(-1, +1)$.

But the following is a remedy. Replace $r_n(\mathbf{x}, \mathbf{y})$ by $r_n(\mathbf{u}, \mathbf{v})$, where $\mathbf{u}$ and $\mathbf{v}$ are computed from $\mathbf{x}$ and $\mathbf{y}$, respectively, according to certain quite general rules given below. The first two of the following five requirements are essential; the others add some convenience:

(1) $\mathbf{u}$ is computed from $\mathbf{x}$ and $\mathbf{v}$ from $\mathbf{y}$: $\mathbf{u} = \Psi(\mathbf{x})$, $\mathbf{v} = \Xi(\mathbf{y})$.

(2) Ψ and Ξ commute with permutations of the components of $\mathbf{x}, \mathbf{u}$ and $\mathbf{y}, \mathbf{v}$.

(3) Ψ and Ξ preserve a monotone ordering of the components of $\mathbf{x}$ and $\mathbf{y}$.

(4) $\Psi = \Xi$.

(5) $\forall a > 0$, $\forall b$, $\exists a_1 > 0$, $\exists b_1$, $\forall \mathbf{x}$ $\Psi(a\mathbf{x} + b) = a_1 \Psi(\mathbf{x}) + b_1$.

Of these requirements, (1) and (2) ensure that $\mathbf{u}$ and $\mathbf{v}$ still satisfy the assumptions of Theorem 3.1 if $\mathbf{x}$ and $\mathbf{y}$ do. If (3) holds, then perfect rank correlations are preserved. Finally (4) and (5) together imply that correlations ± 1 are preserved. In the following two examples, all five requirements hold.

Example 3.1 Let

$$u_i = a(R_i) \qquad (3.2)$$

where R_i is the rank of x_i in $(x_1, \ldots, x_n)$ and $a(\cdot)$ is some monotone scores function. The choice $a(i) = i$ gives the classical Spearman rank correlation between **x** and **y**.

Example 3.2 Let T and S be arbitrary estimates of location and scale satisfying

$$T(a\mathbf{x} + b) = aT(\mathbf{x}) + b, \qquad (3.3)$$

$$S(a\mathbf{x} + b) = |a| S(\mathbf{x}), \qquad (3.4)$$

let ψ be a monotone function, and put

$$u_i = \psi\left(\frac{x_i - T}{S}\right). \qquad (3.5)$$

For example, S could be the median absolute deviation, and T the M-estimate determined by

$$\sum \psi\left(\frac{x_i - T}{S}\right) = 0. \qquad (3.6)$$

The choice $\psi(x) = \text{sign}(x)$ and $T = \text{med}\{x_i\}$ gives the so-called *quadrant correlation*.

Some Properties of Such Modified Correlations

Minimax Bias

Let G and H be centrosymmetric distributions in $\mathbb{R}^2$, and assume that (X, Y) is distributed according to the mixture

$$F = (1 - \varepsilon)G + \varepsilon H.$$

Then the (ordinary) correlation coefficient ρ_F of $\psi(X)$ and $\psi(Y)$ satisfies

$$(1 - \eta)\rho_G - \eta \leqslant \rho_F \leqslant (1 - \eta)\rho_G + \eta. \qquad (3.7)$$

The bounds are sharp, with

$$\frac{\eta}{1-\eta} = \frac{\varepsilon}{1-\varepsilon} \cdot \frac{\sup \psi^2}{E_G(\psi^2)}. \tag{3.8}$$

Thus η is made smallest if $\psi(x) = \text{sign}(x)$; in other words the quadrant correlation is asymptotically minimax with respect to bias. This is analogous to the minimax property of the sample median (Section 4.2).

Tests for Independence

Take the following test problem. Hypothesis: the probability law behind (X^*, Y^*) is

$$X^* = X + \delta \cdot Z,$$
$$Y^* = Y + \delta \cdot Z_1, \tag{3.9}$$

where X, Y, Z, and Z_1 are independent symmetric random variables, with Z and Z_1 being bounded and having the same distribution. Assume $\text{var}(Z) = \text{var}(Z_1) = 1$; δ is a small number.

The alternative is the same, except that $Z = Z_1$.

According to the Neyman-Pearson lemma, the most powerful tests are based on the test statistic

$$\prod_i \frac{h_A(x_i, y_i)}{h_H(x_i, y_i)} \tag{3.10}$$

where h_H and h_A are the densities of (X^*, Y^*) under the hypothesis and the alternative, respectively. If f and g are the densities of X and Y, respectively, we have

$$h_H(x, y) = E[f(x - \delta Z) g(y - \delta Z_1)],$$
$$h_A(x, y) = E[f(x - \delta Z) g(y - \delta Z)], \tag{3.11}$$

and thus

$$\frac{h_A(x, y)}{h_H(x, y)} = 1 + \frac{\text{cov}[f(x - \delta Z), g(y - \delta Z)]}{E[f(x - \delta Z)] E[g(y - \delta Z)]}. \tag{3.12}$$

If f and g can be expanded into a Taylor series

$$f(x - \delta Z) = f(x) - \delta Z f'(x) + \tfrac{1}{2} \delta^2 Z^2 f''(x) - \cdots, \tag{3.13}$$

we obtain that

$$\frac{h_A(x,y)}{h_H(x,y)} = 1 + \delta^2 \frac{f'(x)}{f(x)} \frac{g'(y)}{g(y)} + O(\delta^4),\qquad(3.14)$$

so, asymptotically for $\delta \to 0$, the most powerful test will be based on the test statistic

$$T_n = \sum \psi(x_i)\chi(y_i)\qquad(3.15)$$

where

$$\psi(x) = \frac{f'(x)}{f(x)},\qquad(3.16)$$

$$\chi(x) = \frac{g'(x)}{g(x)}.\qquad(3.17)$$

If we standardize (3.15) by dividing it by its (estimated) standard deviation, then we obtain a robust correlation of the form suggested in Example 3.2.

Under the hypothesis the test statistic (3.15) has expectation 0 and variance

$$E_H(T_n^2) = nE(\psi^2)E(\chi^2).\qquad(3.18)$$

Under the alternative the expectation is

$$E_A(T_n) = \delta^2 nE(\psi')E(\chi'),\qquad(3.19)$$

while the variance stays the same (neglecting higher order terms in δ). It follows that the asymptotic power of the test can be measured by the variance ratio

$$\frac{[E_A(T_n)]^2}{\mathrm{var}_A(T_n)} \approx n\delta^4 \frac{[E(\psi')]^2[E(\chi')]^2}{E(\psi^2)E(\chi^2)}.\qquad(3.20)$$

This holds also if ψ and χ are not related to f and g by (3.16) and (3.17). [For a rigorous treatment of such problems under less stringent regularity conditions, see Hájek and Šidák (1967), p. 75ff.].

A glance at (3.20) shows that there is a close formal analogy to problems in estimation of location. For instance, if the distributions of X^* and Y^*

vary over some sets and we would like to maximize the minimum asymptotic power of a test for independence, we have to find the distributions f and g minimizing Fisher information for location (!).

Of course, this also bears directly on correlation estimates, since in most cases it will be desirable to optimize the estimates so that they are best for nearly independent variables.

A Particular Choice for ψ

Let

$$\psi_c(x) = 2\Phi\left(\frac{x}{c}\right) - 1, \qquad \text{for } c > 0,$$

$$\psi_0(x) = \text{sign}(x),$$

where Φ is the standard normal cumulative.

PROPOSITION 3.2 If (X, Y) is bivariate normal with mean 0 and covariance matrix

$$\begin{pmatrix} 1 & \beta \\ \beta & 1 \end{pmatrix},$$

then

$$E[\psi_c(X)\psi_c(Y)] = \frac{2}{\pi} \arcsin\left(\frac{\beta}{1+c^2}\right).$$

Proof We first treat the special case $c = 0$. We may represent Y as $Y = \beta X - \sqrt{1-\beta^2}\, Z$ where X and Z are independent standard normal. We have

$$E[\psi_0(X)\psi_0(Y)] = 4P\{X > 0, Y > 0\} - 1.$$

Now

$$P\{X > 0, Y > 0\} = P\{X > 0, \beta X - \sqrt{1-\beta^2}\, Z > 0\}$$

$$= \iint_{\substack{x > 0 \\ \beta x - \sqrt{1-\beta^2}\, z > 0}} \frac{1}{2\pi} e^{-(x^2 + z^2)/2} \, dx \, dz$$

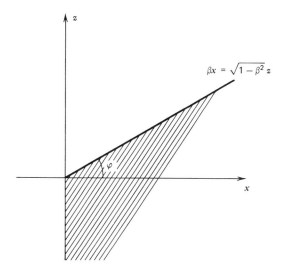

Exhibit 8.3.1

is the integral of the bivariate normal density over the shaded sector in Exhibit 8.3.1. The slope of the boundary line is $\beta/\sqrt{1-\beta^2}$; thus

$$\varphi = \arcsin \beta,$$

and it follows that

$$P(X>0, Y>0) = \frac{1}{4} + \frac{1}{2\pi} \arcsin \beta.$$

Thus the special case is proved.

With regard to the general case, we first note that

$$E\left[\psi_c(X)\psi_c(Y)\right] = 4E\left[\Phi\left(\frac{X}{c}\right)\Phi\left(\frac{Y}{c}\right)\right] - 1.$$

If we introduce two auxiliary random variables Z_1 and Z_2 that are standard normal and independent of X and Y, we can write

$$E\left[\Phi\left(\frac{X}{c}\right)\Phi\left(\frac{Y}{c}\right)\right] = E\left[P^X\{X-cZ_1>0\}P^Y\{Y-cZ_2>0\}\right]$$

$$= P\{X-cZ_1>0, Y-cZ_2>0\},$$

where P^X and P^Y denote conditional probabilities, given X and Y, respectively. But since the correlation of $X - cZ_1$ and $Y - cZ_2$ is $\beta/(1+c^2)$, the general case now follows from the special one. ■

NOTE 1   This theorem exhibits a choice of ψ for which we can recapture the original correlation of X and Y from that of $\psi(X)$ and $\psi(Y)$ in a particularly simple way. However, if this transformation is applied to the elements of a sample covariance/correlation matrix, it in general destroys positive definiteness. So we may prefer to work with the covariances of $\psi(X)$ and $\psi(Y)$, even though they are biased.

NOTE 2   If $T_{X,n}$ and $T_{Y,n}$ are the location estimates determined through

$$\sum \psi(x_i - T_X) = 0,$$

$$\sum \psi(y_i - T_Y) = 0,$$

then the correlation $\rho(\psi(X), \psi(Y))$ can be interpreted as the (asymptotic) correlation between the two location estimates $T_{X,n}$ and $T_{Y,n}$. {Heuristic argument: use the influence function to write

$$T_{X,n} \cong \frac{1}{n} \sum \frac{\psi(x_i)}{E(\psi')},$$

$$T_{Y,n} \cong \frac{1}{n} \sum \frac{\psi(y_i)}{E(\psi')},$$

assuming without loss of generality that the limiting values of $T_{X,n}$ and $T_{Y,n}$ are 0. Thus

$$\left. \text{cov}(T_{X,n}, T_{Y,n}) \cong \frac{1}{n} \frac{E[\psi(X)\psi(Y)]}{[E(\psi')]^2} . \right\}$$

The (relative) efficiency of these covariance/correlation estimates is the square of that of the corresponding location estimates, so the efficiency loss may be quite severe. For instance, assume that the correlation ρ in Proposition 3.2 is small. Then

$$\rho(\psi_c(X), \psi_c(Y)) \approx \beta \frac{1}{(1+c^2)\arcsin[1/(1+c^2)]}$$

and

$$\rho(\psi_0(X), \psi_0(Y)) \approx \beta \cdot \frac{2}{\pi}.$$

Thus if we are testing $\rho(X, Y) = 0$ against $\rho(X, Y) = \beta = \beta_0/\sqrt{n}$, for sample size n, then the asymptotic efficiency of $r_n(\psi_c(X), \psi_c(Y))$ relative to $r_n(X, Y)$ is

$$\left[(1 + c^2) \arcsin\left(\frac{1}{1 + c^2} \right) \right]^{-2}.$$

For $c = 0$ this amounts to $4/\pi^2 \approx 0.41$.

8.4 AN AFFINELY INVARIANT APPROACH

Maximum Likelihood Estimates

Let $f(\mathbf{x}) = f(|\mathbf{x}|)$ be a spherically symmetric probability density in $\mathbb{R}^p$. We apply general nondegenerate affine transformations $\mathbf{x} \rightarrow V(\mathbf{x} - \mathbf{t})$ to obtain a p-dimensional location and scale family of "elliptic" densities

$$f(\mathbf{x}; \mathbf{t}, V) = |\det V| f(|V(\mathbf{x} - \mathbf{t})|). \tag{4.1}$$

The problem is to estimate the vector $\mathbf{t}$ and the matrix V from n observations of $\mathbf{x}$.

Evidently, V is not uniquely identifiable (it can be multiplied by an arbitrary orthogonal matrix from the left), but $V^T V$ is. We can enforce uniqueness of V by suitable side conditions, for example by requiring that it be positive definite symmetric, or that it be lower triangular with a positive diagonal. Mostly, we adopt the latter convention; it is the most convenient one for numerical calculations, but the other is more convenient for some proofs.

The maximum likelihood estimate of $(\mathbf{t}, V)$ is obtained by maximizing

$$\log(\det V) + \text{ave}\{\log f(|V(\mathbf{x} - \mathbf{t})|)\}, \tag{4.2}$$

where $\text{ave}\{\cdot\}$ denotes the average taken over the sample. A necessary condition for a maximum is that (4.2) remain stationary under infinitesimal variations of $\mathbf{t}$ and V. So we let $\mathbf{t}$ and V depend differentiably on a dummy parameter and take the derivative (denoted by a superscribed dot). We obtain the condition

$$\text{tr}(S) + \text{ave}\left\{ \frac{f'(|\mathbf{y}|)}{f(|\mathbf{y}|)} \cdot \frac{\mathbf{y}^T S \mathbf{y}}{|\mathbf{y}|} \right\} - \text{ave}\left\{ \frac{f'(|\mathbf{y}|)}{f(|\mathbf{y}|)} \frac{\dot{\mathbf{t}}^T V^T \mathbf{y}}{|\mathbf{y}|} \right\} = 0, \tag{4.3}$$

with the abbreviations

$$y = V(x - t), \tag{4.4}$$

$$S = \dot{V} V^{-1}. \tag{4.5}$$

Since this should hold for arbitrary infinitesimal variations $\dot{t}$ and $\dot{V}$, (4.3) can be rewritten into the set of simultaneous matrix equations

$$\text{ave}\{w(|y|)y\} = 0, \tag{4.6}$$

$$\text{ave}\{w(|y|)yy^T - I\} = 0, \tag{4.7}$$

where I is the $p \times p$ identity matrix, and

$$w(|y|) = -\frac{f'(|y|)}{|y| f(|y|)}. \tag{4.8}$$

Example 4.1   Let

$$f(|x|) = (2\pi)^{-p/2} \exp\left(\frac{-|x|^2}{2}\right)$$

be the standard normal density. Then $w \equiv 1$, and (4.6) and (4.7) can equivalently be written as

$$t = \text{ave}\{x\}, \tag{4.9}$$

$$(V^T V)^{-1} = \text{ave}\{(x - t)(x - t)^T\}. \tag{4.10}$$

In this case $(V^T V)^{-1}$ is the ordinary covariance matrix of x (the sample one if the average is taken over the sample, the true one if the average is taken over the distribution).

More generally, we call $(V^T V)^{-1}$ a *pseudo-covariance matrix* of x, if t and V are determined from any set of equations

$$\text{ave}\{w(|y|)y\} = 0, \tag{4.11}$$

$$\text{ave}\left\{u(|y|)\frac{yy^T}{|y|^2} - v(|y|)I\right\} = 0, \tag{4.12}$$

with $y = V(x - t)$, and where u, v and w are arbitrary functions.

Note that (4.11) determines location t as a weighted mean

$$t = \frac{\text{ave}\{w(|\mathbf{y}|)\mathbf{x}\}}{\text{ave}\{w(|\mathbf{y}|)\}} \tag{4.13}$$

with weights $w(|\mathbf{y}|)$ depending on the sample.

Similarly, the pseudo-covariance can be written as a kind of scaled weighted covariance

$$(V^T V)^{-1} = \frac{\text{ave}\{s(|\mathbf{y}|)(\mathbf{x}-\mathbf{t})(\mathbf{x}-\mathbf{t})^T\}}{\text{ave}\{s(|\mathbf{y}|)\}} \cdot \frac{\text{ave}\{s(|\mathbf{y}|)\}}{\text{ave}\{v(|\mathbf{y}|)\}} \tag{4.14}$$

with weights

$$s(|\mathbf{y}|) = \frac{u(|\mathbf{y}|)}{|\mathbf{y}|^2} \tag{4.15}$$

depending on the sample. The choice $v = s$ then looks particularly attractive, since it make the scale factor in (4.14) disappear.

8.5   ESTIMATES DETERMINED BY IMPLICIT EQUATIONS

This section shows that (4.12), with arbitrary functions u and v, is in some sense the most general form of an implicit equation determining $(V^T V)^{-1}$.

In order to simplify the discussion, we assume that location is known and fixed to be $\mathbf{t} = 0$. Then we can write (4.12) in the form

$$\text{ave}\{\Psi(V\mathbf{x})\} = 0 \tag{5.1}$$

with

$$\Psi(\mathbf{x}) = s(|\mathbf{x}|)\mathbf{x}\mathbf{x}^T - v(|\mathbf{x}|)I, \tag{5.2}$$

where s is as in (4.15). Is this the most general form of Ψ?

Let us take a sufficiently smooth, but otherwise arbitrary function Ψ from $\mathbb{R}^p$ into the space of symmetric $p \times p$ matrices. This gives us the proper number of equations for the $p(p+1)/2$ unique components of $(V^T V)^{-1}$.

We determine a matrix V such that

$$\text{ave}\{\Psi(V\mathbf{x})\} = 0, \tag{5.3}$$

where the average is taken with respect to a fixed (true or sample) distribution of $\mathbf{x}$.

Let us assume that Ψ and the distribution of $\mathbf{x}$ are such that (5.3) has at least one solution V, that if S is an arbitrary orthogonal matrix, SV is also a solution, and that all solutions lead to the same pseudo-covariance matrix

$$C_{\mathbf{x}} = (V^T V)^{-1} \tag{5.4}$$

This uniqueness assumption implies at once that $C_{\mathbf{x}}$ transforms in the same way under linear transformations B as the classical covariance matrix

$$C_{B\mathbf{x}} = BC_{\mathbf{x}} B^T. \tag{5.5}$$

Now let S be an arbitrary orthogonal transformation and define

$$\Psi_S(\mathbf{x}) = S^T \Psi(S\mathbf{x}) S. \tag{5.6}$$

The transformed function Ψ_S determines a new pseudo-covariance $(W^T W)^{-1}$ through the solution W of

$$\text{ave}\{\Psi_S(W\mathbf{x})\} = \text{ave}\{S^T \Psi(SW\mathbf{x})S\} = 0.$$

Evidently, this is solved by $W = S^T V$, where V is any solution of (5.3), and thus

$$W^T W = V^T S S^T V = V^T V.$$

It follows that Ψ and Ψ_S determine the same pseudo-covariances.

We now form

$$\overline{\Psi}(\mathbf{x}) = \underset{S}{\text{ave}} \{\Psi_S(\mathbf{x})\}, \tag{5.7}$$

by averaging over S (using the invariant measure on the orthogonal group). Evidently, every solution of (5.3) still solves $\text{ave}\{\overline{\Psi}(V\mathbf{x})\} = 0$, but, of course, the uniqueness postulated in (5.4) may have been destroyed by the averaging process.

Clearly, $\overline{\Psi}$ is invariant under orthogonal transformations in the sense that

$$\overline{\Psi}_S(\mathbf{x}) = S^T \overline{\Psi}(S\mathbf{x})S = \overline{\Psi}(\mathbf{x}), \tag{5.8}$$

or equivalently,

$$\overline{\Psi}(S\mathbf{x})S = S\overline{\Psi}(\mathbf{x}). \tag{5.9}$$

Now let $\mathbf{x} \neq 0$ be an arbitrary fixed vector, then (5.9) shows that the matrix $\overline{\Psi}(\mathbf{x})$ commutes with all orthogonal matrices S that keep $\mathbf{x}$ fixed. This implies that the restriction of $\overline{\Psi}(\mathbf{x})$ to the subspace of $\mathbb{R}^p$ orthogonal to $\mathbf{x}$ must be a multiple of the identity. Moreover, for every S that keeps $\mathbf{x}$ fixed, we have

$$S\overline{\Psi}(\mathbf{x})\mathbf{x} = \overline{\Psi}(\mathbf{x})\mathbf{x};$$

hence S also keeps $\overline{\Psi}(\mathbf{x})\mathbf{x}$ fixed, which therefore must be a multiple of $\mathbf{x}$. It follows that $\overline{\Psi}(\mathbf{x})$ can be written in the form

$$\overline{\Psi}(\mathbf{x}) = s(\mathbf{x})\mathbf{x}\mathbf{x}^T - v(\mathbf{x})I$$

with some scalar-valued functions s and v. Because of (5.8) s and v depend on $\mathbf{x}$ only through $|\mathbf{x}|$, and we conclude that $\overline{\Psi}$ is of the form (5.2).

Global uniqueness, as postulated in (5.4), is a rather severe requirement. The arguments carry through in all essential respects with the much weaker local uniqueness requirement that there be a neighborhood of $C_\mathbf{x}$ that contains no other solutions besides $C_\mathbf{x}$. For the symmetrized version (5.2) a set of sufficient conditions for local uniqueness is outlined at the end of Section 8.7.

8.6 EXISTENCE AND UNIQUENESS OF SOLUTIONS

The following existence and uniqueness results are due to Maronna (1976) and Schönholzer (1979). The results are not entirely satisfactory with regard to joint estimation of $\mathbf{t}$ and V.

The Scatter Estimate V

Assume first that location is fixed $\mathbf{t} = 0$. Existence is proved constructively by defining an iterative process converging to a solution V of (4.12). The iteration step from V_m to $V_{m+1} = h(V_m)$ is defined as follows:

$$\left(V_{m+1}^T V_{m+1}\right)^{-1} = \frac{\text{ave}\{s(|V_m\mathbf{x}|)\mathbf{x}\mathbf{x}^T\}}{\text{ave}\{v(|V_m\mathbf{x}|)\}}. \tag{6.1}$$

If the process converges, then the limit V satisfies (4.14) and hence solves (4.12). If (6.1) is used for actual computation, then it is convenient to assume that the matrices V_m are lower triangular with a positive diagonal; for the proofs below, it is more convenient to take them as positive definite symmetric. Clearly, the choice does not matter—both sides of (6.1) are unchanged if V_m and V_{m+1} are multiplied by arbitrary orthogonal matrices from the left.

ASSUMPTIONS

(E-1)   The function $s(r)$ is monotone decreasing and $s(r)>0$ for $r>0$.

(E-2)   The function $v(r)$ is monotone increasing, and $v(r)>0$ for $r\geqslant0$.

(E-3) $u(r)=r^2s(r)$, and $v(r)$ are bounded and continuous.

(E-4) $u(0)/v(0)<p$.

For any hyperplane H in the sample space (i.e., $\dim(H)=p-1$), let

$$P(H)=\text{ave}\{1_{[\mathbf{x}\in H]}\}$$

be the probability of H, or the fraction of observations lying in H, respectively (depending on whether we work with the true or with the sample distribution).

(E-5)   (i)   For all hyperplanes H, $P(H)<1-pv(\infty)/u(\infty)$.

      (ii)   For all hyperplanes H, $P(H)\leqslant1/p$.

LEMMA 6.1   If (E-1), (E-2), (E-3), and (E-5)(i) are satisfied, and if there is an $r_0>0$ such that

$$\frac{\text{ave}\{u(r_0|\mathbf{x}|)\}}{\text{ave}\{v(r_0|\mathbf{x}|)\}}<1,\tag{6.2}$$

then h has a fixed point V.

Proof   Let $\mathbf{z}$ be an arbitrary vector. Then with $V_0=r_0I$ we obtain, from (6.1) and (6.2),

$$\mathbf{z}^T(V_1^TV_1)^{-1}\mathbf{z}=\frac{\text{ave}\{s(r_0|\mathbf{x}|)(\mathbf{z}^T\mathbf{x})^2\}}{\text{ave}\{v(r_0|\mathbf{x}|)\}}$$

$$\leqslant\frac{|\mathbf{z}|^2}{r_0^2}\frac{\text{ave}\{u(r_0|\mathbf{x}|)\}}{\text{ave}\{v(r_0|\mathbf{x}|)\}}\leqslant\frac{|\mathbf{z}|^2}{r_0^2}.$$

Hence

$$(V_1^T V_1)^{-1} < \frac{1}{r_0^2} I$$

(where $A < B$ means that $B - A$ is positive semidefinite). Thus

$$r_0 I < V_1 = h(r_0 I).$$

It follows from (E-1) and (E-2) that $V_{m+1} = h(V_m)$ defines an increasing sequence

$$r_0 I = V_0 < V_1 < V_2 < \cdots.$$

So it suffices to show that the sequence V_m is bounded from above in order to prove convergence $V_m \to V$. Continuity (E-3) then implies that V satisfies (4.14).

Let

$$H = \{ z | \lim | V_m z | < \infty \}.$$

H is a vector space. Assume first that H is a proper subspace of $\mathbb{R}^p$.

Since $V_m < V_{m+1}$, we have

$$I > V_m V_{m+1}^{-2} V_m = \frac{\text{ave}\{ s(|V_m x|)(V_m x)(V_m x)^T \}}{\text{ave}\{ v(|V_m x|) \}}. \tag{6.3}$$

Taking the trace on both sides gives

$$p \geqslant \frac{\text{ave}\{ u(|V_m x|) \}}{\text{ave}\{ v(|V_m x|) \}} \geqslant \frac{\text{ave}\{ u(|V_m x|) \}}{v(\infty)}. \tag{6.4}$$

Since $|V_m x| \uparrow \infty$ for all $x \notin H$, we obtain from the monotone convergence theorem

$$p \geqslant [1 - P(H)] \frac{u(\infty)}{v(\infty)},$$

which contradicts (E-5)(i).

Hence $H = \mathbb{R}^p$, but this is only possible if V_m stays bounded (note that the trace must converge). ∎

Remark   Assumption (6.2) serves to guarantee the existence of a starting matrix V_0, such that $h(V_0) > V_0$. Assume, for instance, that $s(0) > 0$, then (6.2) is satisfied for all sufficiently small r_0. In the limit $r_0 \to 0$, we obtain that $(V_1^T V_1)^{-1}$ then is a multiple of the ordinary covariance matrix.

PROPOSITION 6.2   Assume (E-1) to (E-5). Then h has a fixed point V.

Proof   If s is bounded, the existence of a fixed point follows from Lemma 6.1. If s is unbounded, we choose an $r_1 > 0$ and replace s by $\tilde{s}$:

$$\tilde{s}(r) = s(r_1), \qquad \text{for } r \leqslant r_1$$

$$= s(r), \qquad \text{for } r \geqslant r_1.$$

Let $\tilde{h}$ be the function defined by (6.1) with $\tilde{s}$ in place of s. Lemma 6.1 then implies that $\tilde{h}$ has a fixed point $\tilde{V}$. Since $s \geqslant \tilde{s}$, it follows that, for all V, $h(V) < \tilde{h}(V)$. Hence $h(\tilde{V}) < \tilde{h}(\tilde{V}) = \tilde{V}$, and it follows from (E-1) and (E-2) that $V_{m+1} = h(V_m)$ defines a decreasing sequence

$$\tilde{V} = V_0 > V_1 > V_2 > \cdots.$$

So it suffices to show that $V = \lim V_m$ is nonsingular in order to prove that it is a fixed point of h.

As in the proof of Lemma 6.1, we find

$$I < \frac{\text{ave}\{s(|V_m\mathbf{x}|)(V_m\mathbf{x})(V_m\mathbf{x})^T\}}{\text{ave}\{v(|V_m\mathbf{x}|)\}}, \tag{6.5}$$

and, taking the trace,

$$p \leqslant \frac{\text{ave}\{u(|V_m\mathbf{x}|)\}}{\text{ave}\{v(|V_m\mathbf{x}|)\}} \leqslant \frac{\text{ave}\{u(|V_m\mathbf{x}|)\}}{v(0)}. \tag{6.6}$$

We conclude that not all eigenvalues of V_m can converge to 0, because otherwise

$$p \leqslant \frac{\lim \text{ave}\{u(|V_m\mathbf{x}|)\}}{v(0)} = \frac{u(0)}{v(0)}$$

by the monotone convergence theorem, which would contradict (E-4).

Now assume that $\mathbf{q}_m$ and $\mathbf{z}_m$ are unit-length eigenvectors of V_m belonging to the largest and smallest eigenvalues λ_m and μ_m of V_m, respectively. If we multiply (6.5) from the left and right with $\mathbf{z}_m^T$ and $\mathbf{z}_m$, respectively, we obtain

$$1 \leqslant \frac{\text{ave}\left\{s(|V_m\mathbf{x}|)\mu_m^2(\mathbf{z}_m^T\mathbf{x})^2\right\}}{\text{ave}\,v(|V_m\mathbf{x}|)}. \tag{6.7}$$

Since the largest eigenvalues converge monotonely $\lambda_m \downarrow \lambda > 0$, we obtain

$$G_{m,r} = \left\{\mathbf{x}\,\big|\,|V_m\mathbf{x}| \leqslant r\right\} \subset \left\{\mathbf{x}\,\big|\,\lambda_m|\mathbf{h}_m^T\mathbf{x}| \leqslant r\right\}$$

$$\subset \left\{\mathbf{x}\,\big|\,|\mathbf{h}_m^T\mathbf{x}| \leqslant \frac{r}{\lambda}\right\} = H_{m,r}, \tag{6.8}$$

with $G_{m,r}$ and $H_{m,r}$ defined by (6.8).

Assumption (E-5)(ii) implies that, for each $\varepsilon > 0$, there is a $r_1 > 0$ such that

$$P\{H_{m,r}\} \leqslant \frac{1}{p} + \varepsilon, \qquad \text{for } r \leqslant r_1. \tag{6.9}$$

Furthermore (E-4) implies that we can choose $b > 0$ and $\varepsilon > 0$ such that

$$\frac{u(0) + b}{v(0)}\left(\frac{1}{p} + \varepsilon\right) < 1. \tag{6.10}$$

If $r_0 < r_1$ is chosen such that $u(r) \leqslant u(0) + b$ for $r \leqslant r_0$, then (6.7) to (6.9) imply

$$1 \leqslant \frac{1}{v(0)}\left[\text{ave}\left\{1_{G_{m,r}}(\mathbf{x})u(|V_m\mathbf{x}|) + 1_{G_{m,r}^c}(\mathbf{x})s(|V_m\mathbf{x}|)\mu_m^2(\mathbf{z}_m^T\mathbf{x})^2\right\}\right]$$

$$\leqslant \frac{1}{v(0)}[u(0) + b]\left(\frac{1}{p} + \varepsilon\right) + \text{ave}\left\{\frac{1}{v(0)}\min\left[s(r_0)\mu_m^2|\mathbf{x}|^2, u(\infty)\right]\right\}$$

If $\lim \mu_m = 0$, then the last summand tends to 0 by the dominated convergence theorem; this leads to a contradiction in view of (6.10). Hence $\lim \mu_m > 0$ and the proposition is proved. ■

Uniqueness of the fixed point can be proved under the following assumptions.

ASSUMPTIONS

(U-1) $s(r)$ is decreasing.

(U-2) $u(r) = r^2 s(r)$ is continuous and increasing, and $u(r) > 0$ for $r > 0$.

(U-3) $v(r)$ is continuous and decreasing, and $v(r) \geq 0$, $v(r) > 0$ for $0 \leq r < r_0$.

(U-4) For all hyperplanes $H \subset \mathbb{R}^p$, $P(H) < \frac{1}{2}$.

Remark In view of (E-3), we can prove simultaneous existence and uniqueness only if $v = \text{constant}$ (as in the ML case).

PROPOSITION 6.3 Assume (U-1) to (U-4). If V and V_1 are two fixed points of h, then there is a real number λ such that $V_1 = \lambda V$, and

$$u(|V\mathbf{x}|) = u(\lambda|V\mathbf{x}|), \qquad v(|V\mathbf{x}|) = v(\lambda|V\mathbf{x}|)$$

for almost all $\mathbf{x}$. In particular, $\lambda = 1$ if either u or v is strictly monotone.

We first prove a special case:

LEMMA 6.4 Proposition 6.3 holds if either $V > V_1$ or $V < V_1$.

Proof of the Lemma We may assume without loss of generality that $V_1 = I$. Assume $V > I$ (the case $V < I$ is proved in the same way). Then

$$\frac{\text{ave}\left\{ u(|V\mathbf{x}|) \dfrac{(V\mathbf{x})(V\mathbf{x})^T}{|V\mathbf{x}|^2} \right\}}{\text{ave}\{ v(|V\mathbf{x}|) \}} = \frac{\text{ave}\left\{ u(|\mathbf{x}|) \dfrac{\mathbf{x}\mathbf{x}^T}{|\mathbf{x}|^2} \right\}}{\text{ave}\{ v(|\mathbf{x}|) \}} = I. \qquad (6.11)$$

If we take the trace, we obtain

$$\frac{\text{ave}\{ u(|V\mathbf{x}|) \}}{\text{ave}\{ v(|V\mathbf{x}|) \}} = \frac{\text{ave}\{ u(|\mathbf{x}|) \}}{\text{ave}\{ v(|\mathbf{x}|) \}} = p. \qquad (6.12)$$

In view of (U-2) and (U-3), we must have

$$\text{ave}\{ u(|V\mathbf{x}|) \} = \text{ave}\{ u(|\mathbf{x}|) \},$$

$$\text{ave}\{ v(|V\mathbf{x}|) \} = \text{ave}\{ v(|\mathbf{x}|) \}. \qquad (6.13)$$

Because of $V > I$ this implies

$$u(|V\mathbf{x}|) = u(|\mathbf{x}|),$$

$$v(|V\mathbf{x}|) = v(|\mathbf{x}|), \qquad (6.14)$$

for almost all $\mathbf{x}$.

If either u or v is strictly monotone, this already forces $V=I$.
In view of (6.14) it follows from (6.11) that

$$\text{ave}\left\{ u(|\mathbf{x}|)\left[\frac{(V\mathbf{x})(V\mathbf{x})^T}{|V\mathbf{x}|^2} - \frac{\mathbf{x}\mathbf{x}^T}{|\mathbf{x}|^2} \right] \right\} = 0. \tag{6.15}$$

Now let $\mathbf{z}$ be an eigenvector belonging to the largest eigenvalue λ of V. Then (6.15) implies

$$\text{ave}\left\{ u(|\mathbf{x}|)\left(\frac{\lambda^2}{|V\mathbf{x}|^2} - \frac{1}{|\mathbf{x}|^2} \right)(\mathbf{z}^T\mathbf{x})^2 \right\} = 0. \tag{6.16}$$

The expression inside the curly parentheses of (6.16) is >0 unless either:
(1) $\mathbf{x}$ is an eigenvector of V, to the eigenvalue λ; or (2) $\mathbf{x}$ is orthogonal to $\mathbf{z}$.

If $V=\lambda I$, the lemma is proved. If $V\neq\lambda I$, then (U-4) implies that the union of the $\mathbf{x}$-sets (1) and (2) has a total mass less than 1, which leads to a contradiction. ■

Proof of the Proposition Assume the fixed points are V and I, and neither $V<I$ nor $V>I$. Choose $0<r<1$ so that $rI<V$. Because of (U-2) and (U-3) we have

$$h(rI)^{-2} = \frac{\text{ave}\left\{ u(r|\mathbf{x}|)\dfrac{\mathbf{x}\mathbf{x}^T}{|\mathbf{x}|^2} \right\}}{\text{ave}\{v(r|\mathbf{x}|)\}} \cdot \frac{1}{r^2}$$

$$\leqslant \frac{\text{ave}\left\{ u(|\mathbf{x}|)\dfrac{\mathbf{x}\mathbf{x}^T}{|\mathbf{x}|^2} \right\}}{\text{ave}\{v(|\mathbf{x}|)\}} \cdot \frac{1}{r^2} = \frac{1}{r^2}I,$$

or

$$rI < h(rI).$$

It follows from $rI<I$ and $rI<V$ that $V_1=\lim h^m(rI)$ is a fixed point with $V_1<I$ and $V_1<V$. Then both pairs V_1, I and V_1, V satisfy the assumptions of Lemma 6.4, so V_1, I, and V are scalar multiples of each others. This contradicts the assumption that neither $V<I$ nor $V>I$, and the proposition is proved. ■

The Location Estimate t

If V is kept fixed, say $V=I$, existence and uniqueness of the location estimate **t** are easy to establish, provided $\psi(r)=w(r)r$ is monotone increasing for positive r. Then there is a convex function

$$\rho(\mathbf{x})=\rho(|\mathbf{x}|)=\int_0^{|\mathbf{x}|} \psi(r)\,dr$$

such that **t** can equivalently be defined as minimizing

$$Q(\mathbf{t})=\text{ave}\{\rho(|\mathbf{x}-\mathbf{t}|)\}.$$

We only treat the sample case, so we do not have to worry about the possible nonexistence of the distribution average. Thus the set of solutions **t** is nonempty and convex, and if there is at least one observation **x** such that $\rho''(|\mathbf{x}-\mathbf{t}|)>0$, the solution is in fact unique.

Proof We shall show that Q is strictly convex. Assume that $\mathbf{z}\in\mathbb{R}^p$ depends linearly on a parameter s, and take derivatives with respect to s (denoted by a superscript dot). Then

$$\rho(|\mathbf{z}|)^{\cdot}=\frac{\rho'(|\mathbf{z}|)}{|\mathbf{z}|}\mathbf{z}^T\dot{\mathbf{z}}$$

and

$$\rho(|\mathbf{z}|)^{\cdot\cdot}=\frac{\rho''(|\mathbf{z}|)}{|\mathbf{z}|^2}(\mathbf{z}^T\dot{\mathbf{z}})^2+\frac{\rho'(|\mathbf{z}|)}{|\mathbf{z}|^3}\left[(\mathbf{z}^T\mathbf{z})(\dot{\mathbf{z}}^T\dot{\mathbf{z}})-(\mathbf{z}^T\dot{\mathbf{z}})^2\right]\geq 0,$$

since $\rho'\geq 0$, $(\mathbf{z}^T\dot{\mathbf{z}})^2\leq(\mathbf{z}^T\mathbf{z})(\dot{\mathbf{z}}^T\dot{\mathbf{z}})$, and $\rho''(r)=\psi'(r)\geq 0$. Hence $\rho(|\mathbf{z}|)$ is convex as a function of **z**. Moreover, if $\rho''(|\mathbf{z}|)>0$ and $\rho'(|\mathbf{z}|)>0$, then ρ is strictly convex at the point **z**: if the variation $\dot{\mathbf{z}}$ is orthogonal to **z**, then

$$\rho(|\mathbf{z}|)^{\cdot\cdot}=\frac{\rho'(|\mathbf{z}|)}{|\mathbf{z}|}(\dot{\mathbf{z}}^T\dot{\mathbf{z}})>0,$$

and otherwise

$$\rho(|\mathbf{z}|)^{\cdot\cdot}\geq\frac{\rho''(|\mathbf{z}|)}{|\mathbf{z}|^2}(\mathbf{z}^T\dot{\mathbf{z}})^2>0.$$

In fact $\rho''(r) > 0$, $\rho'(r) = 0$ can only happen at $r = 0$, and $\mathbf{z} = 0$ is a point of strict convexity, as we verify easily by a separate argument. Hence Q is strictly convex, which implies uniqueness. ∎

Joint Estimation of t and V

Joint existence of solutions $\mathbf{t}$ and V is then also easy to establish, if we do not mind somewhat restrictive regularity conditions. Assume that, for each fixed $\mathbf{t}$, there is a unique solution V_t of (4.12), which depends continuously on $\mathbf{t}$, and that for each fixed V there is a unique solution $\mathbf{t}(V)$ of (4.11), which depends continuously on V. It follows from (4.13) that $\mathbf{t}(V)$ is always contained in the convex hull H of the observations. Thus the continuous function $\mathbf{t} \rightarrow \mathbf{t}(V_t)$ maps H into itself and hence has a fixed point by Brouwer's theorem. The corresponding pair $(\mathbf{t}, V_t)$ obviously solves (4.11) and (4.12). Uniqueness of the fixed point so far has been proved only under the assumption that the distribution of the $\mathbf{x}$ has a center of symmetry; in the sample distribution case this is of course very unrealistic [cf. Maronna (1976)].

8.7 INFLUENCE FUNCTIONS AND QUALITATIVE ROBUSTNESS

Our estimates $\mathbf{t}$ and V, defined through (4.11) and (4.12) with the help of averages over the sample distribution, clearly can be regarded as functionals $\mathbf{t}(F)$ and $V(F)$ of some underlying distribution F. The estimates are vector and matrix valued; the influence functions, measuring changes of $\mathbf{t}$ and V under infinitesimal changes of F, clearly are vector and matrix valued too.

Without loss of generality we can choose the coordinate system such that $\mathbf{t}(F) = 0$ and $V(F) = I$. We assume that F is (at least) centrosymmetric. In order to find the influence functions, we have to insert $F_s = (1-s)F + s\delta_\mathbf{x}$ into the defining equations and take the derivative with respect to s at $s = 0$; we denote it by a superscript dot.

We first take (4.11). The procedure just outlined gives

$$-\operatorname{ave}_F\left\{ \frac{w'(|\mathbf{y}|)}{|\mathbf{y}|}(\mathbf{y}^T\dot{\mathbf{t}})\mathbf{y} + w(|\mathbf{y}|)\dot{\mathbf{t}} \right\}$$

$$+\operatorname{ave}_F\left\{ \frac{w'(|\mathbf{y}|)}{|\mathbf{y}|}(\mathbf{y}^T\dot{V}\mathbf{y})\mathbf{y} + w(|\mathbf{y}|)\dot{V}\mathbf{y} \right\} + w(|\mathbf{x}|)\mathbf{x} = 0. \qquad (7.1)$$

The second term (involving $\dot{V}$) averages to 0 if F is centrosymmetric. There is a considerable further simplification if F is spherically symmetric [or, at least, if the conditional covariance matrix of $y/|y|$, given $|y|$, equals $(1/p)I$ for all $|y|$], since then $E\{(y^T\dot{t})y|\ |y|\} = (1/p)|y|^2\dot{t}$. So (7.1) becomes

$$-\text{ave}_F\left\{\frac{1}{p}w'(|y|)|y| + w(|y|)\right\}\dot{t} + w(|x|)x = 0.$$

Hence the influence function for location is

$$IC(x; F, t) = \frac{w(|x|)x}{\text{ave}_F\left\{w(|y|) + \dfrac{1}{p}w'(|y|)|y|\right\}}. \qquad (7.2)$$

Similarly, differentiation of (4.12) gives,

$$\text{ave}_F\left\{\frac{u'(|y|)|y| - 2u(|y|)}{|y|^4}(y^T\dot{V}y)(yy^T) + \frac{u(|y|)}{|y|^2}(\dot{V}yy^T + yy^T\dot{V}^T)\right.$$

$$\left.-\frac{v'(|y|)}{|y|}(y^T\dot{V}y)I\right\}$$

$$-\text{ave}_F\left\{\frac{u'(|y|)|y| - 2u(|y|)}{|y|^4}(y^T\dot{t})(yy^T) + \frac{u(|y|)}{|y|^2}(\dot{t}y^T + y\dot{t}^T)\right.$$

$$\left.-\frac{v'(|y|)}{|y|}(y^T\dot{t})I\right\} + \left\{u(|x|)\frac{xx^T}{|x|^2} - v(|x|)I\right\} = 0. \qquad (7.3)$$

The second term (involving $\dot{t}$) averages to 0 if F is centrosymmetric.

It is convenient to split (7.3) into two equations. We first take the trace of (7.3) and divide it by p. This gives

$$\text{ave}_F\left\{\left[\frac{1}{p}u'(|y|) - v'(|y|)\right]\frac{y^T\dot{V}y}{|y|}\right\} + \left\{\frac{1}{p}u(|x|) - v(|x|)\right\} = 0. \qquad (7.4)$$

If we now subtract (7.4) from the diagonal of (7.3), we obtain

$$
\text{ave}_F \left\{ \frac{u'(|\mathbf{y}|)}{|\mathbf{y}|} \left[\frac{\mathbf{y}\mathbf{y}^T}{|\mathbf{y}|^2} - \frac{1}{p} I \right] (\mathbf{y}^T \dot{V} \mathbf{y}) \right.
$$

$$
+ \frac{u(|\mathbf{y}|)}{|\mathbf{y}|^4} \left[|\mathbf{y}|^2 (\dot{V}\mathbf{y}\mathbf{y}^T + \mathbf{y}\mathbf{y}^T \dot{V}^T) - 2(\mathbf{y}^T \dot{V} \mathbf{y})\mathbf{y}\mathbf{y}^T \right] \Bigg\}
$$

$$
+ u(|\mathbf{x}|)\left(\frac{\mathbf{x}\mathbf{x}^T}{|\mathbf{x}|^2} - \frac{1}{p} I \right) = 0. \tag{7.5}
$$

If F is spherically symmetric, the averaging process can be carried one step further. From (7.4) we then obtain [with $\dot{W} = \frac{1}{2}(\dot{V} + \dot{V}^T)$]

$$
\text{ave}_F \left\{ \left[\frac{1}{p} u'(|\mathbf{y}|) - v'(|\mathbf{y}|) \right] |\mathbf{y}| \right\} \frac{1}{p} \text{tr}(\dot{W}) + \left[\frac{1}{p} u(|\mathbf{x}|) - v(|\mathbf{x}|) \right] = 0, \tag{7.6}
$$

and from (7.5)

$$
\frac{2}{p+2} \text{ave}_F \left\{ u(|\mathbf{y}|) + \frac{1}{p} u'(|\mathbf{y}|)|\mathbf{y}| \right\} \left[\dot{W} - \frac{1}{p} \text{tr}(\dot{W}) \right]
$$

$$
+ u(|\mathbf{x}|)\left(\frac{\mathbf{x}\mathbf{x}^T}{|\mathbf{x}|^2} - \frac{1}{p} I \right) = 0. \tag{7.7}
$$

[Cf. Section 8.10, after (10.15), for this averaging process.] Clearly, only the symmetric part $\dot{W}$ of the influence function $\dot{V} = IC(\mathbf{x}; F, V)$ matters and is determinable. We obtain it in explicit form from (7.6) and (7.7) as

$$
\frac{1}{p} \text{tr}(\dot{W}) = - \frac{\dfrac{1}{p} u(|\mathbf{x}|) - v(|\mathbf{x}|)}{\text{ave}_F \left\{ \left[\dfrac{1}{p} u'(|\mathbf{y}|) - v'(|\mathbf{y}|) \right] |\mathbf{y}| \right\}}
$$

$$
\dot{W} - \frac{1}{p} \text{tr}(\dot{W})I = - \frac{p+2}{2} \frac{u(|\mathbf{x}|)\left(\dfrac{\mathbf{x}\mathbf{x}^T}{|\mathbf{x}|^2} - \dfrac{1}{p} \right)}{\text{ave}_F \left\{ u(|\mathbf{y}|) + \dfrac{1}{p} u'(|\mathbf{y}|)|\mathbf{y}| \right\}} \tag{7.8}
$$

The influence function of the pseudo-covariance is, clearly,

$$IC\left(\mathbf{x}; F,(V^T V)^{-1}\right) = -2\dot{W} \tag{7.9}$$

(assuming throughout that the coordinate system is matched so that $V = I$).

It can be seen from (7.2) and (7.8) that the influence functions are bounded if and only if the functions $w(r)r$, $u(r)$, and $v(r)$ are bounded [and the denominators of (7.2) and (7.8) are not equal to 0].

Qualitative robustness, that is, essentially the continuity of the functionals $\mathbf{t}(F)$ and $V(F)$, is difficult to discuss for the simple reason that we do not yet know for which F these functionals are uniquely defined. However, they are so for elliptical distributions of the type (4.1), and by the implicit function theorem we can then conclude that the solutions are still well defined in some neighborhood. This involves a careful discussion of the influence functions, not only at the model distribution (which is spherically symmetric by assumption), but also in some neighborhood of it. That is, we have to argue directly with (7.1) and (7.3), instead of the simpler (7.2) and (7.8).

Thus we are in good shape if the denominators in (7.2) and (7.8) are strictly positive and if w, wr, $w'r$, $w'r^2$, u, u/r, u', $u'r$, v, v', and $v'r$ are bounded and continuous, because then the influence function is stable at the model distribution, and we can use (2.5.8) to conclude that a small change in F induces only a small change in the values of the functionals.

8.8 CONSISTENCY AND ASYMPTOTIC NORMALITY

The estimates $\mathbf{t}$ and V are consistent and asymptotically normal under relatively mild assumptions, and proofs can be found along the lines of Sections 6.2 and 6.3. While the consistency proof is complicated [the main problem being caused by the fact that we have a simultaneous location-scale problem, where assumptions (A-5) or (B-4) fail], asymptotic normality can be proved straightforwardly by verifying assumptions (N-1) to (N-4). Of course, this imposes some regularity conditions on w, u, and v and the underlying distribution. Note in particular that there will be trouble if $u(r)/r$ is unbounded and there is a pointmass at the origin. For details see Maronna (1976) and Schönholzer (1979).

The asymptotic variances and covariances of the estimates coincide with those of their influence functions, and thus can easily be derived from (7.2) and (7.8). For symmetry reasons location and covariance estimates are asymptotically uncorrelated, and hence asymptotically independent.

The location components $\hat{t}_j$ are asymptotically independent, with asymptotic variance

$$n\,\mathrm{var}(\hat{t}_j) = \frac{p^{-1}E\big[\,w(|\mathbf{x}|)|\mathbf{x}|\,\big]^2}{\big\{E\big[\,w(|\mathbf{x}|)+p^{-1}w'(|\mathbf{x}|)|\mathbf{x}|\,\big]\big\}^2}\,. \tag{8.1}$$

The asymptotic variances and covariances of the components of $\hat{V}$ can be described as follows (we assume that V is lower triangular):

$$n\,\mathrm{var}\!\left(\frac{1}{p}\,\mathrm{tr}\,\hat{V}\right) = \frac{E\big\{\big[\,p^{-1}u(|\mathbf{x}|)-v(|\mathbf{x}|)\,\big]^2\big\}}{\big\{E\big[\,p^{-1}u'(|\mathbf{x}|)|\dot{\mathbf{x}}|-v'(|\mathbf{x}|)|\mathbf{x}|\,\big]\big\}^2}, \tag{8.2}$$

$$n\,\mathrm{var}\!\left(\hat{V}_{jj}-p^{-1}\,\mathrm{tr}\,\hat{V}\right) = \frac{(p-1)(p-2)}{2p^2}\lambda, \tag{8.3}$$

$$nE\big[\big(\hat{V}_{jj}-p^{-1}\,\mathrm{tr}\,\hat{V}\big)\big(\hat{V}_{kk}-p^{-1}\,\mathrm{tr}\,\hat{V}\big)\big]=\frac{p+2}{2p^2}\lambda, \qquad \text{for } j\neq k, \tag{8.4}$$

$$n\,\mathrm{var}\!\left(\hat{V}_{jk}\right)=\frac{p+2}{p}\lambda, \qquad \text{for } j>k, \tag{8.5}$$

with

$$\lambda = \frac{E\big[\,u(|\mathbf{x}|)^2\,\big]}{\big\{E\big[(1/p)u'(|\mathbf{x}|)|\mathbf{x}|+u(|\mathbf{x}|)\big]\big\}^2}\,. \tag{8.6}$$

All other asymptotic covariances between $p^{-1}\,\mathrm{tr}(\hat{V})$, $\hat{V}_{jj}-p^{-1}\,\mathrm{tr}\,\hat{V}$, and $\hat{V}_{jk}$ are 0.

8.9 BREAKDOWN POINT

Let us agree that breakdown occurs when at least one solution of (4.12) misbehaves. Then the breakdown point (with regard to centrosymmetric ε-contamination) is always

$$\varepsilon^* \leqslant \frac{1}{p}\,.$$

This bound is conjectured to be sharp [if we allow asymmetric contamination, then the sharp bound is conjectured to be $1/(p+1)$].

The demonstration follows an idea of W. Stahel (personal communication). Let G and H be centrosymmetric, but not spherically symmetric, distributions in $\mathbb{R}^p$, centered at 0, and put

$$F = (1-\varepsilon)G + \varepsilon H.$$

Assume that $|\mathbf{x}|$ has the same distribution under G and H, hence also under F.

We assume that the conditional covariance matrix of $\mathbf{x}/|\mathbf{x}|$, given $|\mathbf{x}|$, is diagonal under both G and H, namely, with diagonal vector

$$\left(0, \frac{1}{p-1}, \ldots, \frac{1}{p-1}\right)$$

under G, with diagonal vector $(1,0,0,\ldots,0)$ under H. For instance, we may take G to be the distribution of $(0, z_2, \ldots, z_p)$, where $z_2, \ldots, z_p$ are independent standard normal, and H to be the distribution of $(z_1, 0, \ldots, 0)$, where z_1 has a χ-distribution with $p-1$ degrees of freedom. For $\varepsilon = 1/p$, the conditional covariance matrix of $\mathbf{x}/|\mathbf{x}|$, given $|\mathbf{x}|$, under F is diagonal with diagonal vector $(1/p, \ldots, 1/p)$.

Now let $\bar{F}$ be the spherically symmetric distribution obtained by averaging F over the orthogonal group. For both F and $\bar{F}$, the radial distribution (i.e. the distribution of $|\mathbf{x}|$) then is a χ-distribution with $p-1$ degrees of freedom. Clearly, any covariance estimate defined by a relation of the form (4.12), viewed as a functional, then will be the same for F and for $\bar{F}$, namely a certain multiple of the identity matrix.

We interpret this result that a symmetric ε-contamination on the x_1-axis, with $\varepsilon = 1/p$, can cause breakdown.

A breakdown point $\varepsilon^* \leqslant 1/p$ is disappointingly low in high dimension. A possible way out may be the following. Estimate first some location $\mathbf{t}$ and pseudo-covariance $(V^T V)^{-1}$. Then make a search for outliers in the space of y-vectors $[\mathbf{y} = V(\mathbf{x} - \mathbf{t})]$. If the type of breakdown encountered in this section is at all typical (and there is little doubt that it is), the "good" points $\mathbf{y}$ will form a flat disk, while the "bad" points that have caused breakdown will stick out roughly in the axial direction of the disk. A univariate robust scale estimate (e.g., the median absolute deviation) should therefore show a well defined minimum in this direction.

8.10 LEAST INFORMATIVE DISTRIBUTIONS

Location

Consider the family of distributions

$$f(\mathbf{x};\mathbf{t},I)=f(|\mathbf{x}-\mathbf{t}|), \qquad \mathbf{x},\mathbf{t}\in\mathbb{R}^p, \tag{10.1}$$

where f belongs to some convex set $\mathcal{F}$ of densities. Assume that $\mathbf{t}$ depends differentiably on some real parameter θ. Then Fisher information with respect to θ is

$$I(f)=E_{\mathbf{t}}\left\{\left[\frac{\partial}{\partial\theta}\log f(|\mathbf{x}-\mathbf{t}|)\right]^2\right\}$$

$$=E\left\{\left[\frac{f'(|\mathbf{x}|)}{f(|\mathbf{x}|)}\frac{\mathbf{i}^T\mathbf{x}}{|\mathbf{x}|}\right]^2\right\}$$

$$=E\left\{\left[\frac{f'(|\mathbf{x}|)}{f(|\mathbf{x}|)}\right]^2\right\}\cdot\frac{|\mathbf{i}|^2}{p}. \tag{10.2}$$

We now intend to find a $f_0\in\mathcal{F}$ minimizing $I(f)$. Clearly, this is done by minimizing

$$E\left\{\left[\frac{f'(|\mathbf{x}|)}{f(|\mathbf{x}|)}\right]^2\right\}=C_p\int_0^\infty\left[\frac{f'(r)}{f(r)}\right]^2 f r^{p-1}\,dr, \tag{10.3}$$

where C_p denotes the surface area of the unit sphere in $\mathbb{R}^p$. This immediately leads to the variational condition

$$\int_0^\infty\left[-\left(\frac{f'}{f}\right)^2 r^{p-1}-2\left(\frac{f'}{f}r^{p-1}\right)'\right]\delta f\,dr\geqslant 0 \tag{10.4}$$

subject to the side condition

$$\int r^{p-1}\delta f\,dr=0, \tag{10.5}$$

or, with some Lagrange multiplier γ,

$$4\gamma r^{p-1}-\left(\frac{f'}{f}\right)^2 r^{p-1}-2\left(\frac{f'}{f}r^{p-1}\right)'=0. \tag{10.6}$$

on the set of r-values where f can be varied freely; the equality sign should be replaced by ≥ 0 on the set where $\delta f \geq 0$.

With $u = \sqrt{f}$ we obtain the linear differential equation

$$u'' + \frac{p-1}{r} u' - \gamma u = 0, \tag{10.7}$$

valid on the set where f can be freely varied.

Example 10.1   Let $\mathcal{F}$ be the set of spherically symmetric ε-contaminated normal distributions in $\mathbb{R}^3$. Then (10.7) has the particular solution

$$u(r) = \frac{e^{-\sqrt{\gamma}\, r}}{r}. \tag{10.8}$$

Since f_0 and f_0'/f_0 should be continuous, we obtain after some calculations

$$f_0(r) = a e^{-r^2/2}, \qquad \text{for } r \leq r_0$$
$$= b \frac{e^{-cr}}{r^2}, \qquad \text{for } r \geq r_0, \tag{10.9}$$

with

$$a = (1-\varepsilon)(2\pi)^{-3/2},$$
$$b = (1-\varepsilon)(2\pi)^{-3/2} r_0^2 e^{(r_0^2/2)-2}, \tag{10.10}$$
$$c = 2\sqrt{\gamma} = r_0 - \frac{2}{r_0},$$

and thus

$$-\frac{f_0'(r)}{f_0(r)} = r, \qquad \text{for } r \leq r_0$$
$$= c + \frac{2}{r}, \qquad \text{for } r \geq r_0. \tag{10.11}$$

The constants r_0 and ε are related by the requirement that f_0 be a probability density:

$$C_p \int f_0(r) r^{p-1} \, dr = 1. \tag{10.12}$$

In particular, we must have $c > 0$, and hence $r_0 > \sqrt{2}$; the limiting case $c = 0$ corresponds to $r_0 = \sqrt{2}$ and $\varepsilon = 1$.

It can be seen from the nonmonotonicity of (10.11) that $-\log f_0(|\mathbf{x}|)$ is *not* a convex function of $\mathbf{x}$. Hence, in general, the maximum likelihood estimate of location need not be unique, and there are some troubles with consistency proofs when ε is large.

For our present purposes location is but a nuisance parameter, and it is hardly worthwhile to bother with complicated estimates of location. We therefore prefer to work with a simple monotone approximation to (10.11), of the form

$$w(r)r = r, \quad \text{for } r \leqslant r_0$$

$$= r_0, \quad \text{for } r \geqslant r_0; \tag{10.13}$$

compare (4.6).

Covariance

We now consider the family of distributions

$$f(\mathbf{x}; 0, V) = |\det V| f(|V\mathbf{x}|), \quad \mathbf{x} \in \mathbb{R}^p. \tag{10.14}$$

We assume that V depends differentiably on some real parameter θ. Then Fisher information with respect to θ at $V = V_0 = I$ is

$$I(f) = E\left\{ \left[\frac{\partial}{\partial \theta} \log f(\mathbf{x}; 0, V) \right]^2 \right\}$$

$$= E\left\{ \left[\operatorname{tr} \dot{V} + \frac{f'(|\mathbf{x}|)}{f(|\mathbf{x}|)} \frac{\mathbf{x}^T \dot{V} \mathbf{x}}{|\mathbf{x}|} \right]^2 \right\}. \tag{10.15}$$

Because of symmetry it suffices to treat this special case.

In order to simplify (10.15) we first take the conditional expectation, given $|\mathbf{x}|$, that is we average over the uniform distribution on the spheres $|\mathbf{x}| = \text{const}$. The conditional averages of $\mathbf{x}^T \dot{V} \mathbf{x}$ and $(\mathbf{x}^T \dot{V} \mathbf{x})^2$ are $\beta |\mathbf{x}|^2$ and $\gamma |\mathbf{x}|^4$, respectively, with $\beta = (1/p) \operatorname{tr} \dot{V}$ and

$$\gamma = \frac{1}{p(p+2)} \left[(\operatorname{tr} \dot{V})^2 + 2 \sum_{j,k} \dot{V}_{jk}^2 \right]$$

if we assume (without loss of generality) that $\dot{V}$ is symmetric. The easiest way to prove this is to show that, for reasons of symmetry and homogeneity, the averages must be proportional to $|\mathbf{x}|^2$ and $|\mathbf{x}|^4$, respectively, and then to determine the proportionality constants in the special case where $\mathbf{x}$ is p-variate standard normal and V is diagonal. Thus if we put

$$u(r) = -\frac{f'(r)}{f(r)} r, \tag{10.16}$$

we have

$$I(f) = E\left\{\left[u(|\mathbf{x}|) \frac{\mathbf{x}^T \dot{V} \mathbf{x}}{|\mathbf{x}|^2} - p\beta \right]^2\right\}$$

$$= E\left[\gamma u(|\mathbf{x}|)^2 - 2p\beta u(|\mathbf{x}|) + p^2\beta^2 \right]$$

$$= \gamma E\left[u(|\mathbf{x}|)^2 \right] - p^2\beta^2. \tag{10.17}$$

Hence in order to minimize $I(f)$ over $\mathscr{F}$, it suffices to minimize

$$J(f) = E\left[u(|\mathbf{x}|)^2 \right] = C_p \int_0^\infty u(r)^2 r^{p-1} f\, dr$$

$$= C_p \int_0^\infty \frac{f'(r)^2}{f(r)} r^{p+1}\, dr. \tag{10.18}$$

A standard variational argument gives

$$\delta J(f) = C_p \int_0^\infty (-u^2 + 2pu + 2ru') r^{p-1} \delta f\, dr. \tag{10.19}$$

Together with the side condition $C_p \int r^{p-1} \delta f\, dr = 0$, we obtain that the u corresponding to the minimizing f_0 should satisfy

$$2ru' + 2pu - u^2 = c \tag{10.20}$$

for those r where f_0 can be varied freely, or

$$-2ru' + (u-p)^2 = p^2 - c = \kappa^2, \tag{10.21}$$

for some constant κ.

For our purposes we only need the constant solutions corresponding to $u' = 0$. Thus

$$u = p \pm \kappa. \tag{10.22}$$

In particular, let

$$\mathcal{F} = \{ f \mid f(r) = (1 - \varepsilon)\varphi(r) + \varepsilon h(r), h \in \mathfrak{M}_s \} \tag{10.23}$$

be the set of all spherically symmetric contaminated normal densities, with

$$\varphi(r) = (2\pi)^{-p/2} e^{-r^2/2}, \tag{10.24}$$

and $\mathfrak{M}_s$ being the set of all spherically symmetric probability densities in $\mathbb{R}^p$.

Then we verify easily that $J(f)$, and thus $I(f)$, are minimized by choosing

$$\begin{aligned}
u(r) = -\frac{f_0'(r)}{f_0(r)} r = a^2, \qquad &\text{for } 0 \leqslant r \leqslant a \\
= r^2, \qquad &\text{for } a \leqslant r \leqslant b \\
= b^2, \qquad &\text{for } b \leqslant r,
\end{aligned} \tag{10.25}$$

and thus

$$\begin{aligned}
f_0(r) = (1 - \varepsilon)\varphi(a)\left(\frac{a}{r}\right)^{a^2}, \qquad &\text{for } 0 \leqslant r \leqslant a \\
= (1 - \varepsilon)\varphi(r), \qquad &\text{for } a \leqslant r \leqslant b \\
= (1 - \varepsilon)\varphi(b)\left(\frac{b}{r}\right)^{b^2} \qquad &\text{for } b \leqslant r.
\end{aligned} \tag{10.26}$$

The constants a and b satisfy

$$a = \sqrt{(p - \kappa)^+},$$

$$b = \sqrt{p + \kappa}, \tag{10.27}$$

and $\kappa > 0$ has to be determined such that the total mass of f_0 is 1, or equivalently, that

$$C_p\left[\varphi(a)\int_0^a \left(\frac{a}{r}\right)^{a^2} r^{p-1}\,dr + \int_a^b \varphi(r)r^{p-1}\,dr + \varphi(b)\int_b^\infty \left(\frac{b}{r}\right)^{b^2} r^{p-1}\,dr \right]$$

$$= \frac{1}{1-\varepsilon}. \tag{10.28}$$

The maximum likelihood estimate of pseudo-covariance for f_0 can be described by (4.12), with u as in (10.25), and $v\equiv 1$. It has the following minimax property. Let $\mathcal{F}_c \subset \mathcal{F}$ be that subset for which it is a consistent estimate of the identity matrix. Then it minimizes the supremum over $\mathcal{F}_c$ of the asymptotic variances (8.2) to (8.6).

If $\kappa < p$ and hence $a > 0$, the least informative density f_0 is highly unrealistic in view of its singularity at the origin. In other words the corresponding minimax estimate appears to protect against an unlikely contingency. Moreover, if the underlying distribution happens to put a pointmass at the origin (or, if in the course of a computation, a sample point happens to coincide with the current trial value $\mathbf{t}$), (4.12) or (4.14) is not well defined.

If we separate the scale aspects (information contained in $|\mathbf{y}|$) from the directional aspects (information contained in $\mathbf{y}/|\mathbf{y}|$), then it appears that values $a > 0$ are beneficial with regard to the former aspects only—they help to prevent breakdown by "implosion," caused by inliers. The limiting scale estimate for $\kappa \to 0$ is, essentially, the median absolute deviation $\text{med}\{|\mathbf{x}|\}$, and we have already commented upon its good robustness properties in the one-dimensional case. Also the indeterminacy of (4.12) at $\mathbf{y} = 0$ only affects the directional, but not the scale aspects.

With regard to the directional aspects, a value $u(0) \neq 0$ is distinctly awkward. To give some intuitive insight into what is going on, we note that, for the maximum likelihood estimates $\hat{\mathbf{t}}$ and $\hat{\mathbf{V}}$, the linearly transformed quantities $\mathbf{y} = \hat{V}(\mathbf{x}-\hat{\mathbf{t}})$ possess the following property (cf. Exhibit 8.10.1): if the sample points with $|\mathbf{y}| < a$ and those with $|\mathbf{y}| > b$ are moved radially outward and inward to the spheres $|\mathbf{y}| = a$ and $|\mathbf{y}| = b$, respectively, while the points with $a \leqslant |\mathbf{y}| \leqslant b$ are left where they are, then the sample thus modified has the (ordinary) covariance matrix I.

A value $\mathbf{y}$ very close to the origin clearly does not give any directional information; in fact $\mathbf{y}/|\mathbf{y}|$ changes randomly under small random changes of $\mathbf{t}$. We should therefore refrain from moving points to the sphere with radius a when they are close to the origin, but we should like to retain the scale information contained in them. This can be achieved by letting u decrease to 0 as $r \to 0$, and simultaneously changing v so that the trace of

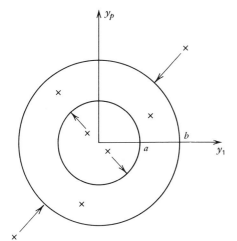

Exhibit 8.10.1   From Huber (1977a), with permission of the publisher.

(4.12) is unchanged. For instance, we might change (10.25) by putting

$$u(r) = \frac{a^2}{r_0} r, \qquad \text{for } r \leqslant r_0 < a \tag{10.29}$$

and

$$v(r) = \left(1 - \frac{a^2}{p}\right) + \frac{a^2}{pr_0} r, \qquad \text{for } r \leqslant r_0$$

$$= 1, \qquad \text{for } r \geqslant r_0. \tag{10.30}$$

Unfortunately, this will destroy the uniqueness proofs of Section 8.6.

It usually is desirable to standardize the scale part of these estimates such that we obtain the correct asymptotic values at normal distributions. This is best done by applying a correction factor τ at the very end, as follows.

Example 10.2   With the u defined in (10.25) we have, for standard normal observations $\mathbf{x}$,

$$E\big[u(\tau|\mathbf{x}|)\big] = a^2 \chi^2\left(p, \frac{a^2}{\tau^2}\right) + b^2\left[1 - \chi^2\left(p, \frac{b^2}{\tau^2}\right)\right]$$

$$+ \tau^2 p\left[\chi^2\left(p+2, \frac{b^2}{\tau^2}\right) - \chi^2\left(p+2, \frac{a^2}{\tau^2}\right)\right], \tag{10.31}$$

			Mass of F_0		
			Below a	Above b	
ε	p	κ	$a=\sqrt{(p-\kappa)^+}$	$b=\sqrt{p+\kappa}$	τ^2
0.01	1	4.1350	0	0.0332	1.0504
	2	5.2573	0	0.0363	1.0305
	3	6.0763	0	0.0380	1.0230
	5	7.3433	0	0.0401	1.0164
	10	9.6307	0.0000	0.0426	1.0105
	20	12.9066	0.0038	0.0440	1.0066
	50	19.7896	0.0133	0.0419	1.0030
	100	27.7370	0.0187	0.0395	1.0016
0.05	1	2.2834	0	0.1165	1.1980
	2	3.0469	0	0.1262	1.1165
	3	3.6045	0	0.1313	1.0873
	5	4.4751	0.0087	0.1367	1.0612
	10	6.2416	0.0454	0.1332	1.0328
	20	8.8237	0.0659	0.1263	1.0166
	50	13.9670	0.0810	0.1185	1.0067
	100	19.7634	0.0877	0.1141	1.0033
0.10	1	1.6086	0	0.1957	1.3812
	2	2.2020	0	0.2101	1.2161
	3	2.6635	0.0445	0.2141	1.1539
	5	3.4835	0.0912	0.2072	1.0908
	10	5.0051	0.1198	0.1965	1.0441
	20	7.1425	0.1352	0.1879	1.0216
	50	11.3576	0.1469	0.1797	1.0086
	100	16.0931	0.1523	0.1754	1.0043
0.25	1	0.8878	0.2135	0.3604	1.9470
	2	1.3748	0.2495	0.3406	1.3598
	3	1.7428	0.2582	0.3311	1.2189
	5	2.3157	0.2657	0.3216	1.1220
	10	3.3484	0.2730	0.3122	1.0577
	20	4.7888	0.2782	0.3059	1.0281
	50	7.6232	0.2829	0.3004	1.0110
	100	10.8052	0.2854	0.2977	1.0055

Exhibit 8.10.2   From Huber (1977a), with permission of the publisher.

where $\chi^2(p, \cdot)$ is the cumulative χ^2-distribution with p degrees of freedom. So we determine τ from $E[u(\tau|\mathbf{x}|)]=p$, and then we multiply the pseudo-covariance $(V^T V)^{-1}$ found from (4.12) by τ^2. Some numerical results are summarized in Exhibit 8.10.2.

Some further remarks are needed on the question of spherical symmetry. First, we should point out that the assumption of spherical symmetry is not needed when minimizing Fisher information. Note that Fisher information is a convex function of f, so by taking averages over the orthogonal group we obtain (by Jensen's inequality)

$$I(\text{ave}\{f\}) \leqslant \text{ave}\{I(f)\},$$

where $\text{ave}\{f\} = \bar{f}$ is a spherically symmetric density. So instead of minimizing $I(f)$ for spherically symmetric f, we might minimize $\text{ave}\{I(f)\}$ for more general f; the minimum will occur at a spherically symmetric f.

Second, we might criticize the approach for being restricted to a framework of elliptic densities (with the exception of Section 8.9).

Such a symmetry assumption is reasonable if we are working with genuinely long-tailed p-variate distributions. But, for instance, in the framework of the gross error model, typical outliers will be generated by a process distinct from that of the main family and hence will have quite a different covariance structure. For example, the main family may consist of a tight and narrow ellipsoid with only a few principal axes significantly different from zero, while there is a diffuse and roughly spherical cloud of outliers. Or it might be the outliers that show a structure and lie along some well-defined lower dimensional subspaces, and so on. Of course, in an affinely invariant framework, the two situations are not really distinguishable.

But we do not seem to have the means to attack such multidimensional separation problems directly, unless we possess some prior information. The estimates developed in Sections 8.4ff. are useful just because they are able to furnish an unprejudiced estimate of the overall shape of the principal part of a pointcloud, from which a more meaningful analysis of its composition might start off.

8.11 SOME NOTES ON COMPUTATION

Unfortunately, so far we have neither a really fast, nor a demonstrably convergent, procedure for calculating simultaneous estimates of location and scatter.

A relatively simple and straightforward approach can be constructed from (4.13) and (4.14):

(1) *Starting values*   For example, let

$$t := \text{ave}\{x\},$$

$$\Sigma := \text{ave}\{(x-t)(x-t)^T\}$$

be the classical estimates. Take the Choleski decomposition $\Sigma = BB^T$, with B lower triangular, and put

$$V := B^{-1}.$$

Then alternate between scatter steps and location steps, as follows.

(2) *Scatter step*   With $y = V(x-t)$ let

$$C := \frac{\text{ave}\{s(|y|)yy^T\}}{\text{ave}\{v(|y|)\}}.$$

Take the Choleski decomposition $C = BB^T$ and put

$$W := B^{-1},$$

$$V := WV.$$

(3) *Location step*   With $y = V(x-t)$ let

$$h := \frac{\text{ave}\{w(|y|)(x-t)\}}{\text{ave}\{w(|y|)\}},$$

$$t := t + h.$$

(4) *Termination rule*   Stop iterating when both $\|W - I\| < \varepsilon$ and $\|Vh\| < \delta$, for some predetermined tolerance levels, for example $\varepsilon = \delta = 10^{-3}$.

Note that this algorithm attempts to improve the numerical properties by avoiding the possibly poorly conditioned matrix $V^T V$.

If either t or V is kept fixed, it is not difficult to show that the algorithm converges under fairly general assumptions. A convergence proof for fixed t is contained in the proof of Lemma 6.1.

For fixed V convergence of the location step can easily be proved if $w(r)$ is monotone decreasing and $w(r)r$ is monotone increasing. Assume for simplicity that $V = I$ and let $\rho(r)$ be an indefinite integral of $w(r)r$. Then

$\rho(|\mathbf{x}-\mathbf{t}|)$ is convex as a function of $\mathbf{t}$, and minimizing $\text{ave}\{\rho(|\mathbf{x}-\mathbf{t}|)\}$ is equivalent to solving (4.11).

As in Section 7.8 we define comparison functions. Let $r_i = |\mathbf{y}_i| = |\mathbf{x}_i - \mathbf{t}^{(m)}|$, where $\mathbf{t}^{(m)}$ is the current trial value and the index i denotes the ith observation. Define the comparison functions u_i such that

$$u_i(r) = a_i + \tfrac{1}{2}b_i r^2,$$

$$u_i(r_i) = \rho(r_i),$$

$$u_i'(r_i) = \rho'(r_i) = w(r_i)r_i.$$

The last condition implies $b_i = w(r_i)$; hence

$$u_i(r) = \rho(r_i) + \tfrac{1}{2}w(r_i)\left(r^2 - r_i^2\right),$$

and, since w is monotone decreasing, we have

$$[u_i(r) - \rho(r)]' = [w(r_i) - w(r)]r \leqslant 0, \qquad \text{for } r \leqslant r_i$$

$$\geqslant 0, \qquad \text{for } r \geqslant r_i.$$

Hence

$$u_i(r) \geqslant \rho(r), \qquad \text{for all } r.$$

Minimizing

$$\text{ave}\{u_i(|\mathbf{x}_i - \mathbf{t}|)\}$$

is equivalent to performing one location step, from $\mathbf{t}^{(m)}$ to $\mathbf{t}^{(m+1)}$; hence $\text{ave}\{\rho(|\mathbf{x}-\mathbf{t}|)\}$ is strictly decreased, unless $\mathbf{t}^{(m)} = \mathbf{t}^{(m+1)}$ already is a solution, and convergence towards the minimum is now easily proved.

Convergence has not been proved yet when $\mathbf{t}$ and V are estimated simultaneously.

The speed of convergence of the location step is satisfactory, but not so that of the more expensive scatter step (most of the work is spent in building up the matrix C).

Some supposedly faster procedures have been proposed by Maronna (1976) and Huber (1977a). The former tried to speed up the scatter step by overrelaxation (in our notation, the Choleski decomposition would be applied to C^2 instead of C, so the step is roughly doubled). The latter proposed using a modified Newton approach instead (with the Hessian matrix replaced by its average over the spheres $|y| = $ const.). But neither of these proposals performed very well in our numerical experiments (Maronna's too often led to oscillatory behavior; Huber's did not really improve the overall speed). A straightforward Newton approach is out of the question because of the high number of variables.

The most successful method so far (with an improvement slightly better than two in overall speed) turned out to be a variant of the conjugate gradient method, using explicit second derivatives. The idea behind it is as follows. Assume that a function $f(\mathbf{z})$, $\mathbf{z} \in \mathbb{R}^n$, is to be minimized, and assume that $\mathbf{z}^{(m)} := \mathbf{z}^{(m-1)} + \mathbf{h}^{(m-1)}$ was the last iteration step. If $\mathbf{g}^{(m)}$ is the gradient of f at $\mathbf{z}^{(m)}$, then approximate the function

$$F(t_1, t_2) = f\left(\mathbf{z}^{(m)} + t_1 \mathbf{g}^{(m)} + t_2 \mathbf{h}^{(m-1)}\right)$$

by a quadratic function $Q(t_1, t_2)$ having the same derivatives up to order two at $t_1 = t_2 = 0$, find the minimum of Q, say at $\hat{t}_1$ and $\hat{t}_2$, and put $\mathbf{h}^{(m)} := \hat{t}_1 \mathbf{g}^{(m)} + \hat{t}_2 \mathbf{h}^{(m-1)}$ and $\mathbf{z}^{(m+1)} := \mathbf{z}^{(m)} + \mathbf{h}^{(m)}$. The first and second derivatives of F should be determined analytically.

If f itself is quadratic, the procedure is algebraically equivalent to the standard descriptions of the conjugate gradient method and reaches the true minimum in n steps (where n is the dimension of $\mathbf{z}$). Its advantage over the more customary versions that determine $\mathbf{h}^{(m)}$ recursively (Fletcher-Powell, etc.) is that it avoids instabilities due to accumulation of errors caused by (1) deviation of f from a quadratic function, and (2) rounding (in essence the usual recursive determination of $\mathbf{h}^{(m)}$ amounts to numerical differentiation).

In our case we start from the maximum likelihood problem (4.2) and assume that we have to minimize

$$Q = -\log(\det V) - \text{ave}\{\log f(|V(\mathbf{x} - \mathbf{t})|).$$

We write $V(\mathbf{x} - \mathbf{t}) = W\mathbf{y}$ with $\mathbf{y} = V_0(\mathbf{x} - \mathbf{t})$; $\mathbf{t}$ and V_0 will correspond to the current trial values. We assume that W is lower triangular and depends linearly on two real parameters s_1 and s_2:

$$W = I + s_1 U_1 + s_2 U_2,$$

where U_1 and U_2 are lower triangular matrices. If

$$Q(W) = -\log(\det W) - \log(\det V_0) - \text{ave}\{\log f(|W\mathbf{y}|)\}$$

is differentiated with respect to a linear parameter in W, we obtain

$$\dot{Q}(W) = -\text{tr}(\dot{W}W^{-1}) + \text{ave}\{s(|W\mathbf{y}|)(W\mathbf{y})^T(\dot{W}\mathbf{y})\},$$

with

$$s(r) = -\frac{f'(r)}{rf(r)}.$$

At $s_1 = s_2 = 0$ this gives

$$\dot{Q}(I) = \text{ave}\{s(|\mathbf{y}|)\mathbf{y}^T\dot{W}\mathbf{y}\} - \text{tr}(\dot{W}),$$

$$\ddot{Q}(I) = \text{ave}\left\{\frac{s'(|\mathbf{y}|)}{|\mathbf{y}|}(\mathbf{y}^T\dot{W}\mathbf{y})(\mathbf{y}^T\dot{W}\mathbf{y}) + s(|\mathbf{y}|)(\dot{W}\mathbf{y})^T(\dot{W}\mathbf{y})\right\} + \text{tr}(\dot{W}\dot{W}).$$

In particular, if we calculate the partial derivatives of Q with respect to the $p(p+1)/2$ elements of W, we obtain from the above that the gradient U_1 can be naturally identified with the lower triangle of

$$U_1 = \text{ave}\{s(|\mathbf{y}|)\mathbf{y}\mathbf{y}^T\} - I.$$

The idea outlined before is now implemented as follows, in such a way that we can always work near the identity matrix and take advantage of the corresponding simpler formulas and better conditioned matrices.

CG-Iteration Step for Scatter

Let **t** and V be the current trial values and write $\mathbf{y} = V(\mathbf{x} - \mathbf{t})$. Let U_1 be lower triangular such that

$$U_1 := \text{ave}\{s(|\mathbf{y}|)\mathbf{y}\mathbf{y}^T\} - I$$

(ignoring the upper triangle of the right-hand side).

In the first iteration step let $j = k = 1$; in all following steps let j and k take the values 1 and 2; let

$$a_{jk} = \text{tr}(U_jU_k) + \text{ave}\left\{\frac{s'(|\mathbf{y}|)}{|\mathbf{y}|}(\mathbf{y}^TU_j\mathbf{y})(\mathbf{y}^TU_k\mathbf{y}) + s(|\mathbf{y}|)(U_j\mathbf{y})^T(U_k\mathbf{y})\right\},$$

$$b_j = -\text{tr}(U_j) + \text{ave}\{s(|\mathbf{y}|)(\mathbf{y}^TU_j\mathbf{y})\}$$

[then $Q(W) \cong Q(I) + \Sigma b_j s_j + \frac{1}{2}\Sigma a_{jk}s_js_k$]. Solve

$$\sum_k a_{jk}s_k + b_j = 0$$

for s_1 and s_2 ($s_2 = 0$ in the first step). Put

$$U_2 := s_1 U_1 + s_2 U_2.$$

Cut U_2 down by a fudge factor if U_2 is too large; for example, let $U_2 := c U_2$, with

$$c = \frac{1}{\max(1, 2d)}$$

where d is the maximal absolute diagonal element of U_2. Put

$$W := I + U_2,$$

$$V := WV.$$

Empirically, with p up to 20 [i.e., up to $p = 20$ parameters for location and $p(p+1)/2 = 210$ parameters for scatter] the procedure showed a smooth convergence down to essentially machine accuracy.

CHAPTER 9

Robustness of Design

9.1 GENERAL REMARKS

We already have encountered two design-related problems. The first was concerned with leverage points (Sections 7.1 and 7.2), the second with subtle questions of bias (Section 7.5). In both cases we had single observations sitting at isolated points in the design space, and the difficulty was, essentially, that these observations were not cross-checkable.

There are many considerations entering into a design. From the point of view of robustness, the most important requirement is to have enough redundancy so that everything can be cross-checked. In this little chapter we give another example of this sort; it illuminates the surprising fact that deviations from linearity that are too small to be detected are already large enough to tip the balance away from the "optimal" designs, which assume exact linearity and put the observations on the extreme points of the observable range, toward the "naive" ones, which distribute the observations more or less evenly over the entire design space (and thus allow us to check for linearity).

One simple example should suffice to illustrate the point; it is taken from Huber (1975). See Sacks and Ylvisaker (1978), as well as Bickel and Herzberg (1979), for interesting further developments.

9.2 MINIMAX GLOBAL FIT

Assume that f is an approximately linear function defined in the interval $I = [-\frac{1}{2}, \frac{1}{2}]$. It should be approximated by a linear function as accurately as possible; we choose mean square error as our measure of disagreement:

$$\int [f(x) - \alpha - \beta x]^2 \, dx. \tag{2.1}$$

All integrals are over the interval I. Clearly, (2.1) is minimized for

$$\alpha_0 = \int f(x)\,dx, \qquad \beta_0 = \frac{\int xf(x)\,dx}{\int x^2\,dx}, \tag{2.2}$$

and the minimum value of (2.1) is denoted by

$$Q_f = \int \left[f(x) - \alpha_0 - \beta_0 x \right]^2 dx. \tag{2.3}$$

Assume now that the values of f are only observable with some measurement errors. Assume that we can observe f at n freely chosen points $x_1, \ldots, x_n$ in the interval I, and that the observed values are

$$y_i = f(x_i) + u_i, \tag{2.4}$$

where the u_i are independent normal $\mathfrak{N}(0, \sigma^2)$.

Our original problem is thus turned into the following: find estimates $\hat{\alpha}$ and $\hat{\beta}$ for the coefficients of a linear function, based on the y_i, such that the expected mean square error

$$Q = E\left\{ \int \left[f(x) - \hat{\alpha} - \hat{\beta} x \right]^2 dx \right\} \tag{2.5}$$

is least possible. Q can be decomposed into a constant part, a bias part, and a variance part:

$$Q = Q_f + Q_b + Q_v, \tag{2.6}$$

where Q_f depends on f alone [see (2.3)], where

$$Q_b = (\alpha_1 - \alpha_0)^2 + \frac{(\beta_1 - \beta_0)^2}{12}, \tag{2.7}$$

with

$$\alpha_1 = E(\hat{\alpha}), \qquad \beta_1 = E(\hat{\beta}), \tag{2.8}$$

and where

$$Q_v = \mathrm{var}(\hat{\alpha}) + \frac{\mathrm{var}(\hat{\beta})}{12}. \tag{2.9}$$

ume now that the response curve f is only approximately linear, say
, where $\eta > 0$ is a small number, and assume that the Statistician
a game against Nature, with loss function $Q(f, \xi)$.

REM 2.1 The game with loss function $Q(f, \xi)$, $f \in \mathcal{F}_\eta = \{f \mid Q_f \leqslant \eta\}$,
saddlepoint (f_0, ξ_0):

$$Q(f, \xi_0) \leqslant Q(f_0, \xi_0) \leqslant Q(f_0, \xi).$$

sign measure ξ_0 has a density of the form $m_0(x) = (ax^2 + b)^+$, and f_0
portional to m_0 (except that an arbitrary linear function can be
to it).

dependence of (f_0, ξ_0) on η can be described in parametric form,
verything depending on the parameter γ. If $\frac{1}{12} \leqslant \gamma \leqslant \frac{3}{20}$, then ξ_0 has
isity

$$m_0(x) = 1 + \tfrac{5}{4}(12\gamma - 1)(12x^2 - 1), \tag{2.16}$$

$$f_0(x) = (12x^2 - 1)\varepsilon, \tag{2.17}$$

$$\varepsilon^2 = \frac{\sigma^2}{n} \cdot \frac{1}{2(12\gamma)^2(12\gamma - 1)}, \tag{2.18}$$

$$\eta = \tfrac{4}{5}\varepsilon^2. \tag{2.19}$$

$\leqslant \gamma < \frac{1}{4}$, the solution is much more complicated, and we better
the parameter to $c \in [0, 1)$, with no direct interpretation of c. Then

$$m_0(x) = \frac{3}{(1 + 2c)(1 - c)^2}(4x^2 - c^2)^+, \tag{2.20}$$

$$\gamma = \frac{3 + 6c + 4c^2 + 2c^3}{20(1 + 2c)}, \tag{2.21}$$

$$f_0(x) = [m_0(x) - 1]\varepsilon, \tag{2.22}$$

$$\varepsilon^2 = \frac{125(1 - c)^3(1 + 2c)^5}{72(3 + 6c + 4c^2 + 2c^3)^2(1 + 3c + 6c^2 + 5c^3)}, \tag{2.23}$$

$$\eta = \frac{25(1 - c)^2(1 + 2c)^3}{18(3 + 6c + 4c^2 + 2c^3)^2}. \tag{2.24}$$

It is convenient to characterize the design by

$$\xi = \frac{1}{n} \sum \delta_{x_i},$$

where δ_x denotes the pointmass 1 at x. We a
measures for ξ [in practice they have to be appr
the form (2.10)].

For the sake of simplicity we consider only
mates

$$\hat{\alpha} = \frac{1}{n} \sum y_i, \qquad \hat{\beta} = \frac{\sum x}{\sum}$$

based on a symmetric design $x_1, \ldots, x_n$. For fix
these are, of course, *the* optimal estimates. Th
designs is inessential and can be removed at
tions; the restriction to linear estimates is r
awkward from a point of view of theoretical p
Then we obtain the following explicit repres

$$Q(f, \xi) = E\left\{ \int \left[f(x) - \hat{\alpha} - \hat{\beta} x \right]^2 dx \right\} =$$

$$= Q_f + \left[(\alpha_1 - \alpha_0)^2 + \frac{(\beta_1 - \beta_0)}{12} \right]$$

with

$$\alpha_1 = E(\hat{\alpha}) \quad = \int f(x) \, d\xi,$$

$$\beta_1 = E(\hat{\beta}) \quad = \frac{\int x f(x) \, d\xi}{\int x^2 \, d\xi},$$

$$\gamma = \int x^2 \, d\xi.$$

If f is exactly linear, then $Q_f = Q_b = 0$,
maximizing γ, that is, by putting all mass of
Note that the uniform design (where ξ has th
to $\gamma = \int x^2 \, dx = \frac{1}{12}$, whereas the "optimal" d
$\gamma = \frac{1}{4}$.

As
$Q_f \leqslant$
plays

THE
has a

The c
is pr
adde

The
with
the d

and

with

and

If
chang

In the limit $\gamma = \frac{1}{4}$, $c = 1$, the solution degenerates and m_0 puts pointmasses $\frac{1}{2}$ at each of the points $\pm \frac{1}{2}$.

Proof We first keep ξ fixed and assume it has a density m. Then $Q(f, \xi)$ is maximized by maximizing the bias term

$$Q_b = (\alpha_1 - \alpha_0)^2 + \frac{(\beta_1 - \beta_0)^2}{12}.$$

Without loss of generality we normalize f such that $\alpha_0 = \beta_0 = 0$. Thus we have to maximize

$$Q_b = \left(\int fm\, dx \right)^2 + \frac{\left(\int xfm\, dx \right)^2}{12\gamma^2} \tag{2.25}$$

under the side conditions

$$\int f\, dx = 0, \tag{2.26}$$

$$\int xf\, dx = 0, \tag{2.27}$$

$$\int f^2\, dx = \eta. \tag{2.28}$$

A standard variational argument now shows that the maximizing f must be of the form

$$f = A \cdot (m - 1) + B \cdot (m - 12\gamma)x \tag{2.29}$$

for some Lagrange multipliers A and B. The multipliers have already been adjusted such that this f satisfies the side conditions (2.26) and (2.27). If we insert f into (2.25) and (2.28), we find that we have to maximize

$$Q_b = A^2 \left[\int (m - 1)^2\, dx \right]^2 + B^2 \frac{\left[\int (m - 12\gamma)^2 x^2\, dx \right]^2}{12\gamma^2} \tag{2.30}$$

under the side condition

$$A^2 \int (m - 1)^2\, dx + B^2 \int (m - 12\gamma)^2 x^2\, dx = \eta. \tag{2.31}$$

This is a linear programming problem (linear in A^2 and B^2), and the maximum is clearly reached on the boundary $A^2 = 0$ or $B^2 = 0$. According as the upper or the lower inequality holds in

$$\int (m-1)^2\, dx \lessgtr \frac{1}{12\gamma^2} \int (m-12\gamma)^2 x^2\, dx, \qquad (2.32)$$

either B or A is zero; it turns out that in all interesting cases the upper inequality applies, so $B = \beta_1 = 0$ (this verification is left to the reader). Thus if we solve for A^2 in (2.31) and insert the solution into (2.30), we obtain an explicit expression for $\sup Q_b$ and hence

$$\sup_f Q(f, \xi) = \eta + \eta \int (m-1)^2\, dx + \frac{\sigma^2}{n}\left(1 + \frac{1}{12\gamma}\right). \qquad (2.33)$$

We now minimize this under the side conditions

$$\int m\, dx = 1, \qquad (2.34)$$

$$\int x^2 m\, dx = \gamma, \qquad (2.35)$$

and obtain that

$$m_0(x) = (ax^2 + b)^+ \qquad (2.36)$$

for some Lagrange multipliers a and b. We verify easily that, for $\frac{1}{12} \leqslant \gamma \leqslant \frac{3}{20}$, both a and b are $\geqslant 0$. For $\frac{3}{20} < \gamma < \frac{1}{4}$ we have $b < 0$. Finally, we minimize over γ, which leads to (2.16) to (2.24). ∎

These results need some interpretation and discussion. First, with any minimax procedure there is the question of whether it is too pessimistic and perhaps safeguards only against some very unlikely contingency. This is not the case here; an approximately quadratic disturbance in f is perhaps the one most likely to occur, so (2.17) makes very good sense. But perhaps f_0 corresponds to such a glaring nonlinearity that nobody in his right mind would want to fit a straight line anyway?

To answer this in an objective fashion, we have to construct a most powerful test for distinguishing f_0 from a straight line.

It is convenient to characterize the design by the design measure

$$\xi = \frac{1}{n} \sum \delta_{x_i}, \tag{2.10}$$

where δ_x denotes the pointmass 1 at x. We allow arbitrary probability measures for ξ [in practice they have to be approximated by a measure of the form (2.10)].

For the sake of simplicity we consider only the traditional linear estimates

$$\hat{\alpha} = \frac{1}{n} \sum y_i, \qquad \hat{\beta} = \frac{\sum x_i y_i}{\sum x_i^2}, \tag{2.11}$$

based on a symmetric design $x_1, \dots, x_n$. For fixed $x_1, \dots, x_n$ and a linear f, these are, of course, *the* optimal estimates. The restriction to symmetric designs is inessential and can be removed at the cost of some complications; the restriction to linear estimates is more serious and certainly awkward from a point of view of theoretical purity.

Then we obtain the following explicit representation of (2.5):

$$Q(f, \xi) = E\left\{ \int \left[f(x) - \hat{\alpha} - \hat{\beta} x \right]^2 dx \right\} = Q_f + Q_b + Q_v$$

$$= Q_f + \left[(\alpha_1 - \alpha_0)^2 + \frac{(\beta_1 - \beta_0)^2}{12} \right] + \frac{\sigma^2}{n}\left(1 + \frac{1}{12\gamma} \right), \tag{2.12}$$

with

$$\alpha_1 = E(\hat{\alpha}) \quad = \int f(x)\, d\xi, \tag{2.13}$$

$$\beta_1 = E(\hat{\beta}) \quad = \frac{\int x f(x)\, d\xi}{\int x^2\, d\xi}, \tag{2.14}$$

$$\gamma = \int x^2\, d\xi. \tag{2.15}$$

If f is exactly linear, then $Q_f = Q_b = 0$, and (2.12) is minimized by maximizing γ, that is, by putting all mass of ξ on the extreme points $\pm\frac{1}{2}$. Note that the uniform design (where ξ has the density $m \equiv 1$) corresponds to $\gamma = \int x^2\, dx = \frac{1}{12}$, whereas the "optimal" design (all mass on $\pm\frac{1}{2}$), has $\gamma = \frac{1}{4}$.

Assume now that the response curve f is only approximately linear, say $Q_f \leqslant \eta$, where $\eta > 0$ is a small number, and assume that the Statistician plays a game against Nature, with loss function $Q(f, \xi)$.

THEOREM 2.1   The game with loss function $Q(f, \xi)$, $f \in \mathcal{F}_\eta = \{f \mid Q_f \leqslant \eta\}$, has a saddlepoint (f_0, ξ_0):

$$Q(f, \xi_0) \leqslant Q(f_0, \xi_0) \leqslant Q(f_0, \xi).$$

The design measure ξ_0 has a density of the form $m_0(x) = (ax^2 + b)^+$, and f_0 is proportional to m_0 (except that an arbitrary linear function can be added to it).

The dependence of (f_0, ξ_0) on η can be described in parametric form, with everything depending on the parameter γ. If $\frac{1}{12} \leqslant \gamma \leqslant \frac{3}{20}$, then ξ_0 has the density

$$m_0(x) = 1 + \tfrac{5}{4}(12\gamma - 1)(12x^2 - 1), \tag{2.16}$$

and

$$f_0(x) = (12x^2 - 1)\varepsilon, \tag{2.17}$$

with

$$\varepsilon^2 = \frac{\sigma^2}{n} \cdot \frac{1}{2(12\gamma)^2(12\gamma - 1)}, \tag{2.18}$$

and

$$\eta = \tfrac{4}{5}\varepsilon^2. \tag{2.19}$$

If $\frac{3}{20} \leqslant \gamma \leqslant \frac{1}{4}$, the solution is much more complicated, and we better change the parameter to $c \in [0, 1)$, with no direct interpretation of c. Then

$$m_0(x) = \frac{3}{(1 + 2c)(1 - c)^2}(4x^2 - c^2)^+, \tag{2.20}$$

$$\gamma = \frac{3 + 6c + 4c^2 + 2c^3}{20(1 + 2c)}, \tag{2.21}$$

$$f_0(x) = [m_0(x) - 1]\varepsilon, \tag{2.22}$$

$$\varepsilon^2 = \frac{125(1 - c)^3(1 + 2c)^5}{72(3 + 6c + 4c^2 + 2c^3)^2(1 + 3c + 6c^2 + 5c^3)}, \tag{2.23}$$

$$\eta = \frac{25(1 - c)^2(1 + 2c)^3}{18(3 + 6c + 4c^2 + 2c^3)^2}. \tag{2.24}$$

In the limit $\gamma = \frac{1}{4}$, $c = 1$, the solution degenerates and m_0 puts pointmasses $\frac{1}{2}$ at each of the points $\pm\frac{1}{2}$.

Proof We first keep ξ fixed and assume it has a density m. Then $Q(f, \xi)$ is maximized by maximizing the bias term

$$Q_b = (\alpha_1 - \alpha_0)^2 + \frac{(\beta_1 - \beta_0)^2}{12}.$$

Without loss of generality we normalize f such that $\alpha_0 = \beta_0 = 0$. Thus we have to maximize

$$Q_b = \left(\int fm \, dx \right)^2 + \frac{\left(\int xfm \, dx \right)^2}{12\gamma^2} \tag{2.25}$$

under the side conditions

$$\int f \, dx = 0, \tag{2.26}$$

$$\int xf \, dx = 0, \tag{2.27}$$

$$\int f^2 \, dx = \eta. \tag{2.28}$$

A standard variational argument now shows that the maximizing f must be of the form

$$f = A \cdot (m - 1) + B \cdot (m - 12\gamma)x \tag{2.29}$$

for some Lagrange multipliers A and B. The multipliers have already been adjusted such that this f satisfies the side conditions (2.26) and (2.27). If we insert f into (2.25) and (2.28), we find that we have to maximize

$$Q_b = A^2 \left[\int (m - 1)^2 \, dx \right]^2 + B^2 \frac{\left[\int (m - 12\gamma)^2 x^2 \, dx \right]^2}{12\gamma^2} \tag{2.30}$$

under the side condition

$$A^2 \int (m - 1)^2 \, dx + B^2 \int (m - 12\gamma)^2 x^2 \, dx = \eta. \tag{2.31}$$

This is a linear programming problem (linear in A^2 and B^2), and the maximum is clearly reached on the boundary $A^2=0$ or $B^2=0$. According as the upper or the lower inequality holds in

$$\int (m-1)^2 \, dx \lesseqgtr \frac{1}{12\gamma^2} \int (m-12\gamma)^2 x^2 \, dx, \tag{2.32}$$

either B or A is zero; it turns out that in all interesting cases the upper inequality applies, so $B=\beta_1=0$ (this verification is left to the reader). Thus if we solve for A^2 in (2.31) and insert the solution into (2.30), we obtain an explicit expression for $\sup Q_b$ and hence

$$\sup_f Q(f, \xi) = \eta + \eta \int (m-1)^2 \, dx + \frac{\sigma^2}{n}\left(1 + \frac{1}{12\gamma}\right). \tag{2.33}$$

We now minimize this under the side conditions

$$\int m \, dx = 1, \tag{2.34}$$

$$\int x^2 m \, dx = \gamma, \tag{2.35}$$

and obtain that

$$m_0(x) = (ax^2 + b)^+ \tag{2.36}$$

for some Lagrange multipliers a and b. We verify easily that, for $\frac{1}{12} \leqslant \gamma \leqslant \frac{3}{20}$, both a and b are $\geqslant 0$. For $\frac{3}{20} < \gamma < \frac{1}{4}$ we have $b<0$. Finally, we minimize over γ, which leads to (2.16) to (2.24). ∎

These results need some interpretation and discussion. First, with any minimax procedure there is the question of whether it is too pessimistic and perhaps safeguards only against some very unlikely contingency. This is not the case here; an approximately quadratic disturbance in f is perhaps the one most likely to occur, so (2.17) makes very good sense. But perhaps f_0 corresponds to such a glaring nonlinearity that nobody in his right mind would want to fit a straight line anyway?

To answer this in an objective fashion, we have to construct a most powerful test for distinguishing f_0 from a straight line.

If ξ is an arbitrary fixed symmetric design, then the most powerful test is based on the test statistic

$$Z = \sum y_i [\, f_0(x_i) - \overline{f}_0 \,],\qquad(2.37)$$

where

$$\overline{f}_0 = \frac{1}{n} \sum f_0(x_i),\qquad(2.38)$$

with f_0 as in (2.17). Under the hypothesis, $E(Z)=0$; $\mathrm{var}(Z)$ is the same under the hypothesis and the alternative. We then obtain the signal-to-noise or variance ratio

$$\frac{(EZ)^2}{\mathrm{var}(Z)} = \frac{1}{\sigma^2} \sum [\, f_0(x_i) - \overline{f}_0 \,]^2 = \frac{n}{\sigma^2} \int [\, f_0(x) - \alpha_1 \,]^2 \, d\xi.\qquad(2.39)$$

Proof (2.37) We test the hypothesis that $f(x) \equiv \overline{f}_0$ against the alternative that $f(x)=f_0(x)$. The most powerful test is given by the Neyman-Pearson lemma; the logarithm of the likelihood ratio $\prod[\, p_1(x_i)/p_0(x_i)\,]$ is

$$-\frac{1}{2} \sum \left[\frac{y_i - f_0(x_i)}{\sigma} \right]^2 + \frac{1}{2} \sum \left(\frac{y_i - \overline{f}_0}{\sigma} \right)^2$$

$$= \frac{1}{\sigma^2} \left\{ \sum y_i [\, f_0(x_i) - \overline{f}_0 \,] - \tfrac{1}{2} \sum [\, f_0(x_i) - \overline{f}_0 \,]^2 \right\}. \quad \blacksquare$$

In particular, the best design for such a test, giving the highest variance ratio, puts one-half of the observations at $x=0$, and one-quarter at each of the endpoints $x= \pm \frac{1}{2}$. The variance ratio is then

$$\frac{(EZ)^2}{\mathrm{var}(Z)} = \frac{9}{4} \frac{n\varepsilon^2}{\sigma^2}.\qquad(2.40)$$

The uniform design $(m \equiv 1)$ gives a variance ratio

$$\frac{(EZ)^2}{\mathrm{var}(Z)} = \frac{4}{5} \frac{n\varepsilon^2}{\sigma^2},\qquad(2.41)$$

and, finally, the minimax design ξ_0 yields

$$\frac{(EZ)^2}{\mathrm{var}(Z)} = \left[\frac{4}{5} + \frac{4}{7}(12\gamma - 1) - (12\gamma - 1)^2 \right] \frac{n\varepsilon^2}{\sigma^2}.\qquad(2.42)$$

Exhibit 9.2.1 gives some numerical values for these variance ratios. Note that: (1) according to (2.18), $n\varepsilon^2/\sigma^2$ is a function of γ alone; and (2) the

γ	$\frac{n\varepsilon^2}{\sigma^2}$	Variance Ratios "Best" (2.40)	"Uniform" (2.41)	"Minimax ξ_0" (2.42)	Quotient (2.42)/(2.41)	$m_0(0)$
0.085	24.029	54.066	19.223	19.488	1.014	0.975
0.090	5.358	12.056	4.287	4.497	1.049	0.900
0.095	2.748	6.183	2.198	2.364	1.076	0.825
0.100	1.736	3.906	1.389	1.518	1.093	0.750
0.105	1.211	2.725	0.969	1.067	1.101	0.675
0.110	0.897	2.018	0.717	0.790	1.101	0.600
0.115	0.691	1.555	0.553	0.603	1.091	0.525
0.120	0.548	1.233	0.438	0.470	1.072	0.450
0.125	0.444	1.000	0.356	0.371	1.045	0.375
0.130	0.367	0.825	0.294	0.296	1.008	0.300
0.135	0.307	0.691	0.246	0.237	0.962	0.225
0.140	0.261	0.586	0.208	0.189	0.908	0.150
0.145	0.223	0.502	0.179	0.151	0.844	0.075
0.150	0.193	0.434	0.154	0.119	0.771	0.

Exhibit 9.2.1 Variance ratios for tests of linearity against a quadratic alternative.

minimax and the uniform design have very similar variance ratios. To give an idea of the shape of the minimax design, its minimal density $m_0(0)$ is also shown.

From this exhibit we can, for instance, infer that, if $\gamma \geq 0.095$ and if we use either the uniform or the minimax design, we are not able to see the nonlinearity of f_0 with any degree of certainty, since the two-sided Neyman -Pearson test with level 10% does not even achieve 50% power (see Exhibit 9.2.2).

To give another illustration let us now take that value of ε for which the uniform design ($m \equiv 1$), minimizing the bias term Q_b, and the "optimal" design, minimizing the variance term Q_v by putting all mass on the extreme points of I, have the same efficiency. As

$$Q(f_0, \text{uni}) = \int f_0^2\, dx + 2\frac{\sigma^2}{n}, \tag{2.43}$$

$$Q(f_0, \text{opt}) = \int f_0^2\, dx + (2\varepsilon)^2 + \frac{4}{3}\frac{\sigma^2}{n}, \tag{2.44}$$

we obtain equality for

$$\varepsilon^2 = \frac{1}{6}\frac{\sigma^2}{n}, \tag{2.45}$$

and the variance ratio (2.41) then is

$$\frac{(EZ)^2}{\text{var}(Z)} = \frac{2}{15}. \tag{2.46}$$

			Variance Ratio			
Level α	1.0	2.0	3.0	4.0	5.0	6.0	9.0
0.01	0.058	0.123	0.199	0.282	0.367	0.450	0.664
0.02	0.093	0.181	0.276	0.372	0.464	0.549	0.750
0.05	0.170	0.293	0.410	0.516	0.609	0.688	0.851
0.10	0.264	0.410	0.535	0.639	0.723	0.790	0.912
0.20	0.400	0.556	0.675	0.764	0.830	0.879	0.957

Exhibit 9.2.2 Power of two-sided tests, in function of the level and the variance ratio.

A variance ratio of about 4 is needed to obtain approximate power 50% with a 5% test (see Exhibit 9.2.2). Hence (2.46) can be interpreted as follows. Even if the pooled evidence of up to 30 experiments similar to the one under consideration suggests that f_0 is linear, the uniform design may still be better than the "optimal" one and may lead to a smaller expected mean square error!

9.3 MINIMAX SLOPE

Conceivably, the situation might be different when we are only interested in estimating the slope β. The expected square error in this case is

$$Q(f, \xi) = E(\hat{\beta} - \beta_0)^2 = (\beta_1 - \beta_0)^2 + \text{var}(\hat{\beta})$$

$$= \left[\frac{\int xf(x)\,dx}{\gamma} \right]^2 + \frac{1}{\gamma} \frac{\sigma^2}{n}, \tag{3.1}$$

if we standardize f such that $\alpha_0 = \beta_0 = 0$ (using the notation of the preceding section).

The game with loss function (3.1) is easy to solve by variational methods similar to those used in the preceding section. For the Statistician the minimax design ξ_0 has density

$$m_0(x) = \frac{1}{(1-2a)^2} \left(1 - \frac{a^2}{x^2} \right)^+, \tag{3.2}$$

for some $0 \leqslant a < \frac{1}{2}$, and for Nature the minimax strategy is

$$f_0(x) \sim [m_0(x) - 12\gamma] x. \tag{3.3}$$

We do not work out the details, but we note that f_0 is crudely similar to a cubic function.

For the following heuristics we therefore use a more manageable, and perhaps even more realistic, cubic f:

$$f(x) = (20x^3 - 3x)\varepsilon. \tag{3.4}$$

This f satisfies $\int f\,dx = \int xf\,dx = 0$ and

$$\int f(x)^2\,dx = \tfrac{1}{7}\varepsilon^2. \tag{3.5}$$

We now repeat the argumentation used in the last paragraphs of Section 9.2.

How large should ε be in order that the uniform design and the "optimal" design are equally efficient in terms of the risk function (3.1)? As

$$Q(f, \text{uni}) = 12\frac{\sigma^2}{n}, \tag{3.6}$$

$$Q(f, \text{opt}) = (4\varepsilon)^2 + 4\frac{\sigma^2}{n}, \tag{3.7}$$

we obtain equality if

$$\varepsilon^2 = \frac{\sigma^2}{2n}. \tag{3.8}$$

The most powerful test between a linear f and (3.4) has the variance ratio

$$\frac{(EZ)^2}{\text{var}(Z)} = \frac{1}{7}\frac{n\varepsilon^2}{\sigma^2}. \tag{3.9}$$

If we insert (3.8) this becomes equal to $\frac{1}{14}$. Thus the situation is even worse than at the end of Section 9.2: even if the pooled evidence of up to 50 experiments similar to the one under consideration suggests that f_0 is linear, the uniform design (which minimizes bias for a not necessarily linear f) may still be better than the "optimal" design (which minimizes variance, assuming that f is exactly linear)!

We conclude from these examples that the so-called optimum design theory (minimizing variance, assuming that the model is exactly correct) is meaningless in a robustness context; we should try rather to minimize bias, assuming that the model is only approximately correct. This had already been recognized by Box and Draper (1959), p. 622: "The optimal design in typical situations in which both variance and bias occur is very nearly the same as would be obtained if *variance were ignored completely* and the experiment designed so as to *minimize bias alone*."

CHAPTER 10

Exact Finite Sample Results

10.1 GENERAL REMARKS

Assume that our data contain 1% gross errors. Then it makes a tremendous conceptual difference whether the sample size is 1000 or 5. In the former case each sample will contain around 10 grossly erroneous values, while in the latter, 19 out of 20 samples are good. In particular, it is not at all clear whether conclusions derived from an asymptotic theory remain valid for small samples. Many people are willing to take a 5% risk (remember the customary levels of statistical tests and confidence intervals!), and possibly, if we are applying a nonrobust optimal procedure, the gains on the good samples might more than offset the losses caused by an occasional bad sample, especially if we are using a realistic (i.e., bounded) loss function.

The main purpose of this chapter is to show that this is not so. We shall find exact, finite sample minimax estimates of location, which, surprisingly, have the same structure as the asymptotically minimax *M*-estimates found in Chapter 4, and they are even quantitatively comparable.

These estimates are derived from minimax robust tests, and thus we have to develop a theory of robust tests.

We begin with a discussion of the structure of some of the neighborhoods used to describe approximately specified probabilities; the goal would be to ultimately develop a kind of interval arithmetics for probability measures (e.g., in Bayesian framework, how we step from an approximate prior to an approximate posterior distribution). It appears that alternating capacities of order two, and occasionally of infinite order, are the appropriate tools in these contexts.

If (and essentially only if) the inaccuracies can be formulated in terms of alternating capacities of order two, the minimax tests have a very simple structure.

10.2   LOWER AND UPPER PROBABILITIES AND CAPACITIES

Let $\mathfrak{M}$ be the set of all probability measures on some measurable space $(\Omega, \mathcal{C})$. We single out four classes of subsets $\mathcal{P} \subset \mathfrak{M}$: those representable through (1) upper expectations, (2) upper probabilities, (3) alternating capacities of order two, (4) alternating capacities of infinite order. Each class contains the following one.

Formally, our treatment is restricted to *finite sets* Ω, even though all the concepts and a majority of the results are valid for much more general spaces. But if we consider the more general spaces, the important conceptual aspects are buried under a mass of technical complications of a measure theoretic and topological nature.

Let $\mathcal{P} \subset \mathfrak{M}$ be an arbitrary nonempty subset. We define the *lower* and the *upper expectation* induced by $\mathcal{P}$ as

$$E_*(X) = \inf_{\mathcal{P}} \int X \, dP, \qquad E^*(X) = \sup_{\mathcal{P}} \int X \, dP, \tag{2.1}$$

and similarly, the *lower* and the *upper probability* induced by $\mathcal{P}$ as

$$v_*(A) = \inf_{\mathcal{P}} P(A), \qquad v^*(A) = \sup_{\mathcal{P}} P(A). \tag{2.2}$$

E_* and E^* are nonlinear functionals conjugate to each other in the sense that

$$E_*(X) = -E^*(-X) \tag{2.3}$$

and

$$v_*(A) = 1 - v^*(A^c). \tag{2.4}$$

Conversely, we may start with an arbitrary pair of conjugate functionals (E_*, E^*) or set functions (v_*, v^*) satisfying (2.3) or (2.4), respectively, and define sets $\mathcal{P}$ by

$$\mathcal{P} = \left\{ P \in \mathfrak{M} \,\middle|\, \int X \, dP \geqslant E_*(X) \quad \text{for all } X \right\}$$

$$= \left\{ P \in \mathfrak{M} \,\middle|\, \int X \, dP \leqslant E^*(X) \quad \text{for all } X \right\} \tag{2.5}$$

or

$$\mathcal{P} = \{P \in \mathfrak{M} \mid P(A) \geqslant v_*(A) \quad \text{for all } A\}$$

$$= \{P \in \mathfrak{M} \mid P(A) \leqslant v^*(A) \quad \text{for all } A\}, \tag{2.6}$$

respectively.

We note that (2.1), followed by (2.5), does not in general restore $\mathcal{P}$; nor does (2.5), followed by (2.1), restore (E_*, E^*). But from the second round on, matters stabilize. We say that $\mathcal{P}$ and (E_*, E^*) *represent* each other if they mutually induce each other through (2.1) and (2.5).

Similarly, we say that $\mathcal{P}$ and (v_*, v^*) *represent* each other if they mutually induce each other through (2.2) and (2.6).

Obviously, it suffices to look at one member of the respective pairs (E_*, E^*) and (v_*, v^*), say E^* and v^*.

These notions immediately provoke a few questions:

(1) What conditions must (E_*, E^*) satisfy so that it is representable by some $\mathcal{P}$? What conditions must $\mathcal{P}$ satisfy so that it is representable by some (E_*, E^*)?

(2) What conditions must (v_*, v^*) satisfy so that it is representable by some $\mathcal{P}$? What conditions must $\mathcal{P}$ satisfy so that it is representable by some (v_*, v^*)?

The answer to (1) is very simple. We first note that every representable $\mathcal{P}$ is closed and convex (since we are working with finite sets Ω, $\mathcal{P}$ can be identified with a subset of the simplex $\{(p_1, \ldots, p_n) \mid \sum p_i = 1, p_i \geqslant 0\}$, so there is a unique natural topology). On the other hand every representable E^* is monotone,

$$X \leqslant Y \Rightarrow E^*(X) \leqslant E^*(Y), \tag{2.7}$$

positively affinely homogeneous,

$$E^*(aX + b) = aE^*(X) + b, \qquad a, b \in \mathbf{R}, \qquad a \geqslant 0, \tag{2.8}$$

and subadditive,

$$E^*(X + Y) \leqslant E^*(X) + E^*(Y). \tag{2.9}$$

E_* satisfies the same conditions (2.7) and (2.8), but is superadditive,

$$E_*(X + Y) \geqslant E_*(X) + E_*(Y). \tag{2.10}$$

PROPOSITION 2.1 $\mathscr{P}$ is representable by an upper expectation E^* iff it is closed and convex. Conversely, (2.7), (2.8) and (2.9) are necessary and sufficient for representability of E^*.

Proof   Assume that $\mathscr{P}$ is convex and closed, and define E^* by (2.1). E^* represents $\mathscr{P}$ if we can show that, for every $Q \notin \mathscr{P}$, there is an X and a real number c such that, for all $P \in \mathscr{P}$, $\int X \, dP \leqslant c < \int X \, dQ$; their existence is in fact guaranteed by one of the well-known separation theorems for convex sets.

Now assume that E^* is monotone, positively affinely homogeneous, and subadditive. It suffices to show that for every X_0 there is a probability measure P such that, for all X, $\int X \, dP \leqslant E^*(X)$, and $\int X_0 \, dP = E^*(X_0)$. Because of (2.8) we can assume without any loss of generality that $E^*(X_0) = 1$. Let $U = \{X \mid E^*(X) < 1\}$. It follows from (2.7) and (2.8) that U is open: with X it also contains all Y such that $Y < X + \varepsilon$, for $\varepsilon = 1 - E^*(X)$. Moreover, (2.9) implies that U is convex. Since $X_0 \notin U$, there is a linear functional λ separating X_0 from U:

$$\lambda(X) < \lambda(X_0), \qquad \text{for all } X \in U. \tag{2.11}$$

With $X = 0$ this implies in particular that $\lambda(X_0)$ is strictly positive, and we may normalize λ such that $\lambda(X_0) = 1 = E^*(X_0)$. Thus we may write (2.11) as

$$E^*(X) < 1 \Rightarrow \lambda(X) < 1. \tag{2.12}$$

In view of (2.7) and (2.8), we have

$$X \leqslant 0 \Rightarrow E^*(X) \leqslant E^*(0) = 0;$$

hence (2.12) implies that, for all $c > 0$, $X \geqslant 0$, we have

$$c\lambda(X) = -\lambda(-cX) > -1;$$

thus $\lambda(X) \geqslant -1/c$. Hence λ is a positive functional. Moreover, we claim that $\lambda(1) = 1$. First, it follows from (2.12) that $\lambda(c) < 1$ for $c < 1$; hence $\lambda(1) \leqslant 1$. On the other hand with $c > 1$ we have $E^*(2X_0 - c) = 2 - c < 1$; hence $\lambda(2X_0 - c) = 2 - c\lambda(1) < 1$, or $\lambda(1) > 1/c$ for all $c > 1$; hence $\lambda(1) = 1$. It follows now from (2.8) and (2.12) that, for all c,

$$E^*(X) < c \Rightarrow \lambda(X) < c;$$

hence $\lambda(X) \leqslant E^*(X)$ for all X, and the probability measure $P(A) = \lambda(1_A)$ is the one we are looking for.   ■

Question (2) is trickier. We note first that every representable (v_*, v^*) will satisfy

$$v_*(\phi) = v^*(\phi) = 0, \qquad v_*(\Omega) = v^*(\Omega) = 1, \tag{2.13}$$

$$A \subset B \quad \Rightarrow \quad v_*(A) \leq v_*(B), \quad v^*(A) \leq v^*(B), \tag{2.14}$$

$$v_*(a \cup B) \geq v_*(A) + v_*(B), \qquad \text{for } A \cap B = \phi, \tag{2.15}$$

$$v^*(A \cup B) \leq v^*(A) + v^*(B). \tag{2.16}$$

But these conditions are not sufficient for (v_*, v^*) to be representable, as the following counterexample shows.

Example 2.1 Let Ω have cardinality $|\Omega| = 4$, and assume that $v_*(A)$ and $v^*(A)$ depend only on the cardinality of A, according to the following table:

$\lvert A \rvert$	0	1	2	3	4
v_*	0	0	$\frac{1}{2}$	$\frac{1}{2}$	1
v^*	0	$\frac{1}{2}$	$\frac{1}{2}$	1	1

Then (v_*, v^*) satisfies the above necessary conditions, but there is only a single additive set function between v_* and v^*, namely $P(A) = |A|/4$; hence (v_*, v^*) is not representable.

Let $\mathcal{D}$ be any collection of subsets of Ω, and let $v_*: \mathcal{D} \to \mathbb{R}_+$ be an arbitrary nonnegative set function. Let

$$\mathcal{P} = \{ P \in \mathfrak{M} \mid P(A) \geq v_*(A) \quad \text{for all } A \in \mathcal{D} \}. \tag{2.17}$$

Dually, $\mathcal{P}$ can also be characterized as

$$\mathcal{P} = \{ P \in \mathfrak{M} \mid P(B) \leq v^*(B) \quad \text{for all } B \text{ with } B^c \in \mathcal{D} \}, \tag{2.18}$$

where $v^*(B) = 1 - v_*(B^c)$.

LEMMA 2.2 The set $\mathcal{P}$ of (2.17) is not empty iff the following condition holds: whenever

$$\sum a_i 1_{A_i} \leq 1, \qquad a_i \geq 0, A_i \in \mathcal{D},$$

then

$$\sum a_i v_*(A_i) \leq 1.$$

Proof   The necessity of the condition is obvious. The sufficiency follows from the next lemma.   ■

We define functionals

$$E_*(X) = \sup\left\{ \sum a_i v_*(A_i) - a \,\middle|\, \sum a_i 1_{A_i} - a \leqslant X, \, a_i \geqslant 0, \, A_i \in \mathcal{D} \right\} \quad (2.19)$$

and $E^*(X) = -E_*(-X)$, or

$$E^*(X) = \inf\left\{ \sum b_i v^*(B_i) - b \,\middle|\, \sum b_i 1_{B_i} - b \geqslant X, \, b_i \geqslant 0, \, B_i^c \in \mathcal{D} \right\}. \quad (2.20)$$

Put

$$v_{*0}(A) = E_*(1_A), \qquad \text{for } A \subset \Omega,$$
$$v^{*0}(A) = E^*(1_A), \qquad \text{for } A \subset \Omega. \quad (2.21)$$

Clearly, $v_* \leqslant v_{*0}$ and $v^{*0} \leqslant v^*$; we verify easily that we obtain the same functionals E_* and E^* if we replace v_* and v^* by v_{*0} and v^{*0} and $\mathcal{D}$ by 2^{Ω} in (2.19) and (2.20).

LEMMA 2.3   Let $\mathcal{P}$ be given by (2.17). If $\mathcal{P}$ is empty, then $E_*(X) = \infty$ and $E^*(X) = -\infty$ identically for all X. Otherwise E_* and E^* coincide with the lower/upper expectations (2.1) defined by $\mathcal{P}$, and v_{*0} and v^{*0} with the lower/upper probabilities (2.2).

Proof   We note first that $E_*(X) \geqslant 0$ if $X \geqslant 0$, and that either $E_*(0) = 0$, or else $E_*(X) = \infty$ for all X. In the latter case $\mathcal{P}$ is empty (this follows from the necessity part of Lemma 2.2 which has already been proved). In the former case we verify easily that E_* (E^*) is monotone, positively affinely homogeneous, and superadditive (subadditive, respectively). The definitions imply at once that $\mathcal{P}$ is contained in the nonempty set $\tilde{\mathcal{P}}$ induced by (E_*, E^*):

$$\mathcal{P} \subset \tilde{\mathcal{P}} = \left\{ P \in \mathcal{M} \,\middle|\, E_*(X) \leqslant \int X \, dP \leqslant E^*(X) \text{ for all } X \right\}. \quad (2.22)$$

But on the other hand it follows from $v_*(A) \leqslant v_{*0}(A)$ and $v^{*0}(A) \leqslant v^*(A)$ that $\mathcal{P} \supset \tilde{\mathcal{P}}$; hence $\mathcal{P} = \tilde{\mathcal{P}}$. The assertion of the lemma follows.   ■

The sufficiency of the condition in Lemma 2.2 follows at once from the remark that it is equivalent to $E_*(0) \leqslant 0$.

PROPOSITION 2.4 (Wolf 1977) A set function v^* on $\mathcal{D} = 2^\Omega$ is representable by some $\mathcal{P}$ iff it has the following property: whenever

$$1_A \leqslant \sum a_i 1_{A_i} - a, \quad \text{with } a_i \geqslant 0, \tag{2.23}$$

then

$$v^*(A) \leqslant \sum a_i v^*(A_i) - a. \tag{2.24}$$

The following weaker set of conditions is in fact sufficient: v^* is monotone $v^*(\phi) = 0$, $v^*(\Omega) = 1$, and (2.24) holds for all decompositions

$$1_A = \sum a_i 1_{A_i} \tag{2.25}$$

where $a_i > 0$ when $A_i \neq \Omega$, and where $(1_{A_1}, \ldots, 1_{A_k})$ is linearly independent.

Proof If $\mathcal{D} = 2^\Omega$, then $v^* = v^{*0}$ is a necessary and sufficient condition for v^* to be representable; this follows immediately from Lemma 2.3. If we spell this out, we obtain (2.23) and (2.24). As (2.23) involves an uncountable infinity of conditions, it is not easy to verify; in the second version (2.25) the number of conditions is still uncomfortably large, but finite [the a_i are uniquely determined if the system $(1_{A_1}, \ldots, 1_{A_k})$ is linearly independent].

To prove the sufficiency of the second set of conditions, assume to the contrary that (2.24) holds for all decompositions (2.25), but fails for some (2.23). We may assume that we have equality in (2.23)—if not, we can achieve it by decreasing some a_i or A_i, or increasing a, on the right-hand side of (2.23). We thus can write (2.23) in the form (2.25), but $(1_{A_1}, \ldots, 1_{A_k})$ then must be linearly dependent. Let k be least possible; then all $a_i \neq 0$, $A_i \neq \phi$, and $a_i > 0$ if $A_i \neq \Omega$. Assume that $\sum c_i 1_{A_i} = 0$, not all $c_i = 0$; then $1_A = \sum (a_i + \lambda c_i) A_i$, for all λ. Let $[\lambda_0, \lambda_1]$ be the interval of λ-values for which $a_i + \lambda c_i \geqslant 0$ for all $A_i \neq \Omega$; clearly, it contains 0 in its interior. Evidently $\sum (a_i + \lambda c_i) v^*(A_i)$ is a linear function of λ, and thus reaches its minimum at one of the endpoints λ_0 or λ_1. There, (2.24) is also violated, but k is decreased by at least one. But k was minimal, which leads to a contradiction. ∎

This proposition gives at least a partial answer to question (2). Note that, in general, several distinct closed convex sets $\mathcal{P}$ induce the same v_* and v^*. The set given by (2.6) is the largest among them. Correspondingly, there will be several upper expectations E^* inducing v^* through $v^*(A) = E^*(1_A)$; (2.20) is the largest one of them, and (2.19) is the smallest lower expectation inducing v_*.

For a given v_* and v^*, there is no simple way to construct the corresponding (extremal) pair E_* and E^*; we can do it either through (2.6) and (2.1) or through (2.19) and (2.20), but either way some awkward suprema and infima are involved.

2-Monotone and 2-Alternating Capacities

The situation is simplified if v_* and v^* are a monotone capacity of order two and an alternating capacity of order two, respectively (or short: 2-*monotone*, 2-*alternating*), that is, if v_* and v^*, apart from the obvious conditions

$$v_*(\phi)=v^*(\phi)=0, \qquad v_*(\Omega)=v^*(\Omega)=1, \tag{2.26}$$

$$A \subset B \quad \Rightarrow \quad v_*(A) \leqslant v_*(B), \quad v^*(A) \leqslant v^*(B), \tag{2.27}$$

satisfy

$$v_*(A \cup B)+v_*(A \cap B) \geqslant v_*(A)+v_*(B), \tag{2.28}$$

$$v^*(A \cup B)+v^*(A \cap B) \leqslant v^*(A)+v^*(B). \tag{2.29}$$

This seemingly slight strengthening of the assumptions (2.13) to (2.16) has dramatic effects.

Assume v^* satisfies (2.26) and (2.27), and define a functional E^* through

$$E^*(X)=\int_0^\infty v^*\{X>t\}\,dt, \qquad \text{for } X \geqslant 0. \tag{2.30}$$

Then E^* is monotone and positively affinely homogeneous, as we verify easily; with the help of (2.8) it can be extended to all X. [Note that, if the construction (2.30) is applied to a probability measure, we obtain the expectation:

$$\int_0^\infty P\{X>t\}\,dt=\int X\,dP, \qquad \text{for } X \geqslant 0.]$$

Similarly, define E_*, with v_* in place of v^*.

PROPOSITION 2.5 The functional E^*, defined by (2.30), is subadditive iff v^* satisfies (2.29). [Similarly, E_* is superadditive iff v_* satisfies (2.28)].

Proof Assume that E^* is subadditive, then

$$E^*(1_A+1_B)=v^*(A \cup B)+v^*(A \cap B),$$

and

$$E^*(1_A) + E^*(1_B) = v^*(A) + v^*(B).$$

Hence if E^* is subadditive, (2.29) holds. The other direction is more difficult to establish. We first note that (2.29) is equivalent to

$$E^*(X \vee Y) + E^*(X \wedge Y) \leqslant E^*(X) + E^*(Y), \qquad \text{for } X, Y \geqslant 0, \quad (2.31)$$

where $X \vee Y$ and $X \wedge Y$ stand for the pointwise supremum and infimum of the two functions X and Y. This follows at once from

$$\{X > t\} \cup \{Y > t\} = \{X \vee Y > t\},$$

$$\{X > t\} \cap \{Y > t\} = \{X \wedge Y > t\}.$$

Since Ω is a finite set, X is a vector $\mathbf{x} = (x_1, \ldots, x_n)$, and E^* is a function of n real variables. The proposition now follows from the following lemma. ∎

LEMMA 2.6 (Choquet) If f is a positively homogeneous function on $\mathbb{R}^n_+$

$$f(c\mathbf{x}) = cf(\mathbf{x}), \qquad \text{for } c \geqslant 0, \qquad (2.32)$$

satisfying

$$f(\mathbf{x} \vee \mathbf{y}) + f(\mathbf{x} \wedge \mathbf{y}) \leqslant f(\mathbf{x}) + f(\mathbf{y}), \qquad (2.33)$$

then f is subadditive:

$$f(\mathbf{x} + \mathbf{y}) \leqslant f(\mathbf{x}) + f(\mathbf{y}). \qquad (2.34)$$

Proof Assume that f is twice continuously differentiable for $\mathbf{x} \neq 0$. Let

$$\mathbf{a} = (x_1 + h_1, x_2, \ldots, x_n),$$

$$\mathbf{b} = (x_1, x_2 + h_2, \ldots, x_n + h_n),$$

with $h_i \geqslant 0$; then $\mathbf{a} \vee \mathbf{b} = \mathbf{x} + \mathbf{h}$, $\mathbf{a} \wedge \mathbf{b} = \mathbf{x}$. If we expand (2.33) into a power series in the h_i, we find that the second order terms must satisfy

$$\sum_{j \neq 1} f_{x_1 x_j} h_1 h_j \leqslant 0;$$

hence

$$f_{x_1 x_j} \leqslant 0, \qquad \text{for} \quad j \neq 1,$$

and more generally

$$f_{x_i x_j} \leqslant 0, \qquad \text{for} \quad i \neq j.$$

Differentiate (2.32) with respect to x_j:

$$cf_{x_j}(c\mathbf{x}) = cf_{x_j}(\mathbf{x}),$$

divide by c, and then differentiate with respect to c:

$$\sum_i x_i f_{x_i x_j} = 0.$$

If F denotes the sum of the second order terms in the Taylor expansion of f at $\mathbf{x}$, we obtain thus

$$2F = - \sum_{i \neq j} x_i x_j f_{x_i x_j} \left(\frac{dx_i}{x_i} - \frac{dx_j}{x_j} \right)^2 \geqslant 0.$$

It follows that f is convex, and because of (2.32), this is equivalent with being subadditive.

If f is not twice continuously differentiable, we must approximate it in a suitable fashion. ∎

In view of Proposition 2.1, we thus obtain that E^* is the upper expectation induced by the set

$$\mathcal{P} = \left\{ P \in \mathfrak{M} \,\Big|\, \int X \, dP \leqslant E^*(X) \text{ for all } X \right\}$$

$$= \{ P \in \mathfrak{M} \,|\, P(A) \leqslant v^*(A) \text{ for all } A \}.$$

Hence every 2-alternating v^* is representable, and the corresponding maximal upper expectation is given by (2.30). In particular, (2.30) implies that, for any monotone sequence $A_1 \subset A_2 \subset \cdots \subset A_k$, it is possible to find a probability $Q \leqslant v^*$ such that, for all i, simultaneously $Q(A_i) = v^*(A_i)$.

Monotone and Alternating Capacities of Infinite Order

Consider the following generalized gross error model: let $(\Omega', \mathcal{C}', P')$ be some probability space, assign to each $\omega' \in \Omega'$ a nonempty subset $T(\omega') \subset \Omega$, and put

$$v_*(A) = P'\{\omega' | T(\omega') \subset A\}, \tag{2.35}$$

$$v^*(A) = P'\{\omega' | T(\omega') \cap A \neq \phi\}. \tag{2.36}$$

We can easily check that v_* and v^* are conjugate set functions. The interpretation is that, instead of the ideal but unobservable outcome ω' of the random experiment, the statistician is shown an arbitrary (not necessarily randomly chosen) element of $T(\omega')$. Clearly, $v_*(A)$ and $v^*(A)$ are lower and upper bounds for the probability that the statistician is shown an element of A.

It is intuitively clear that v_* and v^* are representable; it is easy to check that they are 2-monotone and 2-alternating, respectively. In fact a much stronger statement is true: they are monotone (alternating) of infinite order. We do not define this notion here, but refer the reader to Choquet's fundamental paper (1953/54); by a theorem of Choquet, a capacity is monotone/alternating of infinite order iff it can be generated in the forms (2.35) and (2.36), respectively.

Example 2.2 Let Y and U be two independent real random variables; the first has the idealized distribution P_0, and the second takes two values $\delta \geqslant 0$ and $+\infty$ with probability $1 - \varepsilon$ and ε, respectively. Let T be the interval-valued set function defined by

$$T(\omega') = [Y(\omega') - U(\omega'), Y(\omega') + U(\omega')].$$

Then with probability $\geqslant 1 - \varepsilon$, the statistician is shown a value x that is accurate within δ, that is, $|x - Y(\omega')| \leqslant \delta$, and with probability $\leqslant \varepsilon$, he is shown a value containing a gross error.

The generalized gross error model, using monotone and alternating set functions of infinite order, was introduced by Strassen (1964). There has been a considerable literature on set valued stochastic processes $T(\omega')$ in recent years; in particular, see Harding and Kendall (1974) and Matheron (1975). In a statistical context monotone/alternating capacities of infinite order were used by Dempster (1968) and Shafer (1976). The following example shows another application of such capacities [taken from Huber (1973b)].

Example 2.3 Let α_0 be a probability distribution (the idealized prior) on a finite parameter space Θ. The gross error or ε-contamination model

$$\mathcal{P} = \{\alpha | \alpha = (1 - \varepsilon)\alpha_0 + \varepsilon\alpha_1, \alpha_1 \in \mathfrak{M}\}$$

can be described by an alternating capacity of infinite order, namely,

$$\sup_{\alpha \in \mathscr{P}} \alpha(A) = (1-\varepsilon)\alpha_0(A) + \varepsilon, \qquad \text{for } A \neq \phi$$

$$= 0, \qquad \text{for } A = \phi.$$

Let $p(x|\theta)$ be the conditional probability of observing x, given that θ is true; $p(x|\theta)$ is assumed to be accurately known. Let

$$\beta(\theta|x) = \frac{p(x|\theta)\alpha(\theta)}{\sum_{\theta} p(x|\theta)\alpha(\theta)}$$

be the posterior distribution of θ, given that x has been observed; let $\beta_0(\theta|x)$ be the posterior calculated with the prior α_0.

The inaccuracy in the prior is transmitted to the posterior:

$$v^*(A) = \sup_{\alpha \in \mathscr{P}} \beta(A|x) = \frac{\beta_0(A|x) + s(A)}{1 + s(A)},$$

$$v_*(A) = \inf_{\alpha \in \mathscr{P}} \beta(A|x) = \frac{\beta_0(A|x)}{1 + s(A^c)},$$

where

$$s(A) = \frac{\varepsilon}{1-\varepsilon} \frac{\sup_{\theta \in A} p(x|\theta)}{\sum_{\theta} p(x|\theta)\alpha_0(\theta)}, \qquad \text{for } A \neq \phi$$

$$= 0, \qquad \text{for } A = \phi.$$

Then s satisfies $s(A \cup B) = \max(s(A), s(B))$ and is alternating of infinite order. I do not know the exact order of v^* (it is at least 2-alternating).

10.3 ROBUST TESTS

The classical probability ratio test between two simple hypotheses P_0 and P_1 is not robust: a single factor $p_1(x_i)/p_0(x_i)$, equal or almost equal to 0 or ∞, may upset the test statistic $\prod_1^n p_1(x_i)/p_0(x_i)$. This danger can be averted by censoring the single factors, that is by replacing the test statistic by $\prod_1^n \pi(x_i)$, where $\pi(x_i) = \max\{c', \min[c'', p_1(x_i)/p_0(x_i)]\}$, with $0 < c' < c'' < \infty$.

Somewhat surprisingly, it turns out that this test possesses exact finite sample minimax properties for a wide variety of models: tests of the above structure are minimax for testing between composite hypotheses $\mathcal{P}_0$ and $\mathcal{P}_1$, where $\mathcal{P}_j$ is a neighborhood of P_j in ε-contamination, or total variation, and so on.

In principle P_0 and P_1 can be arbitrary probability measures on arbitrary measurable spaces [cf. Huber (1965)]. But in order to prepare the ground for Section 10.5, from now on we assume that they are probability distributions on the real line. In fact very little generality is lost this way, since almost everything admits a reinterpretation in terms of the real random variable $p_1(X)/p_0(X)$, under various distributions of X.

Let P_0 and P_1, $P_0 \neq P_1$, be two probability measures on the real line. Let p_0 and p_1 be their densities with respect to some measure μ (e.g., $\mu = P_0 + P_1$), and assume that the likelihood ratio $p_1(x)/p_0(x)$ is almost surely (with respect to μ) equal to a monotone function $c(x)$.

Let $\mathfrak{M}$ be the set of all probability measures on the real line, let $0 \leq \varepsilon_0, \varepsilon_1, \delta_0, \delta_1 < 1$ be some given numbers, and let

$$\mathcal{P}_0 = \{Q \in \mathfrak{M} \,|\, Q\{X < x\} \geq (1 - \varepsilon_0)P_0\{X < x\} - \delta_0 \text{ for all } x\},$$

$$\mathcal{P}_1 = \{Q \in \mathfrak{M} \,|\, Q\{X > x\} \geq (1 - \varepsilon_1)P_1\{X > x\} - \delta_1 \text{ for all } x\}. \quad (3.1)$$

We assume that $\mathcal{P}_0$ and $\mathcal{P}_1$ are disjoint (i.e., that ε_j and δ_j are sufficiently small).

It may help to visualize $\mathcal{P}_0$ as the set of distribution functions lying above the solid line $(1 - \varepsilon_0)P_0(x) - \delta_0$ in Exhibit 10.3.1 and $\mathcal{P}_1$ as the set of distribution functions lying below the dotted line $(1 - \varepsilon_1)P_1(x) + \varepsilon_1 + \delta_1$. As before $P\{\cdot\}$ denotes the set function, $P(\cdot)$ the corresponding distribution function: $P(x) = P\{(-\infty, x)\}$.

Now let φ be any (randomized) test between $\mathcal{P}_0$ and $\mathcal{P}_1$, rejecting $\mathcal{P}_j$ with conditional probability $\varphi_j(x)$ given that $\mathbf{x} = (x_1, \ldots, x_n)$ has been observed.

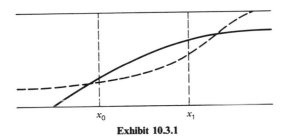

Exhibit 10.3.1

Assume that a loss $L_j > 0$ is incurred if $\mathscr{P}_j$ is falsely rejected; then the expected loss, or risk, is

$$R(Q'_j, \varphi) = L_j E_{Q'_j}(\varphi_j)$$

if $Q'_j \in \mathscr{P}_j$ is the true underlying distribution. The problem is to find a minimax test, that is, to minimize

$$\max_{j=0,1} \sup_{Q'_i \in \mathscr{P}} R(Q'_i, \varphi).$$

These minimax tests happen to have quite a simple structure in our case. There is a least favorable pair $Q_0 \in \mathscr{P}_0$, $Q_1 \in \mathscr{P}_1$, such that, for all sample sizes, the probability ratio tests φ between Q_0 and Q_1 satisfy

$$R(Q'_j, \varphi) \leqslant R(Q_j, \varphi), \qquad \text{for } Q'_j \in \mathscr{P}_j.$$

Thus in view of the Neyman-Pearson lemma, the probability ratio tests between Q_0 and Q_1 form an essentially complete class of minimax tests between $\mathscr{P}_0$ and $\mathscr{P}_1$. The pair Q_0, Q_1 is not unique, in general, but the probability ratio dQ_1/dQ_0 is essentially unique; as already mentioned it will be a censored version of dP_1/dP_0.

It is in fact quite easy to guess such a pair Q_0, Q_1. The successful conjecture is that there are two numbers $x_0 < x_1$, such that the $Q_j(\cdot)$ coincide with the respective boundaries between x_0 and x_1; in particular, their densities thus will satisfy

$$q_j(x) = (1 - \varepsilon_j) p_j(x), \qquad \text{for } x_0 \leqslant x \leqslant x_1. \tag{3.2}$$

On $(-\infty, x_0)$ and on (x_1, ∞), we expect the likelihood ratios to be constant, and we try densities of the form

$$q_j(x) = a p_0(x) + b p_1(x). \tag{3.3}$$

The various internal consistency requirements, in particular that

$$Q_0(x) = (1 - \varepsilon_0) P_0(x) - \delta_0, \qquad \text{for } x_0 \leqslant x \leqslant x_1,$$
$$Q_1(x) = (1 - \varepsilon_1) P_1(x) + \varepsilon_1 + \delta_1, \qquad \text{for } x_0 \leqslant x \leqslant x_1, \tag{3.4}$$

now lead easily to the following explicit formulas (we skip the step-by-step derivation, just stating the final results and then checking them).

Put

$$v' = \frac{\varepsilon_1 + \delta_1}{1 - \varepsilon_1}, \qquad v'' = \frac{\varepsilon_0 + \delta_0}{1 - \varepsilon_0},$$

$$w' = \frac{\delta_0}{1 - \varepsilon_0}, \qquad w'' = \frac{\delta_1}{1 - \varepsilon_1}. \tag{3.5}$$

It turns out to be somewhat more convenient to characterize the middle interval between x_0 and x_1 in terms of $c(x)$ than in terms of the x themselves: $c' < c(x) < 1/c''$ for some constants c' and c'', which are determined later. Since $c(x)$ need not be continuous or strictly monotone, the two variants are not entirely equivalent.

If both $v' > 0$ and $v'' > 0$, we define Q_0 and Q_1 by their densities, as follows. Denote the three regions $c(x) \leqslant c'$, $c' < c(x) < 1/c''$, and $1/c'' \leqslant c(x)$ by I_-, I_0, and I_+, respectively. Then

$$q_0(x) = \frac{1 - \varepsilon_0}{v' + w'c'} [v' p_0(x) + w' p_1(x)], \qquad \text{on } I_-$$

$$= (1 - \varepsilon_0) p_0(x), \qquad \text{on } I_0$$

$$= \frac{(1 - \varepsilon_0) c''}{v'' + w''c''} [w'' p_0(x) + v'' p_1(x)], \qquad \text{on } I_+,$$

$$q_1(x) = \frac{(1 - \varepsilon_1) c'}{v' + w'c'} [v' p_0(x) + w' p_1(x)], \qquad \text{on } I_-$$

$$= (1 - \varepsilon_1) p_1(x), \qquad \text{on } I_0$$

$$= \frac{(1 - \varepsilon_1)}{v'' + w''c''} [w'' p_0(x) = v'' p_1(x)], \qquad \text{on } I_+. \tag{3.6}$$

If, say, $v' = 0$, then $w'' = 0$, and the above formulas simplify to

$$q_0(x) = (1 - \varepsilon_0) \frac{1}{c'} p_1(x), \qquad \text{on } I_-$$

$$= (1 - \varepsilon_0) p_0(x), \qquad \text{on } I_0$$

$$= (1 - \varepsilon_0) c'' p_1(x), \qquad \text{on } I_+,$$

$$q_1(x) = p_1(x), \qquad \text{for all } x. \tag{3.7}$$

It is evident from (3.6) [and (3.7)] that the likelihood ratio has the postulated form

$$\pi(x) = \frac{q_1(x)}{q_0(x)} = \frac{1-\varepsilon_1}{1-\varepsilon_0} c', \qquad \text{on } I_-,$$

$$= \frac{1-\varepsilon_1}{1-\varepsilon_0} c(x), \qquad \text{on } I_0$$

$$= \frac{1-\varepsilon_1}{1-\varepsilon_0} \cdot \frac{1}{c''}, \qquad \text{on } I_+. \qquad (3.8)$$

Moreover, since $p_1(x)/p_0(x) = c(x)$ is monotone, (3.6) implies

$$q_0(x) \leqslant (1-\varepsilon_0)p_0(x), \qquad \text{on } I_-,$$

$$q_0(x) \geqslant (1-\varepsilon_0)p_0(x), \qquad \text{on } I_+, \qquad (3.9)$$

and dual relations hold for q_1.

In view of (3.9) we have $Q_j \in \mathscr{P}_j$, with $Q_j(\cdot)$ touching the boundary between x_0 and x_1 if four relations hold, the first of which is

$$\int_{c(x)<c'} \left[(1-\varepsilon_0)p_0(x) - q_0(x) \right] d\mu = \delta_0. \qquad (3.10)$$

The other three are obtained by interchanging left and right, and the roles of P_0 and P_1.

If we insert (3.6) into (3.10), we obtain the equivalent condition

$$\int \left[c'p_0(x) - p_1(x) \right]^+ d\mu = v' + w'c'. \qquad (3.11)$$

Of the other three relations, one coincides with (3.11), the other two with

$$\int \left[c''p_1(x) - p_0(x) \right]^+ d\mu = v'' + w''c''. \qquad (3.12)$$

We must now show that (3.11) and (3.12) have solutions c' and c'', respectively. Evidently, it suffices to discuss (3.11).

If $v' = 0$, we have the trivial solution $c' = 0$ (and perhaps also some others). Let us exclude this case and put

$$f(z) = \frac{\int (zp_0 - p_1)^+ \, d\mu}{v' + w'z}, \qquad z \geqslant 0. \qquad (3.13)$$

We have to find a z such that $f(z) = 1$. Let $\Delta \geqslant 0$, then

$$f(z+\Delta) - f(z) = \frac{\Delta \int_E (v' + w'c) p_0 \, d\mu + \int_{E'} (v' + wz)(z + \Delta - c) p_0 \, d\mu}{(v' + w'z)[v' + w'(z+\Delta)]}$$

(3.14)

with

$$E = \{x \mid c(x) \leqslant z\}, \qquad E' = \{x \mid z < c(x) \leqslant z + \Delta\}.$$

Hence

$$0 \leqslant f(z+\Delta) - f(z) \leqslant \frac{\Delta}{v' + w'(z+\Delta)},$$

and it follows that f is monotone increasing and continuous.

As $z \to \infty$, $f(z) \to 1/w'$, and as $z \to 0$, $f(z) \to 0$. Thus there is a solution c' for which $f(c') = 1$, provided $w' < 1$. (Note that $w' \geqslant 1$ implies $\mathcal{P}_0 = \mathfrak{M}$; hence $\mathcal{P}_0 \cap \mathcal{P}_1 = \phi$ ensures $w' < 1$.)

It can be seen from (3.11) and (3.14) that $f(z)$ is strictly monotone for

$$z > c_1 = \text{ess. inf } c(x).$$

Since $f(z) = 0$ for $0 \leqslant z \leqslant c_1$, the solution c' is unique.

We can write the likelihood ratio between Q_0 and Q_1 in the form

$$\frac{q_1(x)}{q_0(x)} = \frac{1 - \varepsilon_1}{1 - \varepsilon_0} \tilde{\pi}(x)$$

with

$$\tilde{\pi}(x) = c', \qquad \text{on } I_-$$

$$= c(x), \qquad \text{on } I_0$$

$$= 1/c'', \qquad \text{on } I_+ \, .$$

(assuming that $c' < 1/c''$).

LEMMA 3.1

$$Q_0'\{\tilde{\pi} < t\} \geqslant Q_0\{\tilde{\pi} < t\}, \qquad \text{for } Q_0' \in \mathcal{P}_0$$

$$Q_1'\{\tilde{\pi} < t\} \leqslant Q_1\{\tilde{\pi} < t\}, \qquad \text{for } Q_1' \in \mathcal{P}_1.$$

Proof   These relations are trivially true for $t \le c'$ and for $t > 1/c''$. For $c' < t \le 1/c''$ they boil down to the inequalities in (3.1). ∎

In other words among all distributions in $\mathcal{P}_0$, $\tilde{\pi}$ is stochastically largest for Q_0, and among all distributions in $\mathcal{P}_1$, $\tilde{\pi}$ is stochastically smallest for Q_1.

THEOREM 3.2   For any sample size n and any level α, the Neyman-Pearson test of level α between Q_0 and Q_1, namely

$$\varphi(\mathbf{x}) = 1, \qquad \text{for} \quad \prod_1^n \tilde{\pi}(x_i) > C$$

$$= \gamma, \qquad \text{for} \quad \prod_1^n \tilde{\pi}(x_i) = C$$

$$= 0, \qquad \text{for} \quad \prod_1^n \tilde{\pi}(x_i) < C,$$

where C and γ are chosen such that $E_{Q_0}\varphi = \alpha$, is a minimax test between $\mathcal{P}_0$ and $\mathcal{P}_1$, with the same level

$$\sup_{\mathcal{P}_0} E\varphi = \alpha$$

and the same minimum power

$$\inf_{\mathcal{P}_1} E\varphi = E_{Q_1}\varphi.$$

Proof   This is an immediate consequence of Lemma 3.1 and of the following well-known Lemma 3.3 [putting $U_i = \log \tilde{\pi}(X_i)$, $\mathcal{L}(X_i) = Q$, etc.]. ∎

LEMMA 3.3   Let (U_i) and (V_i), $i = 1, 2, \ldots$, be two sequences of random variables, such that the U_i are independent among themselves, the V_i are independent among themselves, and U_i is stochastically larger than V_i, for all i. Then, for all n, $\sum_1^n U_i$ is stochastically larger than $\sum_1^n V_i$.

Proof   Let (Z_i) be a sequence of independent random variables with uniform distribution in $(0, 1)$, and let $F_i \le G_i$ be the distribution functions of U_i and V_i, respectively. Then $F_i^{-1}(Z_i)$ has the same distribution as U_i, $G_i^{-1}(Z_i)$ has the same distribution as V_i, and the conclusion follows easily from $F_i^{-1}(Z_i) \ge G_i^{-1}(Z_i)$. ∎

For the above we have assumed that $c' < 1/c''$. We now show that this is equivalent to our initial assumption that $\mathcal{P}_0$ and $\mathcal{P}_1$ are disjoint. If $c' = 1/c''$, then $Q_0 = Q_1$, and the sets $\mathscr{P}_0$ and $\mathscr{P}_1$ overlap. Since the solutions c' and c'' of (3.11) and (3.12) are monotone increasing in the ε_j, δ_j, the overlap is even worse if $c' > 1/c''$. On the other hand if $c' < 1/c''$, then $Q_0 \neq Q_1$, and $Q_0\{\tilde{\pi} < t\} \geqslant Q_1\{\tilde{\pi} < t\}$ with strict inequality for some $t = t_0$ [the power of a Neyman-Pearson test exceeds its size, cf. Lehmann (1959), p. 67, Corollary 1]. In view of Lemma 3.1, then $Q'_0\{\tilde{\pi} < t_0\} > Q'_1\{\tilde{\pi} < t_0\}$; hence $\mathscr{P}_0$ and $\mathscr{P}_1$ do not overlap.

The limiting test for the case $c' = 1/c''$ is of some interest; it is a kind of sign test, based on the number of observations for which $p_1(x)/p_0(x) \geqslant c'$. Incidentally, if $\varepsilon_0 = \varepsilon_1$, the limiting value is $c' = 1$.

Particular Cases

In the following we assume that either $\delta_j = 0$ or $\varepsilon_j = 0$. Note that the set $\mathscr{P}_0$, defined in (3.1), contains each of the following five sets (1) to (5), and that Q_0 is contained in each of them. *It follows that the minimax tests of Theorem 3.2 are also minimax for testing between neighborhoods specified in terms of ε-contamination, total variation, Prohorov distance, Kolmogorov distance, and Lévy distance,* assuming only that $p_1(x)/p_0(x)$ is monotone for the pair of idealized model distributions.

(1) *ε-contamination*   With $\delta_0 = 0$,

$$\{Q \in \mathfrak{M} \mid Q = (1 - \varepsilon_0) P_0 + \varepsilon_0 H, \ H \in \mathfrak{M}\}.$$

(2) *Total variation*   With $\varepsilon_0 = 0$,

$$\{Q \in \mathfrak{M} \mid \forall A \mid Q\{A\} - P_0\{A\}\mid \leqslant \delta_0\}.$$

(3) *Prohorov*   With $\varepsilon_0 = 0$ and $P_{0,\eta}(x) = P_0(x - \eta)$,

$$\{Q \in \mathfrak{M} \mid \forall A\, Q\{A\} \leqslant P_{0,\eta}\{A^\eta\} + \delta_0\}.$$

(4) *Kolmogorov*   With $\varepsilon_0 = 0$,

$$\{Q \in \mathfrak{M} \mid \forall x \mid Q(x) - P_0(x)\mid \leqslant \delta_0\}.$$

(5) *Lévy*   With $\varepsilon_0 = 0$ and $P_{0,\eta}(x) = P_0(x - \eta)$,

$$\{Q \in \mathfrak{M} \mid P_{0,\eta}(x - \eta) - \delta_0 \leqslant Q(x) \leqslant P_{0,\eta}(x + \eta) + \delta_0 \text{ for all } x\}.$$

Note that the gross error model (1) and the total variation model (2) make sense in arbitrary probability spaces; a closer look at the above proof shows that monotonicity of $p_1(x)/p_0(x)$ then is not needed and that the proof carries through in arbitrary probability spaces.

Furthermore note that the hypothesis $\mathcal{P}_0$ of (3.1) is such that it contains with every Q also all Q' stochastically smaller than Q; similarly, $\mathcal{P}_1$ contains with every Q also all Q' stochastically larger than Q. This has the important consequence that, if $(P_\theta)_{\theta \in \mathbf{R}}$ is a monotone likelihood ratio family, that is, if $p_{\theta_1}(x)/p_{\theta_0}(x)$ is monotone increasing in x if $\theta_0 < \theta_1$, then the test of Theorem 3.2 constructed for neighborhoods $\mathcal{P}_j$ of P_{θ_j}, $j=0,1$, is not only a minimax test for testing θ_0 against θ_1, but also for testing $\theta \leqslant \theta_0$ against $\theta \geqslant \theta_1$.

Example 3.1 *Normal Distribution* Let P_0 and P_1 be normal distributions with variance 1 and mean $-a$ and $+a$, respectively. Then $g(x) = p_1(x)/p_0(x) = e^{2ax}$. Assume that $\varepsilon_0 = \varepsilon_1 = \varepsilon$, and $\delta_0 = \delta_1 = \delta$; then for reasons of symmetry $c' = c''$. Write the common value in the form $c' = e^{-2ak}$; then (3.11) reduces to

$$e^{-2ak}\Phi(a-k) - \Phi(-a-k) = \frac{\varepsilon + \delta + \delta e^{-2ak}}{1-\varepsilon}. \qquad (3.15)$$

Assume that k has been determined from this equation. Then the logarithm of the test statistic in Theorem 3.2 is, apart from a constant factor,

$$h(\mathbf{x}) = \sum_{1}^{n} \psi(x_i), \qquad (3.16)$$

with

$$\psi(x) = \max(-k, \min(k, x)). \qquad (3.17)$$

Exhibit 10.3.2 shows some numerical results. Note that the values of k are surprisingly small: if $\delta \geqslant 0.0005$, then $k \leqslant 2.5$, and if $\delta \geqslant 0.01$, then $k \leqslant 1.5$, for all choices of a.

a	$k=0$	0.5	1.0	1.5	2.0	2.5
0.05	0.020	0.010	0.004	0.0014	0.0004	0.00010
0.1	0.040	0.020	0.008	0.0029	0.0008	0.00019
0.2	0.079	0.039	0.016	0.0055	0.0015	0.00035
0.5	0.191	0.090	0.034	0.0103	0.0025	0.00048
1.0	0.341	0.162	0.040	0.0087	0.0016	0.00022
1.5	0.433	0.135	0.027	0.0042	0.0005	0.00006
2.0	0.477	0.111	0.014	0.0015	0.0001	0.00001

Exhibit 10.3.2 Values of δ in function of a and k ($\varepsilon = 0$). From Huber (1968), with permission of the publisher.

Example 3.2 *Binomial Distributions* Let $\Omega = \{0, 1\}$, and let $b(x|p) = p^x(1-p)^{1-x}$, $x = 0, 1$. The problem is to test between $p = \pi_0$ and $p = \pi_1$, $0 \leqslant \pi_0 < \pi_1 \leqslant 1$, when there is uncertainty in terms of total variation. This means that

$$\mathscr{P}_j = \{b(\cdot|p)|0 \leqslant p \leqslant 1, \pi_j - \delta_j \leqslant p \leqslant \pi_j + \delta_j\}.$$

It is evident that the minimax tests between $\mathscr{P}_0$ and $\mathscr{P}_1$ coincide with the Neyman-Pearson tests of the same level between $b(\cdot|\pi_0 + \delta_0)$ and $b(\cdot|\pi_1 - \delta_1)$, provided $\pi_0 + \delta_0 < \pi_1 - \delta_1$. (This trivial example is used to construct a counterexample in the following section).

In general, the level and power of these robust tests are not easy to determine. It is, however, possible to attack such problems asymptotically, assuming that, simultaneously, the hypotheses approach each other at a rate $\theta_1 - \theta_0 \sim n^{-1/2}$, while the neighborhood parameters ε and δ shrink at the same rate. For details, see Section 11.2.

10.4 SEQUENTIAL TESTS

Let $\mathscr{P}_0$ and $\mathscr{P}_1$ be two composite hypotheses as in the preceding section, and let Q_0 and Q_1 be a least favorable pair with probability ratio $\pi(x) = q_1(x)/q_0(x)$. We saw that this pair is least favorable for all fixed sample sizes. What happens if we use the sequential probability ratio test (SPRT) between Q_0 and Q_1 to discriminate between $\mathscr{P}_0$ and $\mathscr{P}_1$?

Put $\gamma(x) = \log \pi(x)$ and let us agree that the SPRT terminates as soon as

$$K' < \sum_{i \leqslant n} \gamma(x_i) < K'' \tag{4.1}$$

is violated for the first time $n = N(\mathbf{x})$, and that we decide in favor of $\mathscr{P}_0$ or $\mathscr{P}_1$, respectively, according as the left or right inequality in (4.1) is violated, respectively. Somewhat more generally, we may allow randomization on the boundary, but we leave this to the reader.

Assume, for example, that Q_0' is true. We have to compare the stochastic behavior of the cumulative sums $\sum \gamma(x_i)$ under Q_0' and Q_0. According to the proof of Lemma 3.3, there are functions $f \geqslant g$ and independent random variables Z_i such that $f(Z_i)$ and $g(Z_i)$ have the same distribution as $\gamma(X_i)$ under Q_0 and Q_0', respectively. Thus if the cumulative sum $\sum g(Z_i)$ leaves the interval (K', K'') first at K'', $\sum f(Z_i)$ will do the same, but even earlier. Therefore the probability of falsely rejecting $\mathscr{P}_0$ is at least as large under Q_0 as under Q_0'. A similar argument applies to the other hypothesis $\mathscr{P}_1$, and we conclude that the pair (Q_0, Q_1) *is also least favorable in the sequential case, as far as the probabilities of error are concerned.*

It need *not be least favorable for the expected sample size*, as the following example shows.

Example 4.1   Assume that $X_1, X_2, \ldots$ are independent Bernoulli variables

$$P\{X_i = 1\} = 1 - P\{X_i = 0\} = p,$$

and that we are testing the hypothesis $\mathcal{P}_0 = \{p \leqslant \alpha\}$ against the alternative $\mathcal{P}_1 = \{p \geqslant \frac{1}{2}\}$, where $0 < \alpha < \frac{1}{2}$. There is a least favorable pair Q_0, Q_1, corresponding to $p = \alpha$ and $p = \frac{1}{2}$, respectively (cf. Example 3.2). Then

$$\gamma(x) = \log \frac{q_1(x)}{q_0(x)} = -\log 2(1-\alpha), \qquad \text{for } x = 0$$

$$= -\log 2\alpha, \qquad \text{for } x = 1. \tag{4.2}$$

Assume $\alpha \leqslant 2^{-m-1}$, where m is a positive integer; then

$$\frac{-\log 2\alpha}{\log 2(1-\alpha)} \geqslant \frac{m \log 2}{\log 2 + \log(1-\alpha)} \geqslant m, \tag{4.3}$$

and we verify easily that the SPRT between $p = \alpha$ and $p = \frac{1}{2}$ with boundaries

$$K' = -m \log 2(1-\alpha),$$

$$K'' = -\log 2\alpha - (m-1) \log 2(1-\alpha) \tag{4.4}$$

can also be described by the simple rule:

(1)   Decide for $\mathcal{P}_1$ at the first appearance of a 1.
(2)   But decide for $\mathcal{P}_0$ after m zeros in a row.

The probability of deciding for $\mathcal{P}_0$ is $(1-p)^m$, the probability of deciding for $\mathcal{P}_1$ is $1 - (1-p)^m$, and the expected sample size is

$$E_p(N) = \sum_{k=0}^{m-1} (1-p)^k = \frac{1 - (1-p)^m}{p}. \tag{4.5}$$

Note that the expected sample size reaches its maximum (namely m) for $p = 0$, that is, *outside of the interval* $[\alpha, \frac{1}{2}]$. The probabilities of error of the first and of the second kind are bounded from above by $1 - (1-\alpha)^m \leqslant$

$m\alpha \leqslant m2^{-m-1}$ and by 2^{-m}, respectively, and thus can be made arbitrarily small [this disproves conjecture 8(i) of Huber (1965)].

However, if the boundaries K' and K'' are so far away that the behavior of the cumulative sums is essentially determined by their nonrandom drift

$$\sum \gamma(X_i) \sim nE_{Q'}[\gamma(X)], \qquad (4.6)$$

then the expected sample size is asymptotically equal to

$$E_{Q_0'}(N) \sim \frac{K'}{E_{Q_0'}(\gamma(X))}, \qquad \text{for } Q_0' \in \mathscr{P}_0,$$

$$E_{Q_1'}(N) \sim \frac{K''}{E_{Q_1'}(\gamma(X))}, \qquad \text{for } Q_1' \in \mathscr{P}_1. \qquad (4.7)$$

This heuristic argument can be made precise with the aid of the standard approximations for the expected sample sizes [cf., e.g., Lehmann (1959)]. In view of the inequalities of Theorem 3.2, it follows that the right-hand sides of (4.7) are indeed maximized for Q_0 and Q_1, respectively. So the pair (Q_0, Q_1) is in a certain sense, asymptotically least favorable also for the expected sample size if $K' \to -\infty$ and $K'' \to +\infty$.

10.5 THE NEYMAN-PEARSON LEMMA FOR 2-ALTERNATING CAPACITIES

Ordinarily, sample size n minimax tests between two composite alternatives $\mathscr{P}_0$ and $\mathscr{P}_1$ have a fairly complex structure. Setting aside all measure theoretic complications, they are Neyman-Pearson tests based on a likelihood ratio $\bar{q}_1(\mathbf{x})/\bar{q}_0(\mathbf{x})$, where each $\bar{q}_j$ is a mixture of product densities on Ω^n:

$$\bar{q}_j(\mathbf{x}) = \int \prod_{i=1}^{n} q(x_i)\lambda_j(dq).$$

Here, λ_j is a probability measure supported by the set $\mathscr{P}_j$; in general, λ_j depends both on the level and on the sample size.

The simple structure of the minimax tests found in Section 10.3 therefore was a surprise. On closer scrutiny it turned out that this had to do with the fact that all the "usual" neighborhoods $\mathscr{P}$ used in robustness theory could be characterized as $\mathscr{P} = \mathscr{P}_v$ with $v = (v, \bar{v})$ being a pair of conjugated 2-monotone/2-alternating capacities (see Section 10.2).

The following summarizes the main results of Huber and Strassen (1973). Let Ω be a Polish space (complete, separable, metrizable), equipped with its Borel σ-algebra $\mathcal{C}$, and let $\mathfrak{M}$ be the set of all probability measures on $(\Omega, \mathcal{C})$. Let $\bar{v}$ be a real valued set function defined on $\mathcal{C}$, such that

$$\bar{v}(\phi)=0, \qquad \bar{v}(\Omega)=1, \tag{5.1}$$

$$A \subset B \Rightarrow \bar{v}(A) \leqslant \bar{v}(B), \tag{5.2}$$

$$A_n \uparrow A \Rightarrow \bar{v}(A_n) \uparrow \bar{v}(A), \tag{5.3}$$

$$F_n \downarrow F, \; F_n \text{ closed} \Rightarrow \bar{v}(F_n) \downarrow \bar{v}(F), \tag{5.4}$$

$$\bar{v}(A \cup B) + \bar{v}(A \cap B) \leqslant \bar{v}(A) + \bar{v}(B). \tag{5.5}$$

The conjugate set function $\underline{v}$ is defined by

$$\underline{v}(A) = 1 - \bar{v}(A^c). \tag{5.6}$$

A set function $\bar{v}$ satisfying (5.1) to (5.5) is called a 2-*alternating capacity*, and the conjugate function $\underline{v}$ shall be called a 2-*monotone capacity*.

It can be shown that any such capacity is regular in the sense that, for every $A \in \mathcal{C}$,

$$\bar{v}(A) = \sup_K \bar{v}(K) = \inf_G \bar{v}(G), \tag{5.7}$$

where K ranges over the compact sets contained in A and G over the open sets containing A.

Among these requirements (5.4) is equivalent to

$$\mathcal{P}_v = \{ P \in \mathfrak{M} \mid P \leqslant \bar{v} \} = \{ P \in \mathfrak{M} \mid P \geqslant \underline{v} \}$$

being weakly compact, and (5.5) could be replaced by: for any monotone sequence of closed sets $F_1 \subset F_2 \subset \cdots$, $F_i \subset \Omega$, there is a $Q \leqslant \bar{v}$ that simultaneously maximizes the probabilities of the F_i, that is $Q(F_i) = \bar{v}(F_i)$, for all i.

Example 5.1 Let Ω be compact. Define $\bar{v}(A) = (1-\varepsilon)P_0(A) + \varepsilon$ for $A \neq \phi$. Then $\bar{v}$ satisfies (5.1) to (5.5), and

$$\mathcal{P}_v = \{ P \mid P = (1-\varepsilon)P_0 + \varepsilon H, \; H \in \mathfrak{M} \}$$

is the ε-contamination neighborhood of P_0.

Example 5.2 Let Ω be compact metric. Define $\bar{v}(A) = \min[P_0(A^\delta) + \varepsilon, 1]$ for compact sets $A \neq \phi$, and use (5.7) to extend v to $\mathcal{C}$. Then v satisfies (5.1) to (5.5), and

$$\mathcal{P}_v = \left\{ P \in \mathfrak{M} \mid P(A) \leqslant P_0(A^\delta) + \varepsilon \text{ for all } A \in \mathcal{Q} \right\}$$

is a Prohorov neighborhood of P_0.

Now let $\bar{v}_0$ and $\bar{v}_1$ be two 2-alternating capacities on Ω, and let $\underline{v}_0$ and $\underline{v}_1$ be their conjugates.

Let A be a critical region for testing between $\mathcal{P}_0 = \{P \in \mathfrak{M} \mid P \leqslant \bar{v}_0\}$ and $\mathcal{P}_1 = \{P \in \mathfrak{M} \mid P \leqslant \bar{v}_1\}$, that is, reject $\mathcal{P}_0$ if $x \in A$ is observed. Then the upper probability of falsely rejecting $\mathcal{P}_0$ is $\bar{v}_0(A)$, and of falsely accepting $\mathcal{P}_0$ is $\bar{v}_1(A^c) = 1 - \underline{v}_1(A)$.

Assume that $\mathcal{P}_0$ is true with prior probability $t/(1+t)$, $0 \leqslant t \leqslant \infty$, then the upper Bayes risk of the critical region A is by definition

$$\frac{t}{1+t} \bar{v}_0(A) + \frac{1}{1+t}\left[1 - \underline{v}_1(A)\right].$$

This is minimized by minimizing the 2-alternating set function

$$w_t(A) = t\bar{v}_0(A) - \underline{v}_1(A) \tag{5.8}$$

through a suitable choice of A.

It is not very difficult to show that, for each t, there is a critical region A_t minimizing (5.8). Moreover, the sets A_t can be chosen decreasing, that is, $A_t = \cup_{s > t} A_s$.

Define

$$\pi(x) = \inf\{t \mid x \notin A_t\}. \tag{5.9}$$

If $\bar{v}_0 = \underline{v}_0$, $\bar{v}_1 = \underline{v}_1$ are ordinary probability measures, then π is a version of the Radon-Nikodym derivative dv_1/dv_0, so the above constitutes a natural generalization of this notion to 2-alternating capacities.

The crucial result now is given in the following theorem.

THEOREM 5.1 (Neyman-Pearson lemma for capacities) There exist two probabilities $Q_0 \in \mathcal{P}_0$ and $Q_1 \in \mathcal{P}_1$ such that, for all t,

$$Q_0\{\pi > t\} = \bar{v}_0\{\pi > t\},$$

$$Q_1\{\pi > t\} = \underline{v}_1\{\pi > t\},$$

and that $\pi = dQ_1/dQ_0$.

Proof   See Huber and Strassen (1973, with correction 1974).   ■

In other words among all distributions in $\mathcal{P}_0$, π is stochastically largest for Q_0, and among all distributions in $\mathcal{P}_1$, π is stochastically smallest for Q_1.

The conclusion of Theorem 5.1 is essentially identical to that of Lemma 3.1, and we conclude, just as there, that the Neyman-Pearson tests between Q_0 and Q_1, based on the test statistic $\Pi_{i=1}^{n}\pi(x_i)$, are minimax tests between $\mathcal{P}_0$ and $\mathcal{P}_1$, and this for arbitrary levels and sample sizes.

10.6   ESTIMATES DERIVED FROM TESTS

In this section we derive a rigorous correspondence between tests and interval estimates of location.

Let $X_1, \ldots, X_n$ be random variables whose joint distribution belongs to a location family, that is,

$$\mathcal{L}_\theta(X_1, \ldots, X_n) = \mathcal{L}_0(X_1 + \theta, \ldots, X_n + \theta); \tag{6.1}$$

the X_i need not be independent.

Let $\theta_1 < \theta_2$, and let φ be a (randomized) test of θ_1 against θ_2, of the form

$$\varphi(\mathbf{x}) = 0, \qquad \text{for } h(\mathbf{x}) < C$$

$$= \gamma, \qquad \text{for } h(\mathbf{x}) = C$$

$$= 1, \qquad \text{for } h(\mathbf{x}) > C. \tag{6.2}$$

The test statistic h is arbitrary, except that $h(\mathbf{x} + \theta) = h(x_1 + \theta, \ldots, x_n + \theta)$ is assumed to be a monotone increasing function of θ. Let

$$\alpha = E_{\theta_1}\varphi$$

and

$$\beta = E_{\theta_2}\varphi$$

be the level and the power of this test.

As $\alpha = E_0\varphi(\mathbf{x} + \theta_1)$, $\beta = E_0\varphi(\mathbf{x} + \theta_2)$, and $\varphi(\mathbf{x} + \theta)$ is monotone increasing in θ, we have $\alpha \leq \beta$.

We define two random variables T^* and T^{**} by

$$T^* = \sup\{\theta \,|\, h(\mathbf{x} - \theta) > C\},$$

$$T^{**} = \inf\{\theta \,|\, h(\mathbf{x} - \theta) < C\}, \tag{6.3}$$

and put

$$T^0 = T^* \text{ with probability } 1 - \gamma$$

$$= T^{**} \text{ with probability } \gamma. \tag{6.4}$$

This randomization should be independent of $(X_1, \ldots, X_n)$; for example, take a uniform $(0,1)$ random variable U that is independent of $(X_1, \ldots, X_n)$ and let T^0 be a deterministic function of $(X_1, \ldots, X_n, U)$, defined in the obvious way: $T^0(\mathbf{X}, U) = T^*$ or T^{**} according as $U \geqslant \gamma$ or $U < \gamma$.

Evidently all three statistics T^*, T^{**}, and T^0 are translation invariant in the sense that $T(\mathbf{x} + \theta) = T(\mathbf{x}) + \theta$.

We note that $T^* \leqslant T^{**}$ and that

$$\{\mathbf{x} | T^* > \theta\} \subset \{\mathbf{x} | h(\mathbf{x} - \theta) > C\} \subset \{\mathbf{x} | T^* \geqslant \theta\},$$

$$\{\mathbf{x} | T^{**} > \theta\} \subset \{\mathbf{x} | h(\mathbf{x} - \theta) \geqslant C\} \subset \{\mathbf{x} | T^{**} \geqslant \theta\}. \tag{6.5}$$

If $h(\mathbf{x} - \theta)$ is continuous as a function of θ, these relations simplify to

$$\{T^* > \theta\} = \{h(\mathbf{x} - \theta) > C\},$$

$$\{T^{**} \geqslant \theta\} = \{h(\mathbf{x} - \theta) \geqslant C\}.$$

In any case we have, for an arbitrary joint distribution of $X_1, \ldots, X_n$ and arbitrary θ,

$$P\{T^0 > \theta\} = (1 - \gamma) P\{T^* > \theta\} + \gamma P\{T^{**} > \theta\}$$

$$\leqslant (1 - \gamma) P\{h(\mathbf{X} - \theta) > C\} + \gamma P\{h(\mathbf{X} - \theta) \geqslant C\}$$

$$= E\varphi(\mathbf{X} - \theta).$$

For $T^0 \geqslant \theta$ the inequality is reversed; thus

$$P\{T^0 > \theta\} \leqslant E\varphi(\mathbf{X} - \theta) \leqslant P\{T^0 \geqslant \theta\}. \tag{6.6}$$

For the translation family (6.1) we have, in particular,

$$E_{\theta_1}\varphi(\mathbf{X}) = E_0\varphi(\mathbf{X} + \theta_1) = \alpha.$$

Since T^0 is translation invariant, this implies

$$P_\theta\{T^0 + \theta_1 > \theta\} \leqslant \alpha \leqslant P_\theta\{T^0 + \theta_1 \geqslant \theta\}, \tag{6.7}$$

and similarly,

$$P_\theta\{T^0 + \theta_2 > \theta\} \leq \beta \leq P_\theta\{T^0 + \theta_2 \geq \theta\}. \qquad (6.8)$$

We conclude that $[T^0 + \theta_1, T^0 + \theta_2]$ is a (*fixed-length*) *confidence interval such that the true value* θ *lies to its left with probability* $\leq \alpha$, *and to its right with probability* $\leq 1 - \beta$. For the open interval $(T^0 + \theta_1, T^0 + \theta_2)$ the inequalities are reversed, and the probabilities of error become $\geq \alpha$ and $\geq 1 - \beta$ respectively.

In particular, if the distribution of T^0 is continuous, then $P_\theta\{T^0 + \theta_1 = \theta\} = P_\theta\{T^0 + \theta_2 = \theta\} = 0$; therefore we have equality in either case, and $(T^0 + \theta_1, T^0 + \theta_2)$ catches the true value with probability $\beta - \alpha$.

The following lemma gives a sufficient condition for the absolute continuity of the distribution of T^0.

LEMMA 6.1 If the joint distribution of $\mathbf{X} = (X_1, \ldots, X_n)$ is absolutely continuous with respect to Lebesgue measure in $\mathbb{R}^n$, then every translation invariant measurable estimate T has an absolutely continuous distribution with respect to Lebesgue measure in $\mathbb{R}$.

Proof We prove the lemma be explicitly writing down the density of T: if the joint density of $\mathbf{X}$ is $f(\mathbf{x})$, then the density of T is

$$g(t) = \int f(y_1 - T(\mathbf{y}) + t, \ldots, y_{n-1} - T(\mathbf{y}) + t, -T(\mathbf{y}) + t) \, dy_1 \ldots dy_{n-1},$$

$$(6.9)$$

where $\mathbf{y}$ is short for $(y_1, \ldots, y_{n-1}, 0)$. In order to prove (6.9) it suffices to verify that, for every bounded measurable function w,

$$\int w(t) g(t) \, dt = \int w(T(\mathbf{x})) f(\mathbf{x}) \, dx_1 \ldots dx_n. \qquad (6.10)$$

By Fubini's theorem we can interchange the order of integrations on the left-hand side:

$$\int w(t) g(t) \, dt = \int \left\{ \int w(t) f(\cdots) \, dt \right\} dy_1 \ldots dy_{n-1},$$

where the argument list of $f(\cdots)$ is the same as in (6.9). We substitute $t = T(\mathbf{y}) + x_n = T(\mathbf{y} + x_n)$ in the inner integral and change the order of integrations again:

$$\int w(t)g(t)\,dt = \int \left\{ \int w(T(\mathbf{y}+x_n)) f(\mathbf{y}+x_n)\,dy_1 \ldots dy_{n-1} \right\} dx_n.$$

Finally, we substitute $x_i = y_i + x_n$ for $i = 1,\ldots,n-1$ and obtain the desired equivalence (6.10). ∎

NOTE 1   The assertion that the distribution of a translation invariant estimate T is continuous, provided the observations X_i are independent with identical continuous distributions, is plausible but false [cf. Torgersen (1971)].

NOTE 2   It is possible to obtain confidence intervals with *exact* one-sided error probabilities α and $1-\beta$ also in the general discontinuous case if we are willing to choose a sometimes open, sometimes closed interval. More precisely, when $U \geqslant \gamma$ and thus $T^0 = T^*$, and if the set $\{\theta \mid h(\mathbf{x}-\theta) > C\}$ is open, choose the interval $[T^0+\theta_1, T^0+\theta_2)$; if it is closed, choose $(T^0+\theta_1, T^0+\theta_2]$. When $T^0 = T^{**}$ and $\{\theta \mid h(\mathbf{x}-\theta) \geqslant C\}$ is open, take $[T^0+\theta_1, T^0+\theta_2)$; if it is closed, take $(T^0+\theta_1, T^0+\theta_2]$.

NOTE 3   The more traditional nonrandomized compromise $T^{00} = \frac{1}{2}(T^* + T^{**})$ between T^* and T^{**} in general does *not* satisfy the crucial relation (6.6).

NOTE 4   Starting from the translation invariant estimate T^0, we can reconstruct a test between θ_1 and θ_2, having the original level α and power β, as follows. In view of (6.6)

$$P_{\theta_1}\{T^0 > 0\} \leqslant \alpha \leqslant P_{\theta_1}\{T^0 \geqslant 0\},$$

$$P_{\theta_2}\{T^0 > 0\} \leqslant \beta \leqslant P_{\theta_2}\{T^0 \geqslant 0\}.$$

Hence if T^0 has a continuous distribution so that $P_\theta\{T^0 = 0\} = 0$ for all θ, we simply take $\{T^0 > 0\}$ as the critical region. In the general case we would have to split the boundary $T^0 = 0$ in the manner of Note 2 (for that, the mere value of T^0 does not quite suffice—we also need to know on which side the confidence intervals are open and closed, respectively).

Rank tests are particularly attractive to derive estimates from, since they are distribution free under the null hypothesis; the sign test is so generally, and the others at least for symmetric distributions. This leads to distribution-free confidence intervals—the probabilities that the true value lies to the left or the right of the interval, respectively, do not depend on the underlying distribution.

Example 6.1 *Sign Test* Assume that the $X_1, \ldots, X_n$ are independent, with common distribution $F_\theta(x) = F(x-\theta)$, where F has median 0 and is continuous at 0. We test $\theta_1 = 0$ against $\theta_2 > 0$, using the test statistic

$$h(\mathbf{x}) = \sum 1_{[x_i > 0]}; \tag{6.11}$$

assume that the level of the test is α. Then there will be an integer c, independent of the special F, such that the test rejects the hypothesis if the cth order statistic $x_{(c)} > 0$, accepts it if $x_{(c+1)} \leqslant 0$, and randomizes it if $x_{(c)} \leqslant 0 < x_{(c+1)}$. The corresponding estimate T^0 randomizes between $x_{(c)}$ and $x_{(c+1)}$, and is a distribution-free lower confidence bound for the true median:

$$P_\theta\{\theta < T^0\} \leqslant \alpha \leqslant P_\theta\{\theta \leqslant T^0\}. \tag{6.12}$$

As F is continuous at its median, $P_\theta\{\theta = T^0\} = P_0\{0 = T^0\} = 0$, we have in fact equality in (6.12). (The upper confidence bound $T^0 + \theta_2$ is uninteresting, since its level depends on F.)

Example 6.2 *Wilcoxon and Similar Tests* Assume that $X_1, \ldots, X_n$ are independent with common distribution $F_\theta(x) = F(x-\theta)$, where F is continuous and symmetric. Rank the absolute values of the observations, and let R_i be the rank of $|x_i|$. Define the test statistic

$$h(\mathbf{x}) = \sum_{x_i > 0} a(R_i).$$

If $a(\cdot)$ is an increasing function [as for the Wilcoxon test: $a(i) = i$], then $h(\mathbf{x} + \theta)$ is increasing in θ. It is easy to see that it is piecewise constant, with jumps possible at the points $\theta = -\frac{1}{2}(x_i + x_j)$. It follows that T^0 randomizes between two (not necessarily adjacent) values of $\frac{1}{2}(x_i + x_j)$.

It is evident from the foregoing results that there is a precise correspondence between optimality properties for tests and estimates. For instance, the theory of locally most powerful rank tests for location leads to locally most efficient R-estimates, that is to estimates T maximizing the probability that $(T - \Delta, T + \Delta)$ catches the true value of the location parameter (i.e., the center of symmetry of F), provided Δ is chosen sufficiently small.

10.7 MINIMAX INTERVAL ESTIMATES

The minimax robust tests of Section 10.3 can be translated in a straightforward fashion into location estimates possessing exact finite sample minimax properties.

Let G be an absolutely continuous distribution on the real line, with a continuous density g such that $-\log g$ is strictly convex on its convex support (which need not be the whole real line).

Let $\mathcal{P}$ be a "blown-up" version of G:

$$\mathcal{P} = \{F \in \mathfrak{M} \mid (1-\varepsilon_0)G(x) - \delta_0 \leqslant F(x) \leqslant (1-\varepsilon_1)G(x) + \varepsilon_1 + \delta_1 \text{ for all } x\}.$$

$$(7.1)$$

Assume that the observations $X_1, \ldots, X_n$ of θ are independent, and that the distributions F_i of the observational errors $X_i - \theta$ lie in $\mathcal{P}$.

We intend to find an estimate T that minimizes the probability of under- or overshooting the true θ by more than a, where $a > 0$ is a constant fixed in advance. That is, we want to minimize

$$\sup_{\mathcal{P}, \theta} \max\left[P\{T < \theta - a\}, P\{T > \theta + a\} \right].$$

$$(7.2)$$

We claim that this problem is essentially equivalent to finding minimax tests between $\mathcal{P}_{-a}$ and $\mathcal{P}_{+a}$, where $\mathcal{P}_{\pm a}$ are obtained by shifting the set $\mathcal{P}$ of distribution functions to the left and right by amounts $\pm a$.

More precisely, define the two distribution functions G_{-a} and G_{+a} by their densities

$$g_{-a}(x) = g(x+a),$$

$$g_{+a}(x) = g(x-a).$$

$$(7.3)$$

Then

$$c(x) = \frac{g(x-a)}{g(x+a)} \qquad (7.4)$$

is strictly monotone increasing wherever it is finite.

Expand $P_0 = G_{-a}$ and $P_1 = G_{+a}$ to composite hypotheses $\mathcal{P}_0$ and $\mathcal{P}_1$ according to (3.1), and determine a least favorable pair $(Q_0, Q_1) \in \mathcal{P}_0 \times \mathcal{P}_1$. Determine the constants C and γ of Theorem 3.2 such that errors of both kinds are equally probable under Q_0 and Q_1:

$$E_{Q_0}\varphi = E_{Q_1}(1-\varphi) = \alpha.$$

$$(7.5)$$

If $\mathcal{P}_{-a}$ and $\mathcal{P}_{+a}$ are the translates of $\mathcal{P}$ to the left and to the right by the amount $a > 0$, then it is easy to verify that

$$Q_0 \in \mathcal{P}_{-a} \subset \mathcal{P}_0,$$

$$Q_1 \in \mathcal{P}_{+a} \subset \mathcal{P}_1.$$

$$(7.6)$$

If we now determine an estimate T^0 according to (6.3) and (6.4) from the test statistic

$$h(\mathbf{x}) = \prod_1^n \tilde{\pi}(x_i)$$

of Theorem 3.2, then (6.6) shows that

$$Q_0'\{T^0 > 0\} \leqslant E_{Q_0'}\varphi(\mathbf{X}) \leqslant \alpha, \qquad \text{for } Q_0' \in \mathscr{P}_0,$$

$$Q_1'\{T^0 < 0\} \leqslant E_{Q_1'}(1 - \varphi(\mathbf{X})) \leqslant \alpha, \qquad \text{for } Q_1' \in \mathscr{P}_1. \tag{7.7}$$

On the other hand for any statistic T satisfying

$$Q_0\{T = 0\} = Q_1\{T = 0\} = 0, \tag{7.8}$$

we must have

$$\max\left[Q_0\{T > 0\}, Q_1\{T < 0\}\right] \geqslant \alpha. \tag{7.9}$$

This follows from the remark that we can view T as a test statistic for testing between Q_0 and Q_1, and the minimax risk is α according to (7.5). Since Q_0 and Q_1 have densities, any translation invariant estimate, in particular T^0, satisfies (7.8) (Lemma 6.1). In view of (7.6) we have proved the following theorem.

THEOREM 7.1 The estimate T^0 minimizes (7.2); more precisely, if the distributions of the errors $X_i - \theta$ are contained in $\mathscr{P}$, then for all θ,

$$P\{T^0 < \theta - a\} \leqslant \alpha,$$

$$P\{T^0 > \theta + a\} \leqslant \alpha,$$

and the bound α is the best possible for translation invariant estimates.

NOTE The restriction to translation invariant estimates can be dropped in view of the Hunt-Stein theorem [Lehmann (1959), p. 335].

It is useful to discuss particular cases of this theorem. Assume that G is symmetric, and that $\varepsilon_0 = \varepsilon_1$ and $\delta_0 = \delta_1$. Then for reasons of symmetry $C = 1$ and $\gamma = \frac{1}{2}$. Put

$$\psi(x) = \log\frac{q_1(x)}{q_0(x)}; \tag{7.10}$$

then

$$\psi(x) = \max\left\{ -k, \min\left[k, \log\frac{g(x-a)}{g(x+a)} \right] \right\}, \qquad (7.11)$$

and T^* and T^{**} are the smallest and the largest solutions of

$$\sum \psi(x_i - T) = 0, \qquad (7.12)$$

respectively, and T^0 randomizes between them with equal probability. Actually, $T^* = T^{**}$ with overwhelming probability; $T^* < T^{**}$ occurs only if the sample size $n = 2m$ is even and the sample has a large gap in the middle [so that all summands in (7.12) have values $\pm k$]. Although, ordinarily, the nonrandomized midpoint estimate $T^{00} = \frac{1}{2}(T^* + T^{**})$ seems to have slightly better properties than the randomized T^0, it does *not* solve the minimax problem; see Huber (1968) for a counterexample.

In the particular case where $G = \Phi$ is the normal distribution, $\log g(x - a)/g(x+a) = 2ax$ is linear, and after dividing through $2a$, we obtain our old acquaintance

$$\psi(x) = \max\left[-k, \min(k, x) \right]. \qquad (7.13)$$

Thus the *M*-estimate T^0, as defined by (7.12) and (7.13), has two quite different minimax robustness properties for approximately normal distributions:

(1) It minimizes the maximal asymptotic variance, for symmetric ε-contamination.

(2) It yields exact, finite sample minimax interval estimates, for not necessarily symmetric ε-contamination (and for indeterminacy in terms of Kolmogorov distance, total variation and other models as well).

In retrospect it strikes us as very remarkable that the ψ defining the finite sample minimax estimate does not depend on the sample size (only on ε, δ, and a), even though, as already mentioned, 1% contamination has conceptionally quite different effects for sample size 5 and for sample size 1000.

The above results assume the scale to be fixed. For the more realistic case, where scale is a nuisance parameter, no exact finite sample results are known.

CHAPTER 11

Miscellaneous Topics

11.1 HAMPEL'S EXTREMAL PROBLEM

The minimax approaches described in Chapters 4 and 10 do not generalize beyond problems possessing a high degree of symmetry. This symmetry (e.g., translation invariance) is essential in their case, since it allows us to extend the parametrization of the idealized model throughout a neighborhood.

Hampel (1968, 1974b) proposed an alternative approach that avoids this problem by strictly staying at the idealized model: minimize the asymptotic variance of the estimate at the model, subject to a bound on the gross error sensitivity. This works for essentially arbitrary one-parameter families (and can even be extended to multiparameter problems). The main drawback is of a conceptual nature: only "infinitesimal" deviations from the model are allowed.

For L- and R-estimates the concept of gross-error sensitivity is of questionable value (compare Examples 3.5.1 and 3.5.2). We therefore restrict our attention to M-estimates.

Let $f_\theta(x) = f(x; \theta)$ be a family of probability densities, relative to some measure μ, indexed by a real parameter θ. We intend to estimate θ by an M-estimate $T = T(F)$, where the functional T is defined through an implicit equation

$$\int \psi(x; T(F))F(dx) = 0. \tag{1.1}$$

The function ψ is to be determined by the following extremal property.
Subject to Fisher consistency

$$T(F_\theta) = \theta \tag{1.2}$$

(where $dF_\theta = f_\theta \, d\mu$), and subject to a prescribed bound $k(\theta)$ on the gross error sensitivity

$$|IC(x; F_\theta, T)| \leqslant k(\theta), \qquad \text{for all } x, \tag{1.3}$$

the resulting estimate should minimize the asymptotic variance

$$\int IC(x; F_\theta, T)^2 \, dF_\theta. \tag{1.4}$$

Hampel showed that the solution is of the form

$$\psi(x; \theta) = [g(x; \theta) - a(\theta)]_{-b(\theta)}^{+b(\theta)}, \tag{1.5}$$

where

$$g(x; \theta) = \frac{\partial}{\partial \theta} \log f(x; \theta), \tag{1.6}$$

and where $a(\theta)$ and $b(\theta) > 0$ are some functions of θ; we are using the notation $[x]_u^v = \max[u, \min(v, x)]$.

How should we choose $k(\theta)$? Hampel had left the choice open, noting that the problem fails to have a solution if $k(\theta)$ is too small, and pointing out that it might be preferable to start with a sensible choice for $b(\theta)$, and then to determine the corresponding values of $a(\theta)$ and $k(\theta)$. We now sketch a somewhat more systematic approach, by proposing that $k(\theta)$ should be an arbitrarily chosen, but fixed, multiple of the "average error sensitivity" [i.e., of the square root of the asymptotic variance (1.4)]. Thus we put

$$k(\theta)^2 = k^2 \int IC(x; F_\theta, T)^2 \, dF_\theta, \tag{1.7}$$

where the constant k clearly must satisfy $k \geqslant 1$, but otherwise can be chosen freely (we would tentatively recommend the range $1 < k \leqslant 2.5$).

We now discuss existence and uniqueness of $a(\theta)$ and $b(\theta)$, when $k(\theta)$ is defined by (1.7).

The influence function of an M-estimate (1.1) at F_θ can be written as

$$IC(x; F_\theta, T) = \frac{\psi(x; \theta)}{\int \psi(x; \theta) g(x; \theta) f(x; \theta) \, d\mu}, \tag{1.8}$$

see (3.2.13). Here, we have used Fisher consistency and have transformed the denominator by an integration by parts.

The side conditions (1.2) and (1.3) now may be rewritten as

$$\int \psi(x;\theta)f(x;\theta)d\mu=0 \tag{1.9}$$

and

$$\psi(x;\theta)^2 \leqslant k^2 \int \psi(x;\theta)^2 f(x;\theta)\,d\mu, \tag{1.10}$$

while the expression to be minimized is

$$\frac{\int \psi(x;\theta)^2 f(x;\theta)\,d\mu}{\left[\int \psi(x;\theta)g(x;\theta)f(x;\theta)\,d\mu\right]^2}. \tag{1.11}$$

This extremal problem can be solved separately for each value of θ. Existence of a minimizing ψ follows in a straightforward way from the fact that ψ is bounded (1.10) and from weak compactness of the unit ball in L_∞.

The explicit form of the minimizing ψ can be found by the standard methods of the calculus of variations.

If we apply a small variation $\delta\psi$ to the ψ in (1.9) to (1.11), we obtain as a necessary condition for the extremum

$$\int [\psi-\lambda g+\nu]\delta\psi f d\mu \geqslant 0,$$

where λ and ν are Lagrange multipliers. Since ψ is only determined up to a multiplicative constant, we may standardize $\lambda=1$, and it follows that $\psi=g-\nu$ for those x where it can be freely varied [i.e., where we have strict inequality in (1.10)]. Hence the solution must be of the form (1.5), apart from an arbitrary multiplicative constant, and excepting a limiting case to be discussed later [corresponding to $b(\theta)=0$].

We first show that $a(\theta)$ and $b(\theta)$ exist, and that under mild conditions they are uniquely determined by (1.9) and by the following relation derived from (1.10):

$$b(\theta)^2=k^2 \int \psi(x;\theta)^2 f(x;\theta)\,d\mu. \tag{1.12}$$

To simplify the writing we work at one fixed θ and drop both arguments x and θ from the notation.

Existence and uniqueness of the solution (a, b) of (1.9) and (1.12) can be established by a method we have used already in Chapter 7. Namely, put

$$\rho(z) = \tfrac{1}{2}(k^{-2} + z^2), \qquad \text{for } |z| \leqslant 1$$

$$= \tfrac{1}{2}(k^{-2} - 1) + |z|, \qquad \text{for } |z| > 1, \tag{1.13}$$

and let

$$Q(a, b) = E\left\{\rho\left(\frac{g-a}{b}\right)b - |g|\right\}. \tag{1.14}$$

We note that Q is a convex function of (a, b) [this is a special case of (7.7.9)ff], and that it is minimized by the solution (a, b) of the two equations

$$E\left[\rho'\left(\frac{g-a}{b}\right)\right] = 0, \tag{1.15}$$

$$E\left[\frac{g-a}{b}\rho'\left(\frac{g-a}{b}\right) - \rho\left(\frac{g-a}{b}\right)\right] = 0, \tag{1.16}$$

obtained from (1.14) by taking partial derivatives with respect to a and b. But these two equations are equivalent to (1.9) and (1.12), respectively.

Note that this amounts to estimating a location parameter a and a scale parameter b for the random variable g by the method of Huber (1964, "Proposal 2"); compare Example 6.4.1. In order to see this let $\psi_0(z) = \rho'(z) = \max(-1, \min(1, z))$, and rewrite (1.15) and (1.16) as

$$E\left[\psi_0\left(\frac{g-a}{b}\right)\right] = 0, \tag{1.15'}$$

$$E\left[\psi_0\left(\frac{g-a}{b}\right)^2\right] = \frac{1}{k^2}. \tag{1.16'}$$

As in Chapter 7 it is easy to show that there is always some pair (a_0, b_0) with $b_0 \geqslant 0$ minimizing $Q(a, b)$.

We first take care of the limiting case $b_0 = 0$. For this it is advisable to scale ψ differently, namely to divide the right-hand side of (1.5) by $b(\theta)$. In the limit $b = 0$ this gives

$$\psi(x; \theta) = \text{sign}(g(x; \theta) - a(\theta)). \tag{1.17}$$

The differential conditions for $(a_0, 0)$ to be a minimum of Q now have a $\leqslant$ sign in (1.16), since we are on the boundary, and they can be written out as

$$\int \text{sign}(g(x; \theta) - a(\theta)) f(x; \theta) \, d\mu = 0, \tag{1.18}$$

$$1 \geqslant k^2 P\{g(x; \theta) \neq a(\theta)\}. \tag{1.19}$$

If $k > 1$, and if the distribution of g under F_θ is such that

$$P\{g(x; \theta) = a\} < 1 - k^{-2} \tag{1.20}$$

for all real a, then (1.19) clearly cannot be satisfied. It follows that (1.20) is a sufficient condition for $b_0 > 0$. Conversely, the choice $k = 1$ forces $b_0 = 0$. In particular, if $g(x; \theta)$ has a continuous distribution under F_θ, $k > 1$ is a necessary and sufficient condition for $b_0 > 0$.

Assume now that $b_0 > 0$. Then, in a way similar to that in Section 7.7, we find that Q is strictly convex at (a_0, b_0) provided the following two assumptions are true:

(1) $|g - a_0| < b_0$ with nonzero probability.
(2) Conditionally on $|g - a_0| < b_0$, g is not constant.
It follows that then (a_0, b_0) is unique.

In other words we now have determined a ψ that satisfies the side conditions (1.9) and (1.10), and for which (1.11) is stationary under infinitesimal variations of ψ, and it is the unique such ψ. Thus we have found the unique solution to the minimum problem.

Unless $a(\theta)$ and $b(\theta)$ can be determined in closed form, the actual calculation of the estimate $T_n = T(F_n)$ through solving (1.1) may still be quite difficult. Also we may encounter the usual problems of ML estimation caused by nonuniqueness of solutions.

The limiting case $b = 0$ is of special interest, since it corresponds to a generalization of the median. In detail this estimate works as follows. We first determine the median $a(\theta)$ of $g(x; \theta) = (\partial/\partial\theta) \log f(x; \theta)$ under the true distribution F_θ. Then we estimate $\hat{\theta}_n$ from a sample of size n such that one-half of the sample values of $g(x_i; \hat{\theta}_n) - a(\hat{\theta}_n)$ are positive, and the other half negative.

11.2 SHRINKING NEIGHBORHOODS

An interesting asymptotic approach to robust testing (and, through the methods of Section 10.6, to estimation) is obtained by letting both the alternative hypotheses and the distance between them shrink with increasing sample size. This approach was first utilized by Huber-Carol (1970); recently, it has been further exploited by Rieder (1978, 1979, 1980a, b). The issues involved here deserve some discussion.

First, we note that the exact finite sample results of Chapter 10 are not easy to deal with; unless the sample size n is very small, the size and minimum power are hard to calculate. This suggests the use of asymptotic approximations. Indeed for large values of n, the test statistics, or more precisely, their logarithms (10.3.16) are approximately normal. But for increasing n, either the size or the power of these tests, or both, tend to 0 or 1, respectively, exponentially fast, which corresponds to a limiting theory we are interested in only very rarely. In order to get limiting sizes and powers that are bounded away from 0 and 1, the hypotheses must approach each other at the rate $n^{-1/2}$ (at least in the nonpathological cases). If the diameters of the composite alternatives are kept constant, while they approach each other until they touch, we typically end up with a limiting sign-test. This may be a very sensible test for extremely large sample sizes (cf. Section 4.2 for a related discussion in an estimation context), but the underlying theory is relatively dull. So we shrink the hypotheses at the same rate $n^{-1/2}$, and then we obtain nontrivial limiting tests.

Now two related questions pose themselves:

(1) Determine the asymptotic behavior of the sequence of exact, finite sample minimax tests.

(2) Find the properties of the limiting test; is it asymptotically equivalent to the sequence of the exact minimax tests?

The appeal of this approach lies in the fact that it does not make any assumptions about symmetry, and we therefore have good chances to obtain a workable theory of asymptotic robustness for tests and estimates without assuming symmetry.

However, there are conceptual drawbacks connected with these shrinking neighborhoods; somewhat pointedly, we may say that these tests are robust with regard to zero contamination only!

It appears that there is an intimate connection between limiting robust tests determined on the basis of shrinking neighborhoods and the robust estimates found through Hampel's extremal problem (Section 11.1), which share the same conceptual drawbacks.

This connection is now sketched very briefly; details can be found in the references mentioned at the beginning of this section; compare, in particular, Theorem 3.7 of Rieder (1978).

Assume that $(P_\theta)_\theta$ is a sufficiently regular family of probability measures, with densities p_θ, indexed by a real parameter θ. To fix the idea consider total variation neighborhoods $\mathcal{P}_{\theta,\delta}$ of P_θ, and assume that we are to test robustly between the two composite hypotheses

$$\mathcal{P}_{\theta \pm n^{-1/2}\tau, \, n^{-1/2}\delta}. \tag{2.1}$$

According to Chapter 10 the minimax tests between these hypotheses will be based on test statistics of the form

$$\sum \psi_n(X_i), \tag{2.2}$$

where $\psi_n(X)$ is a censored version of

$$\log\!\left(\frac{p_{\theta + n^{-1/2}\tau}(X)}{p_{\theta - n^{-1/2}\tau}(X)} \right). \tag{2.3}$$

Clearly, the limiting test will be based on

$$\sum \psi(X_i), \tag{2.4}$$

where $\psi(X)$ is a censored version of

$$\frac{\partial}{\partial \theta}\big[\log p_\theta(X)\big]. \tag{2.5}$$

It can be shown under quite mild regularity conditions that the limiting test is indeed asymptotically equivalent to the sequence of exact minimax tests.

If we standardize ψ by subtracting its expected value, so that

$$\int \psi \, dP_\theta = 0, \tag{2.6}$$

it turns out that the censoring is symmetric:

$$\psi(X) = \left[\frac{\partial}{\partial \theta} \log p_\theta - a_\theta \right]_{-b_\theta}^{+b_\theta}. \tag{2.7}$$

Note that this is formally identical to (1.5) and (1.6). In our case the constants a_θ and b_θ are determined by

$$\int \left(\frac{\partial}{\partial \theta} \log p_\theta - a_\theta - b_\theta\right)^+ dP_\theta = \int \left(\frac{\partial}{\partial \theta} \log p_\theta - a_\theta + b_\theta\right)^- dP_\theta = \frac{\delta}{\tau}. \quad (2.8)$$

In the above case the relations between the exact finite sample tests and the limiting test are straightforward, and the properties of the latter are easy to interpret. In particular, (2.8) shows that it will be very nearly minimax along a whole family of total variation neighborhood alternatives with a constant ratio δ/τ.

Trickier problems arise if such a shrinking sequence is used to describe and characterize the robustness properties of some given test. We noted earlier that some estimates get relatively less robust when the neighborhood shrinks, in the precise sense that the estimate is robust, but $\lim b(\varepsilon)/\varepsilon = \infty$; compare Section 3.5. In particular, the normal scores estimate has this property. It is therefore not surprising that the robustness properties of the normal scores test do not show up in a naive shrinking neighborhood model [cf. Rieder (1979, 1980b)].

References

D. F. Andrews et al. (1972), *Robust Estimates of Location: Survey and Advances*, Princeton University Press, Princeton N.J.

F. J. Anscombe (1960), Rejection of outliers, *Technometrics*, **2**, pp. 123–147.

V. I. Averbukh and O. G. Smolyanov (1967), The theory of differentiation in linear topological spaces, *Russian Math. Surveys*, **22**, pp. 201–258.

———(1968), The various definitions of the derivative in linear topological spaces, *Russian Math. Surveys*, **23**, pp. 67–113.

R. Beran (1974), Asymptotically efficient adaptive rank estimates in location models, *Ann. Statist.*, **2**, pp. 63–74.

———(1977a), Robust location estimates, *Ann. Statist.*, **5**, pp. 431–444.

———(1977b), Minimum Hellinger distance estimates for parametric models, *Ann. Statist.*, **5**, pp. 445–463.

———(1978), An efficient and robust adaptive estimator of location, *Ann. Statist.*, **6**, pp. 292–313.

P. J. Bickel (1973), On some analogues to linear combinations of order statistics in the linear model, *Ann. Statist.*, **1**, pp. 597–616.

———(1975), One-step Huber estimates in the linear model, *J. Amer. Statist. Ass.*, **70**, pp. 428–434.

———(1976), Another look at robustness: A review of reviews and some new developments, *Scand. J. Statist.*, **3**, pp. 145–168.

P. J. Bickel and A. M. Herzberg (1979), Robustness of design against autocorrelation in time I, *Ann. Statist.*, **7**, pp. 77–95.

P. J. Bickel and J. L. Hodges (1967), The asymptotic theory of Galton's test and a related simple estimate of location, *Ann. Math. Statist.*, **4**, pp. 68–85.

P. Billingsley (1968), *Convergence of Probability Measures*, John Wiley, New York.

G. E. P. Box and N. R. Draper (1959), A basis for the selection of a response surface design, *J. Amer. Statist. Ass.*, **54**, pp. 622–654.

N. Bourbaki (1952), *Intégration*, Ch. III, Hermann, Paris.

H. Chen, R. Gnanadesikan, and J. R. Kettenring (1974), Statistical methods for grouping corporations, *Sankhya*, **B36**, pp. 1–28.

H. Chernoff, J. L. Gastwirth, and M. V. Johns (1967), Asymptotic distribution of linear combinations of functions of order statistics with applications to estimation, *Ann. Math. Statist.*, **38**, pp. 52–72.

G. Choquet (1953/54), Theory of capacities, *Ann. Inst. Fourier*, **5**, pp. 131–292.

_____(1959), Forme abstraite du théorème de capacitabilité, *Ann. Inst. Fourier*, **9**, pp. 83–89.

J. R. Collins (1976), Robust estimation of a location parameter in the presence of asymmetry, *Ann. Statist.*, **4**, pp. 68–85.

C. Daniel and F. S. Wood (1971), *Fitting Equations to Data*, John Wiley, New York.

H. E. Daniels (1954), Saddle point approximations in statistics, *Ann. Math. Statist.*, **25**, pp. 631–650.

_____(1976), Paper presented at the Grenoble Statistics Meeting, 1976.

A. P. Dempster (1967), Upper and lower probabilities induced by a multivalued mapping, *Ann. Math. Statist.*, **38**, pp. 325–339.

_____(1968), A generalization of Bayesian inference, *J. Roy. Statist. Soc.*, **B30**, pp. 205–247.

_____(1975), A subjectivist look at robustness, Proc. 40th Session I. S. I., Warsaw, *Bull. Int. Statist. Inst.*, **46**, Book 1, pp. 349–374.

L. Denby and C. L. Mallows (1977), Two diagnostic displays for robust regression analysis, *Technometrics*, **19**, pp. 1–13.

S. J. Devlin, R. Gnanadesikan, and J. R. Kettenring (1975), Robust estimation and outlier detection with correlation coefficients, *Biometrika*, **62**, pp. 531–545.

_____(1979), Robust estimation of dispersion matrices and principal components, submitted to *J. Amer. Statist. Ass.*

J. L. Doob (1953), *Stochastic Processes*, John Wiley, New York.

R. M. Dudley (1969), The speed of mean Glivenko-Cantelli convergence, *Ann. Math. Statist.*, **40**, pp. 40–50.

R. Dutter (1975), Robust regression: Different approaches to numerical solutions and algorithms, Res. Rep. no. 6, Fachgruppe für Statistik, Eidgen. Technische Hochschule, Zurich.

_____(1976), LINWDR: Computer linear robust curve fitting program, Res. Rep. no. 10, Fachgruppe für Statistik, Eidgen. Technische Hochschule, Zurich.

_____(1977a), Numerical solution of robust regression problems: Computational aspects, a comparison. *J. Statist. Comput. Simul.*, **5**, pp. 207–238.

_____(1977b), Algorithms for the Huber estimator in multiple regression, *Computing*, **18**, pp. 167–176.

_____(1978), Robust regression: LINWDR and NLWDR, COMPSTAT 1978, Proc., in *Computational Statistics*, L. C. A. Corsten, Ed., Physica-Verlag, Vienna.

A. S. Eddington (1914), *Stellar Movements and the Structure of the Universe*, Macmillan, London.

W. Feller (1966), *An Introduction to Probability Theory and its Applications*, Vol. II, John Wiley, New York.

C. A. Field and F. R. Hampel (1980), Small sample asymptotic distributions of *M*-estimates of location, submitted to *Biometrika*.

A. A. Filippova (1962), Mises' theorem of the asymptotic behavior of functionals of empirical distribution functions and its statistical applications, *Theor. Prob. Appl.*, **7**, pp. 24–57.

R. A. Fisher (1920), A mathematical examination of the methods of determining the accuracy of an observation by the mean error and the mean square error, *Monthly Not. Roy. Astron. Soc.*, **80**, pp. 758–770.

D. Gale and H. Nikaidô (1965), The Jacobian matrix and global univalence of mappings, *Math. Ann.*, **159**, pp. 81–93.

R. Gnanadesikan and J. R. Kettenring (1972), Robust estimates, residuals and outlier detection with multiresponse data, *Biometrics*, **28**, pp. 81–124.

A. M. Gross (1977), Confidence intervals for bisquare regression estimates, *J. Amer. Statist. Ass.*, **72**, pp. 341–354.

J. Hájek (1968), Asymptotic normality of simple linear rank statistics under alternatives, *Ann. Math. Statist.*, **39**, pp. 325–346.

J. Hájek (1972), Local asymptotic minimax and admissibility in estimation, in: *Proc. Sixth Berkeley Symposium on Mathematical Statistics and Probability*, Vol. 1. University of California Press, Berkeley.

J. Hájek and V. Dupač (1969), Asymptotic normality of simple linear rank statistics under alternatives, II, *Ann. Math. Statist.*, **40**, pp. 1992–2017.

J. Hájek and Z. Šidák (1967), *Theory of Rank Tests*, Academic, New York.

W. C. Hamilton (1970), The revolution in crystallography, *Science*, **169**, pp. 133–141.

F. R. Hampel (1968), Contributions to the theory of robust estimation, Ph. D. Thesis, University of California, Berkeley.

_____(1971), A general qualitative definition of robustness, *Ann. Math. Statist.*, **42**, pp. 1887–1896.

_____(1973a), Robust estimation: A condensed partial survey, *Z. Wahrscheinlichkeitstheorie Verw. Gebiete*, **27**, pp. 87–104.

_____(1973b), Some small sample asymptotics, Proc. Prague Symposium on Asymptotic Statistics, Prague.

_____(1974a), Rejection rules and robust estimates of location: An analysis of some Monte Carlo results, Proc. European Meeting of Statisticians and 7th Prague Conference on Information Theory, Statistical Decision Functions and Random Processes, Prague, 1974.

_____(1974b), The influence curve and its role in robust estimation, *J. Amer. Statist. Ass.*, **62**, pp. 1179–1186.

_____(1975), Beyond location parameters: Robust concepts and methods, Proc. 40th Session I. S. I., Warsaw 1975, *Bull. Int. Statist. Inst.*, **46**, Book 1, pp. 375–382.

_____(1976), On the breakdown point of some rejection rules with mean, Res. Rep. no. 11, Fachgruppe für Statistik, Eidgen. Technische Hochschule, Zurich.

E. F. Harding and D. G. Kendall (1974), *Stochastic Geometry*, Wiley, London.

R. W. Hill (1977), Robust regression when there are outliers in the carriers, Ph. D. Thesis, Harvard University, Cambridge, Mass.

R. V. Hogg (1967), Some observations on robust estimation, *J. Amer. Statist. Ass.*, **62**, pp. 1179–1186.

_____(1972), More light on kurtosis and related statistics, *J. Amer. Statist. Ass.*, **67**, pp. 422–424.

_____(1974), Adaptive robust procedures, *J. Amer. Statist. Ass.*, **69**, pp. 909–927.

D. C. Hoaglin and R. E. Welsch (1978), The hat matrix in regression and ANOVA, *Amer. Statist.*, **32**, pp. 17–22.

P. W. Holland and R. E. Welsch (1977), Robust regression using iteratively reweighted least squares, *Comm. Statist.*, **A6**, pp. 813–827.

P. J. Huber (1964), Robust estimation of a location parameter, *Ann. Math. Statist.*, **35**, pp. 73–101.

_____(1965), A robust version of the probability ratio test, *Ann. Math. Statist.*, **36**, pp. 1753–1758.

_____(1966), Strict efficiency excludes superefficiency, (Abstract), *Ann. Math. Statist.*, **37**, p. 1425.

_____(1967), The behavior of maximum likelihood estimates under nonstandard conditions, in: *Proc. Fifth Berkeley Symposium on Mathematical Statistics and Probability*, Vol. 1, University of California Press, Berkeley.

_____(1968), Robust confidence limits, *Z. Wahrscheinlichkeitstheorie Verw. Gebiete*, **10**, pp. 269–278.

_____(1969), *Théorie de l'Inférence Statistique Robuste*, Presses de l'Université, Montreal.

_____(1970), Studentizing robust estimates, in: *Nonparametric Techniques in Statistical Inference*, M. L. Puri, Ed., Cambridge University Press, Cambridge, England.

_____(1972), Robust statistics: A review, *Ann. Math. Statist.*, **43**, pp. 1041–1067.

_____(1973a), Robust regression: Asymptotics, conjectures and Monte Carlo, *Ann. Statist.*, **1**, pp. 799–821.

_____(1973b), The use of Choquet capacities in statistics, *Bull. Int. Statist. Inst.*, *Proc. 39th Session*, **45**, pp. 181–191.

_____(1975), Robustness and designs, in: *A Survey of Statistical Design and Linear Models*, J. N. Srivastava, Ed., North Holland, Amsterdam.

_____(1976), Kapazitäten statt Wahrscheinlichkeiten? Gedanken zur Grundlegung der Statistik, *Jber. Deutsch. Math.-Verein.*, **78**, H.2, pp. 81–92.

_____(1977a), Robust covariances, in: *Statistical Decision Theory and Related Topics, II*, S. S. Gupta and D. S. Moore, Eds., Academic Press, New York.

_____(1977b), *Robust Statistical Procedures*, Regional Conference Series in Applied Mathematics No. 27, Soc. Industr. Appl. Math, Philadelphia, Penn.

_____(1979), Robust smoothing, *Proc. ARO Workshop on Robustness in Statistics*, April 11–12, 1978, R. L. Launer and G. N. Wilkinson, Eds., Academic Press, New York.

P. J. Huber and R. Dutter (1974), Numerical solutions of robust regression problems, in: *COMPSTAT 1974, Proc. in Computational Statistics*, G. Bruckmann, Ed., Physika Verlag, Vienna.

P. J. Huber and V. Strassen (1973), Minimax tests and the Neyman-Pearson lemma for capacities, *Ann. Statist.*, **1**, pp. 251–263; **2**, pp. 223–224.

C. Huber-Carol (1970), Etude asymptotique de tests robustes, Ph.D. Thesis, Eidgen. Technische Hochschule, Zurich.

L. A. Jaeckel (1971a), Robust estimates of location: Symmetry and asymmetric contamination, *Ann. Math. Statist.*, **42**, pp. 1020–1034.

_____(1971b), Some flexible estimates of location, *Ann. Math. Statist.*, **42**, pp. 1540–1552.

_____(1972), Estimating regression coefficients by minimizing the dispersion of the residuals, *Ann. Math. Statist.*, **43**, pp. 1449–1458.

J. Jurečková (1971), Nonparametric estimates of regression coefficients, *Ann. Math. Statist.*, **42**, pp. 1328–1338.

L. Kantorovič and G. Rubinstein (1958), On a space of completely additive functions, *Vestnik, Leningrad Univ.*, **13**, no. 7 (*Ser. Mat. Astr. 2*), pp. 52–59, in Russian.

J. L. Kelley (1955), *General Topology*, Van Nostrand, New York.

G. D. Kersting (1978), Die Geschwindigkeit der Glivenko-Cantelli-Konvergenz gemessen in der Prohorov-Metrik, Habilitationsschrift, Georg-August-Universität, Göttingen.

B. Kleiner, R. D. Martin, and D. J. Thomson (1979), Robust estimation of power spectra, *J. Roy. Statist. Soc.*, **B41**, No. 3, pp. 313–351.

W. S. Krasker and R. E. Welsch (1980), Efficient bounded influence regression estimation using alternative definitions of sensitivity, unpublished.

H. W. Kuhn and A. W. Tucker (1951), Nonlinear programming, in: *Proc. Second Berkeley Symposium on Mathematical Statistics and Probability*, University of California Press, Berkeley.

C. L. Lawson and R. J. Hanson (1974), *Solving Least Squares Problems*, Prentice Hall, Englewood Cliffs, N.J.

L. LeCam (1953), On some asymptotic properties of maximum likelihood estimates and related Bayes' estimates, *Univ. Calif. Publ. Statist.*, **1**, pp. 277–330.

E. L. Lehmann (1959), *Testing Statistical Hypotheses*, John Wiley, New York.

R. A. Maronna (1976), Robust *M*-estimators of multivariate location and scatter, *Ann. Statist.*, **4**, pp. 51–67.

G. Matheron (1975), *Random Sets and Integral Geometry*, John Wiley, New York.

R. Miller (1964), A trustworthy jackknife, *Ann. Math. Statist.*, **35**, pp. 1594–1605.

_____(1974), The jackknife—A review, *Biometrika*, **61**, pp. 1–15.

F. Mosteller and J. W. Tukey (1977), *Data Analysis and Regression*, Addison-Wesley, Reading, Mass.

J. Neveu (1964), *Bases Mathématiques du Calcul des Probabilités*, Masson, Paris; English translation by A. Feinstein, (1965), *Mathematical Foundations of the Calculus of Probability*, Holden-Day, San Francisco.

Y. V. Prohorov (1956), Convergence of random processes and limit theorems in probability theory, *Theor. Prob. Appl.*, **1**, pp. 157–214.

M. H. Quenouille (1956), Notes on bias in estimation, *Biometrika*, **43**, pp. 353–360.

J. A. Reeds (1976), On the definition of von Mises functionals, Ph. D. thesis, Dept. of Statistics, Harvard University, Cambridge, Mass.

D. A. Relles and W. H. Rogers (1977), Statisticians are fairly robust estimators of location, *J. Amer. Statist. Ass.*, **72**, pp. 107–111.

W. J. J. Rey (1978), *Robust Statistical Methods*, Lecture Notes in Mathematics, 690, Springer-Verlag, Berlin.

H. Rieder (1978), A robust asymptotic testing model, *Ann. Statist.*, **6**, pp. 1080–1094.

_____(1979), Robustness of one and two sample rank tests against gross errors, to appear in *Ann. Statist.*, **9**.

_____(1980a), On local asymptotic minimaxity and admissibility in robust estimation, unpublished.

_____(1980b), Qualitative robustness of rank tests, unpublished.

M. Romanowski and E. Green (1965), Practical applications of the modified normal distribution, *Bull. Géodésique*, **76**, pp. 1–20.

J. Sacks (1975), An asymptotically efficient sequence of estimators of a location parameter, *Ann. Statist.*, **3**, pp. 285–298.

J. Sacks and D. Ylvisaker (1972), A note on Huber's robust estimation of a location parameter, *Ann. Math. Statist.*, **43**, pp. 1068–1075.

_____(1978) Linear estimation for approximately linear models, *Ann. Statist.*, **6**, pp. 1122–1137.

H. Schönholzer (1979), Robuste Kovarianz, Ph. D. Thesis, Eidgen. Technische Hochschule, Zurich.

F. W. Scholz (1971), Comparison of optimal location estimators, Ph. D. Thesis, Dept. of Statistics, University of California, Berkeley.

G. Shafer (1976), *A Mathematical Theory of Evidence*, Princeton University Press, Princeton, N.J.

G. R. Shorack (1976), Robust studentization of location estimates, *Statistica Neerlandica*, **30**, pp. 119–141.

C. Stein (1956), Efficient nonparametric testing and estimation, in: *Proc. Third Berkeley Symposium on Mathematical Statistics and Probability*, Vol. 1, University of California Press, Berkeley.

S. M. Stigler (1969), Linear functions of order statistics, *Ann. Math. Statist.*, **40**, pp. 770–788.

_____(1973), Simon Newcomb, Percy Daniell and the history of robust estimation 1885–1920, *J. Amer. Statist. Assoc.*, **68**, pp. 872–879.

C. J. Stone (1975), Adaptive maximum likelihood estimators of a location parameter., *Ann. Statist.*, **3**, pp. 267–284.

V. Strassen (1964), Messfehler und Information, *Z. Wahrscheinlichkeitstheorie Verw. Gebiete*, **2**, pp. 273–305.

_____(1965), The existence of probability measures with given marginals, *Ann. Math. Statist.*, **36**, pp. 423–439.

K. Takeuchi (1971), A uniformly asymptotically efficient estimator of a location parameter, *J. Amer. Statist. Ass.*, **66**, pp. 292–301.

E. N. Torgerson (1970), Comparison of experiments when the parameter space is finite, *Z. Wahrscheinlichkeitstheorie Verw. Gebiete*, **16**, pp. 219–249.

_____(1971), A counterexample on translation invariant estimators, *Ann. Math. Statist.*, **42**, pp. 1450–1451.

J. W. Tukey (1958), Bias and confidence in not-quite large samples (Abstract), *Ann. Math. Statist.*, **29**, p. 614.

_____(1960), A survey of sampling from contaminated distributions, in: *Contributions to Probability and Statistics*, I. Olkin, Ed., Stanford University Press, Stanford, Calif.

_____(1970), *Exploratory Data Analysis*, mimeographed preliminary edition.

_____(1977), *Exploratory Data Analysis*, Addison-Wesley, Reading, Mass.

R. von Mises (1937), Sur les fonctions statistiques, in: *Conférence de la Réunion Internationale des Mathématiciens*, Gauthier-Villars, Paris; also in: *Selecta R. von Mises*, Vol. II, Amer. Math. Soc., Providence, R.I. 1964.

_____(1947), On the asymptotic distribution of differentiable statistical functions, *Ann. Math. Statist.*, **18**, pp. 309–348.

G. Wolf (1977), Obere und untere Wahrscheinlichkeiten, Ph. D. Thesis, Eidgen, Technische Hochschule, Zurich.

V. J. Yohai and R. A. Maronna (1979), Asymptotic behavior of *M*-estimators for the linear model, *Ann. Statist.*, **7**, pp. 258–268.

Index